The Origin of Metamorphic and Metasomatic Rocks

The Origin of Metamorphic and Metasomatic Rocks

A treatise on recrystallization and replacement in the earth's crust

By
Hans Ramberg

THE UNIVERSITY OF CHICAGO PRESS

THE UNIVERSITY OF CHICAGO PRESS, CHICAGO 37
Cambridge University Press, London, N.W. 1, England

Published 1952. Composed and printed by THE UNIVERSITY OF CHICAGO PRESS, *Chicago, Illinois, U.S.A.*

PREFACE

The study of metamorphism has advanced rapidly throughout the last decade; yet little of this progress had been compiled and interpreted in books when the manuscript for the present volume was begun in 1948. Since then some texts in petrology have appeared which include the more recent advancements in this science. As these modern texts deal chiefly with magmatic and sedimentary petrology, a need for a modern comprehensive book on metamorphism is still felt. It is hoped that the present volume will satisfy a part of that need.

A successful tackling of the complex problems in metamorphic petrology depends heavily upon thermodynamics and crystal chemistry. A liberal portion of the present book is therefore devoted to those aspects of thermodynamics and general crystal chemistry which are especially significant for the clarification of metamorphic and metasomatic phenomena. It cannot be emphasized too strongly, however, that much more knowledge and understanding of these basic sciences are needed than may be gained from the present book.

A few words may be said about the arrangement of the matter in the present volume. Some useful and fundamental principles of thermodynamics and individual cases of equilibrium systems of metamorphic minerals are treated first. Although these discussions are basic for the considerations of the petrogenetic processes to which the last two-thirds of the book is devoted, the petrogenetic sections of the book may well be read first, because references are frequently made to the physicochemical laws discussed in the beginning sections.

The collection of erratic data and descriptions of rocks or field relations are kept at a minimum, whereas the emphasis is put upon the discussion of general principles of petrogenesis. Several of the numerous unsolved problems in petrology are pointed out and discussed. It is hoped that this approach may stimulate students to do creative and independent work in the field of metamorphic and metasomatic petrology.

The writer wishes to express his sincere thanks to the persons who have been of direct or indirect help in preparing this book. Dr. W. H. Newhouse' encouraging interest in the work and his inspiring attitude toward our science in general have been of great value to the author. Drs. N. L. Bowen, C. A. Chapman, and G. E. Goodspeed have kindly read the manuscript and offered valuable suggestions. The first draft of the manuscript was read by Dr. Tom F. Barth, who was permitted, upon his request, to make use of some of its content in the preparation of his recent book, *Theoretical Petrology* (New York: John Wiley & Sons, 1952). The author's wife, Marie-Louise Ramberg, has been a great help in reading proof and in preparing the photographic prints used in the book. Professor Arne Noe-Nygaard has permitted the author to publish photos taken on expeditions for the Geological Survey of Greenland. Dr. F. Laves has generously permitted the publication of his original alkali feldspar diagram (Fig. 25). Dr. H. Yoder has placed experimental data at the writer's disposal (Fig. 38). Dr. F. Pettijohn has kindly allowed the author to publish a photograph, Figure 125. Mr. W. J. Oulten has made most of the drawings.

The following publishers have kindly permitted reproduction of various tables and figures: Addison-Wesley Press, Inc., New York; Cambridge University Press, Cambridge, England; Clarendon Press, Oxford, England; and Harper and Brothers, New York.

HANS RAMBERG

CHICAGO
August 1952

INTRODUCTION

UP UNTIL a few decades ago it was assumed by most geologists that metamorphism and metasomatism involved petrogenetic processes of but secondary rank. The school of Rosenbusch and Niggli in Europe and the experimental school of Bowen in the United States had taught that most crystalline rocks of nonsedimentary origin were emplaced by means of solidifying magmatic melts. Metamorphism and metasomatism were able only to cause secondary recrystallizations in the magmatic or sedmentary rocks and perhaps to cause minor compositional changes in the solid rock bodies. It was widely held that the hot igneous melts were the real cause for the rise in crustal temperature indicated by zones of metamorphism; and it was taken for granted by most that the moving material in metasomatism had a magmatic source (the emanation theory), if not actually being a true magmatic melt (the pore-magma hypothesis). Magmatism was the hen, metamorphism the egg.

This view is essentially correct in the cases of shallow subvolcanic magmatic intrusions, with their external contact zones of metamorphism. For the major portions of the accessible crust of the earth, however, this view is no longer tenable in light of the data gathered in the field and in laboratories, to say nothing about the impact of the results from theoretical analyses of rock-making processes. It has been shown that very little of the rock shell of the earth consists of rocks directly crystallized out of magmas. This conclusion is based upon analyses of the architecture and structure of the heterogeneous lithosphere and upon the temperature of formation of rocks as inferred from geologic thermometers. The temperatures indicated by the mineral assemblages in the gneiss granitic lithosphere are rarely above some 600° C.—more commonly considerably below this value. Such mineral assemblages did not crystallize directly out of a silicate melt, a conclusion also in harmony with the texture and structure of the rocks involved.

This situation may occasionally be explained by secondary recrystallization *in situ*, but in many cases one must assume that the rock matter was emplaced at submagmatic temperatures in the solid state. This happened partly by flowage in the crystalline state, in part by migrations of individual ions or molecules through the solid crust, or by capillary solution flowing through the pore network in rocks. These submagmatic processes belong to the domain of metamorphism in the broadest sense of the word. Hence one realizes that metamorphism has played a dominant role in the evolution of the earth's crust.

These results of the workers in the field of metamorphism have had some interesting impacts upon the magmatic theories. The modern school of igneous rocks and volcanism is no longer so concerned about the crystallization process of magmas and the petrogenetic consequences of this process. This is rather well understood in light of the theory of phase equilibria in closed systems, but it has so far failed to explain convincingly any *major* fraction of the evolution of the earth's crust. The modern magmatists (Van Bemmelen, Rittmann, Verhoogen, and others) are more concerned about the origin of the magmatic melts than about the formation of solid rocks from these melts. They are inclined to believe that magmas are created by refusion within the uppermost some 100-km.-thick crystalline crust of the earth. The magmas are the children of ultra-metamorphism rather than the metamorphism being the child of magmatism.

These conclusions were, in fact, anticipated more than a hundred years ago by the Norwegian geologist Keilhau, and later by several geologists of the French school, on the basis of field observations. But at that time physical and chemical theories were lagging far behind, and it was impossible to present the geological observations and facts in the light of accepted laws of physics and chemistry. It has not been until rather recently that the theory of the basic sciences has reached a stage at which they can be useful for the interpretation of a considerable part of the metasomatic and metamorphic petrology. We had to wait for the modern evolution in crystal chemistry before the happenings in solid rocks could be explained. Even today, however, the present theories of physics and chemistry are unable to offer any detailed explanation of several of the most common field observations from metamorphic and metasomatic areas. The writer is thinking in particular about the difficulties involved in understanding the origin of numerous quartzo-feldspathic rock bodies in gneissic areas and in the explanation of several aspects of metamorphic differentiation and in the co-occurrence and paragenetic grouping of certain metasomatic minerals.

On the other hand, there are a number of metamorphic and metasomatic phenomena which are excellently explained by the known laws of physics and chemistry, as, for example, the occurrence of mineral assemblages after the facies principle and the growth of metasomatic minerals preferably in places of pressure minima, as well as several aspects of the evolution of structures in schists and gneisses. We may anticipate that in not too remote a future several of our largest petrogenetic puzzles may be solved by intimate co-operation between field and laboratory work and theoretical considerations.

CONTENTS

LIST OF ILLUSTRATIONS

Figures

ABBREVIATIONS OF MINERAL NAMES USED ON FIGURES

ab. albite
act. actinolite
alm. almandite (almandine)
an. anorthite
and. andalusite
anth. anthophyllite
antig. antigorite
bi. biotite
cal. calcite
ch. chlorite
cord. cordierite
di. diopside
dol. dolomite
enst. enstatite
ep. epidote
Fe-tr. ferrotremolite
fo. forsterite
gar. garnet
ge. geikielite
gross. grossularite
h'bl. hornblende
hed. hedenbergite
hyp. hypersthene
il. ilmenite
ja. jadeite
ky. kyanite
mu. muscovite
ne. nepheline
ol. olivine
omph. omphacite
or. potash feldspar
per. periclase
phlog. phlogopite
pl. plagioclase
p'op. pyrope
py. pyroxene
q. quartz
sill. sillimanite
spess. spessartite
stau. staurolite
tr. tremolite
zo. zoisite

DEFINITIONS

THE term *metamorphism* was introduced to describe the indisputable field observations which proved that solid rocks were not "dead" bodies but, exposed to the geological forces in the lithosphere, were continuously changing—mineralogically, chemically, and structurally. It was not easy to understand this particular behavior of solid rocks at the beginning of the science of geology. At that time the laws of chemical equilibrium, of chemical kinetics, the phase rule, and the mass-action law were still unknown, and the understanding of crystalline plasticity was but vague.

Gradually geologists realized that the processes of metamorphism were governed by the laws of physics and chemistry and that the directions of the minero-chemical and structural alterations were correlative with changes in the external pressure (stress) and temperature affecting the rock. This simple principle has established itself as the fundamental law of metamorphism.

Metamorphism, in modern language (Eskola, 1939; Harker, 1932; Turner, 1948), is confined to endogenic rock-making processes taking place in a P,T-field situated between that of weathering and that of magmatism.

It is worthy of note that metamorphic alterations are rarely *isochemical*. More commonly, the alteration is *allochemical* (Eskola, 1939), i.e., accompanied by the loss and gain of some elements, or it is combined with *metasomatism*.

Furthermore, it is expedient to distinguish between the following types of metamorphism:

1. *Regional metamorphism* takes place in connection with the formation of folded mountain chains over large orogenic regions. It may be allochemical or isochemical and is usually connected with considerable mechanical motion of the rock complexes (folding, plastic flow, faulting, thrusting, and all kinds of tectonic transportation of solid rocks); so it has been called *dynamo-thermal metamorphism*. It happens that mechanical motion is negligible or lacking, however, and the term *load metamorphism* or *plutonic metamorphism* has been applied. On the other hand, when the mechanical deformation in the rock is more extensive than the chemical, then *kinetic metamorphism* in a restricted sense takes place. Regional metamorphism is usually progressive, i.e., the recrystallization takes place by rising temperature (and pressure). If chemical adjustments take place

by decreasing temperature (and pressure), the term *diaphtoresis* is used.

2. *Contact metamorphism* takes place in rocks adjacent to hot magma masses. It may be isochemical; if it is allochemical, it is called *contact metasomatism*. An extreme type of contact metamorphism and metasomatism is *pyrometamorphism* (affecting, e.g., inclusions in lavas).

3. *Autometamorphism* may be considered a third type of metamorphism. It denotes the alterations occurring in magmatic rocks which cool slowly after their solidification.

Other types of metamorphism may be distinguished (Turner, 1948). The classification is, to a great extent, based on the origin of the metamorphism, and we shall therefore discuss the theory of metamorphic processes before we return to the classification problem.

NATURAL BOUNDARY BETWEEN METAMORPHISM AND MAGMATISM

If a rock is exposed to rising temperature (and pressure), its mineral assemblage changes according to definite rules subsequently to be explained; and, at sufficiently high temperatures, a silicate melt will form, which, with rising temperature, gradually becomes more basic. This refusion process is called *anatexis* or *palingenesis*. When a natural silicate melt cools, the changes occur in the opposite direction: first a basic rock crystallizes from the cooling melt, and gradually the rock as well as the melt changes in composition until the melt is completely solidified.

The processes which take place in the solid rocks at temperatures below which no natural silicate melt can exist are called *metamorphic processes*.

The boundary between metamorphism and magmatism in a P,T-diagram thus simply corresponds to the boundary between the stability fields of natural silicate melts, on the one side, and of crystalline minerals, on the other side, i.e., the solidus surface of the natural rock-making systems. For dry silicate systems, this boundary is well known from numerous experiments. Corresponding to the change of melting temperature with composition of the silicate systems, there must be a change of position in the P,T-field of the boundary between magmatism and metamorphism. Unfortunately, the exact position of the said boundary for natural rocks cannot be determined from the available experimental data.

Some of the most pertinent experiments are those of Goranson (1931–38), who showed that an ordinary granitic melt was unable to adsorb more than some 10 per cent of water even at pressures corresponding to 15-km. depth (see p. 244).

According to Goranson's experiments, water and rock-making silicate melts are only partly miscible, showing that no continuity exists between magmatic melts and hydrothermal solutions (see P. Eskola, 1939, p. 378). It is, therefore, impossible, although it has often been assumed, that a gradually increasing concentration of water can depress the crystallization temperature of a magma down to any value convenient for the interpretation of metamorphism as a secondary effect of magmas. In the field of metamorphism, the phases involved are the crystalline minerals and dilute hydrous fluids, together with the adsorbed intergranular "films"; in the field of magmatism, the phases involved are the concentrated silicate melts, the crystallizing minerals, and the fugitive magmatic gases (Fig. 1).[1]

At the present stage of our discussion, it is not necessary to consider

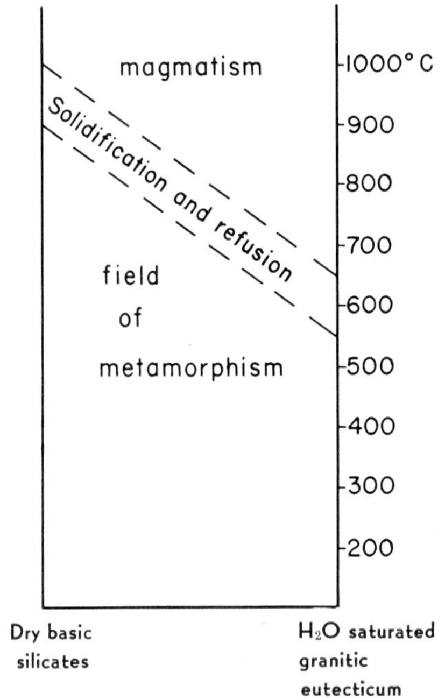

FIG. 1.—The fields of metamorphism and magmatism

more fully the relations between magmatism and metamorphism. The reader is referred to the sections on granitization and pegmatitization (pp. 238 and 248).

1. The fluid phase, which is immiscible with the silicate melt, is usually called the "magmatic rest fluid."

THERMODYNAMICS OF METAMORPHIC PROCESSES

THE laws of thermodynamics give us the conditions for equilibrium in any physical system; these laws also give us the directions and the driving energies of the natural or spontaneous processes occurring in unstable systems. The processes forming metamorphic and metasomatic rocks are certainly also brought about by instabilities of different kinds, the sole aim of the several processes being to bring the mineral assemblages and rock complexes involved toward states of stable thermodynamic equilibrium (Goldschmidt, 1911).

Considering the earth as a whole and disregarding any substantial and energetic communication with the universe, it is evident that the net result of all geological processes is a more stable over-all situation, with an increase in the entropy and a corresponding decrease in the free energy, as explained later. In metamorphic petrology we are chiefly concerned with smaller or greater confined parts of the earth's crust, however; and in such cases energetic and substantial communication between the complex considered and the surroundings may provide changes in entropy and free energy in either direction in the complex under consideration.

It is true that thermodynamic equilibrium is never reached within a greater part of the earth's crust; it is unfortunately also true that thermodynamics is unable to inform us about the velocities of the several processes. Even so, we may say that the laws of thermodynamics are our most important tools in the theoretical handling and interpretation of metamorphic-metasomatic phenomena.

However, the general laws of thermodynamics are of little value if some crucial data are not known. We need to know such data as: the heats of reaction, heat capacities, partial molal volumes, temperature of transformation at given pressures, heat of mixing, etc., of rock-making substances.

Provided that a sufficient number of the above-mentioned data are known, thermodynamics is in a position to evaluate the equilibrium diagrams of the mineral assemblages. One is then able to determine, more or less accurately, the P,T-conditions at which rocks are formed and to calculate the trend of reaction in unstable mineral associations. Furthermore, thermodynamic considerations inform us about the directions and trends of the long-distance transportation of matter involved in metasomatism.

Thermodynamic data from rock-making minerals are deplorably few—

so few, indeed, that it becomes tempting to describe the metamorphic processes without regard to the general laws of thermodynamics. However, this is a delusion, for the fewer empirical data we have, the more necessary it is to make use of the principal rules of physical chemistry in order to guide our thinking and to find the logical connections between the scattered data. To be sure, if our thermodynamic data from rock-making minerals were plentiful, we would not have to make use of the principal laws of thermodynamics in order to connect the data in the proper way.

Some of the well-known dry silicate melt–crystal equilibrium diagrams may serve as good examples of this statement. By making enough runs at different temperatures (and/or pressure) one can construct the equilibrium diagrams properly without the help of any theory. It is true that such empirically determined diagrams will be more accurate than those calculated theoretically from an insufficient number of fixed points. But in the case of metamorphic minerals syntheses are difficult, if not impossible, in many instances. We shall indeed need a rather thorough knowledge of the general laws of thermodynamics in order to get a maximum of understanding out of our scattered data from metamorphic rocks and mineral assemblages. The following pages will therefore be devoted to some of the laws of thermodynamics and their application to metamorphic-metasomatic petrology.

Let us consider a rock during the process of metamorphic recrystallization. The rock consists mainly of solids with negligible amounts of pore fluids. Each mineral is an assemblage of certain ions, atoms, or molecules arranged in a "rigid" crystal lattice. However, the crystal lattices are not so rigid and immobile as they were believed to be some decades ago. Different types of motion of the elementary particles take place in the lattices (Hedvall, 1938; Jost, 1937; Barrer, 1941). Thermal oscillation about energy depressions in the lattices constitutes the main type of motion, although a certain percentage of the elementary particles are charged with enough thermal energy to escape from the energy depressions. Such energy-rich particles may jump into empty spaces elsewhere in the lattice, they may exchange places with other energy-rich particles, or they may simply jump out of the lattice if they are situated at or near the surface. Along the grain boundaries and in the mosaic fissures in single crystals, there are small openings which are preferred by the movable elementary particles (ions, molecules, atoms), because the energy required to move the particles along these openings is usually much less than that required to move them inside the lattices. The particles are skating along the crystal interfaces, as Hedvall likes to express it. If a liquid pore solution is present, it may occupy space between the mineral grains and may also constitute the liquid inclusions in the minerals. The fluid will carry, in solution, all the constituent elements of the rock.

This, roughly, is the picture of a rock capable of undergoing metamorphic changes. Field observations indicate that if the rock is moved from a certain P,T-combination into another, then new minerals are formed in an attempt to conform to the equilibrium conditions of the new P,T-combination. For this reason, Eskola was able to refer the metamorphic rocks to definite mineral facies (Eskola, 1921*a*–39), a concept which was initiated in the excellent work by V. M. Goldschmidt from the Oslo area in 1911.

Let us consider a rock as a closed system (i.e., no interchange of matter permitted between rock and surroundings) in which chemical equilibrium is attained at a given pressure-temperature combination, P_1, T_1, where P denotes confining homogeneous pressure. By transferring the rock to new physical conditions, P_2, T_2, the rock loses or gains energy by communication with the surroundings. If $T_2 > T_1$, the rock absorbs an amount of heat given by the heat capacity of the rock, the difference $T_2 - T_1$, and the heat of reactions absorbed by the chemical processes occurring inside the rock body. The rock also absorbs or loses a comparatively small amount of mechanical energy because of the change of volume brought about by the chemical reactions and by the change of pressure.

In studying the mineral reaction and phase transitions accompanying the changes just mentioned, the concept of *molal free energy* is of the greatest significance. *Equilibrium* between several phases exists when the partial molal free energies of corresponding atomic species in the several phases have the same value, this equilibrium is *stable* if the molal free energies at the same time are at a minimum and any natural isothermal physicochemical process proceeds in such a direction that the total free energy decreases.

The mineral interactions and short-distance diffusions connected therewith are thus set in motion by the free-energy gradients created by the altered pressure and temperature affecting the rock. The partial molal free energies of the several minerals (and pore solution) change differently with pressure and temperature, and they also change with the composition of the liquid and solid solutions. Therefore, by changing pressure and temperature, some minerals will disintegrate, others will grow, and mixed crystals will change their composition in a rock undergoing metamorphism.

When undertaking a thermodynamic analysis of these processes, it is fortunately not necessary to consider the mechanism and kinetics of the processes, nor is it necessary to consider the complicated solubility of minerals in pore fluids. In most treatises on metamorphism the stability of the minerals, their disintegration, and their growth are considered with the help of their solubility in an imaginary pore fluid. The law of solubility product and the mass-action rule are the main tools applied under this analysis. It may be shown, however, that such a procedure is a very indirect one and often extremely complicated and inconvenient. Application of the free energy is much simpler, more direct, and offers better security for a correct trend in our theoretical considerations.

We have, so far, considered a rock as a closed system. The more common cases, in which rocks are to be considered as open systems capable of absorbing and expelling chemical elements, will be treated later (p. 169). We shall see that whether considerations closely parallel to those above may then be applied or not depends largely on the mechanism of the transportation of matter during rock metasomatism.

THE CONCEPT OF FREE ENERGY, A CLUE TO THE EVOLUTION OF MINERAL ASSEMBLAGES IN METAMORPHIC ROCKS

Realizing that the concept of free energy is the best and safest tool in our analysis of chemical processes in rocks, we shall need to consider this quantity in some detail. Let us consider a perfect crystal at 0° K. and zero pressure. In this crystal, all the atoms or ions are at rest in the equilibrium positions which are determined by the integrated interatomic (-ionic) forces of all the particles constituting the crystal unit and, furthermore, by the relative amount and relative radii of the several types of atoms or ions. The constituent particles are kept in their positions by the different bonding forces acting between the particles. Owing to these forces, a certain amount of energy is needed to disintegrate the crystal into its individual constituents and transfer it to a very dilute "gas." This energy of disintegration, when occurring at absolute zero and referring to 1 mol of the crystal, is called the "lattice energy," ϵ_0. Suppose that we consider two modifications of a certain compound, for example, monoclinic and rhombic enstatite, at absolute zero. These two modifications will have different lattice energies because the constituents are arranged in somewhat different ways and perhaps because the distances between the particles are not exactly the same. The modification with the greatest lattice energy is the stablest one, because here the several particles are resting at lower energy levels than in the modification with a small lattice energy.

These relationships are visualized in Figure 2, where the lattice energy of modification *1* is $\epsilon_1 = E_3 - E_1$, and of modification *2*, $\epsilon_2 = E_3 - E_2$ and in which E_3 is the energy level of the dispersed particles.

It is customary to consider the energy level of the gas as the zero level and all other levels as negative, because the energy of the disintegrated dilute gas is the highest. If, at the same time, the gas is said to have zero free energy, then the free energy of modification *1* is $-\epsilon_1$ and of modification *2*, $-\epsilon_2$, and the free-energy difference between modifications *2* and *1* is $-\epsilon_2 - (-\epsilon_1) = \epsilon_1 - \epsilon_2$. We realize that the modification with the lowest free energy is the stable one. These simple relationships between lattice energy and free energy hold only at absolute zero temperature (see below).

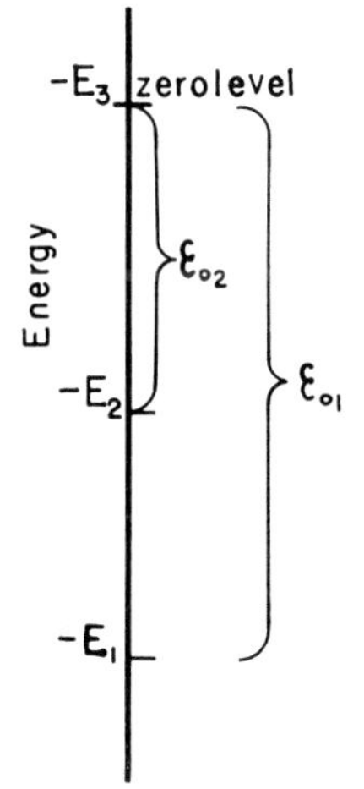

FIG. 2.—Energy levels E_1 and E_2 of two crystalline modifications, and of the corresponding ionic gas, E_3, at 0° K. $E_3 = 0$. $-\epsilon_{01}$ and $-\epsilon_{02}$ are the lattice energies of modifications 1 and 2, respectively.

Let us then assume that the crystal at absolute zero is exposed to a confining pressure P which is put upon the crystal by a semipermeable inclosure, so that the individual vaporized atoms or ions of the crystal are able to move freely through the walls of the container. If the molal volume of the crystal at 0° K. is V, then a mechanical energy equal to PV will now automatically help to disintegrate 1 mol of the crystal (which, for convenience, is supposed to be incompressible). The excess energy needed to disintegrate the crystal to the standard gas state is equal to $\epsilon_0 - PV$, where ϵ_0 is the lattice energy. With the assumption that the free energy of the standard gas is zero, the free energy of the crystal at pressure P is $-(\epsilon_0 - PV) = -\epsilon_0 + PV$.

At pressure P and temperature 0° K. the molal free energies of the two modifications previously considered are: $-\epsilon_1 + PV_1$ and $-\epsilon_2 + PV_2$, respectively, in which V_1 and V_2 are the molal volumes of the incompressible crystalline modifications. Figure 3 shows the free energies of two modifications at absolute zero at different pressures. We shall return to the relationship between pressure and molal free energy shortly, but first we shall consider some other aspects of the free energy.

The relationship between free energy and temperature is controlled by the *entropy* of the considered phase. We shall, therefore, briefly recall the

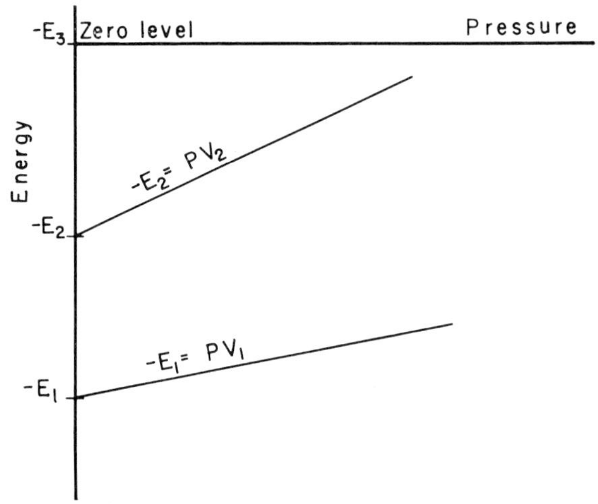

FIG. 3.—Effect of pressure on the energy of two crystalline modifications with different molal volumes at 0° K.

concept of entropy. A perfect crystal has, as a consequence of the so-called "third law" of thermodynamics, zero entropy at absolute zero. That is so because entropy is a measure of randomness or disorder among the particles constituting the phase in such a way that the more varied the states of one and the same type of particles inside the phase, or the greater the number of different types of particles existing in identical states, the greater the entropy. At 0° K., the perfect crystal of stoichiometric composition has identical particles resting at identical lattice points, and identical points are occupied by but one type of particle. There is no disorder or randomness in this situation.

With rising temperature, the entropy increases because of the randomly distributed thermal vibrations of the several constituents of the crystal. At constant pressure, the heat capacity, C_p, controls the change of entropy with changing temperature, inasmuch as we confine ourselves to the temperature range of one modification only.

If one crystalline modification of a compound is stable from 0° K. to T, then its entropy per mol at T is:

$$S_T = \int_0^T \frac{C_p}{T}\, dT\,, \tag{1}$$

where the molal heat capacity, C_p, itself varies with temperature. If phase transitions take place, such as modification changes, melting, and vaporization, the entropies of these transitions must be added; hence:

$$S_T = \int_0^{T_1} \frac{C_p}{T}\, dT + \frac{\Delta H_1}{T_1} + \int_{T_1}^{T_2} \frac{C_{p2}}{T}\, dT + \frac{\Delta H_2}{T_2} \ldots, \text{etc.}\,, \tag{2}$$

where ΔH_1, ΔH_2, etc., are the heats of transition or melting, and S_T is the entropy per mol of the phase stable at T.

Figure 4 shows some of the few known C_p versus T curves of some minerals. Table 1 gives examples of entropies at certain temperatures.

An important relation between Gibb's molal free energy, F, and pressure and temperature is combined in the second law of thermodynamics:

$$F = H - ST \tag{3}$$

or

$$F = U + PV - ST\,, \tag{4}$$

in which H is the heat content of the phase, S the entropy, V the volume, and U the internal energy (all are molal quantities), and P and T are the considered pressure and absolute temperature.

Consider, now, a crystal of n mol of potash feldspar (orthoclase), $n\mathrm{KAlSi_3O_8}$, having a total free energy F at a given P,T combination. This total free energy can be split into several partial molal free energies, $\bar{F}$, which can be assigned to each of the individual constituents in the potash feldspar crystal:

$$F^{\mathrm{or}} = n\bar{F}_{\mathrm{K}}^{\mathrm{or}} + n\bar{F}_{\mathrm{Al}}^{\mathrm{or}} + 3n\bar{F}_{\mathrm{si}}^{\mathrm{or}} + 8n\bar{F}_{\mathrm{o}}^{\mathrm{or}}\,. \tag{5}$$

For an anorthite crystal of m mol $\mathrm{CaAl_2Si_2O_8}$, we have the following relation:

$$F^{\mathrm{an}} = m\bar{F}_{\mathrm{Ca}}^{\mathrm{an}} + 2m\bar{F}_{\mathrm{Al}}^{\mathrm{an}} + 2m\bar{F}_{\mathrm{si}}^{\mathrm{an}} + 8m\bar{F}_{\mathrm{o}}^{\mathrm{an}}\,. \tag{6}$$

At chemical equilibrium between the potash feldspar and the anorthite, the following requirements must be fulfilled:

$$\bar{F}_{\mathrm{Al}}^{\mathrm{an}} = \bar{F}_{\mathrm{Al}}^{\mathrm{or}}\,; \qquad \bar{F}_{\mathrm{si}}^{\mathrm{an}} = \bar{F}_{\mathrm{si}}^{\mathrm{or}}\,; \qquad \bar{F}_{\mathrm{o}}^{\mathrm{or}} = \bar{F}_{\mathrm{o}}^{\mathrm{an}}\,. \tag{7}$$

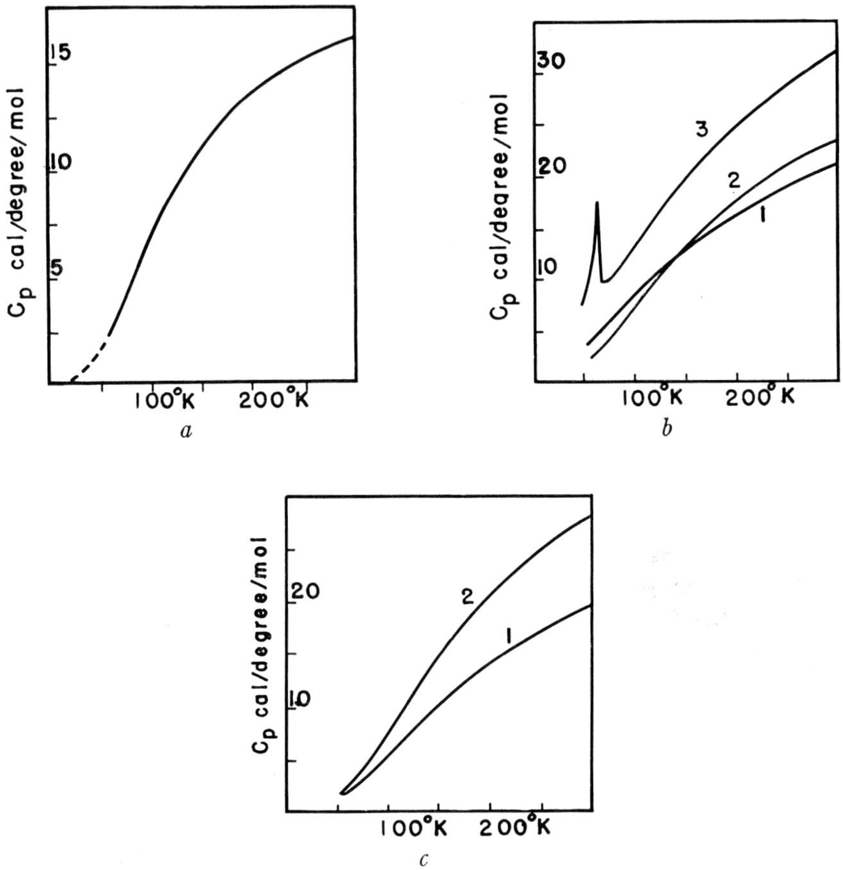

FIG. 4.—Heat capacity, C_p, of some minerals at low temperatures. a = CaF_2; b, curve *1* = $MnSiO_3$, curve *2* = $ZrSiO_4$, curve *3* = Fe_2SiO_4; c, curve *1* = $MgSiO_3$, curve *2* = Mg_2SiO_4. (After K. K. Kelley, 1941, and S. S. Todd, 1949.)

TABLE 1

ENTROPIES OF SOME MINERALS, IN CAL/MOL/DEGREE; TEMPERATURE IN °K.*

Substance	$S_{298.16}$	$S_{800°} - S_{298.16}$	$S_{800°}$	Substance	$S_{298.16}$	$S_{800°} - S_{298.16}$	$S_{800°}$
Quartz, SiO_2	10.00±0.1	13.88	23.88	Fe_2SiO_4.....	35.4 ±1.4	39.04	74.5
Tridymite..	10.36±0.15	14.00	24.35	$MnSiO_3$....	21.3 ±0.3	25.08	46.4
Cristobalite.	10.19±0.1	14.21	24.40	$ZrSiO_4$.....	20.1 ±0.3	....	....
Mg_2SiO_4....	22.7 ±0 2	.	......	CaF_2.......	16.46±0.08	18.01	34.5
$MgSiO_3$....	16.2 ±0.2	24.40	40.6				

* Data after Kelley, 1941, 1948, and 1949; Todd, 1949; Mosesman and Pitzer, 1941.

That is, the partial molal free energies of the elements which the two phases have in common must be the same for the two phases. (It is evident, also, that small amounts of K enter the anorthite lattice and that still smaller amounts of Ca go into the potash feldspar. Therefore, it is also true that $\bar{F}_{\text{K}}^{\text{an}} = \bar{F}_{\text{K}}^{\text{or}}$ and $\bar{F}_{\text{Ca}}^{\text{an}} = \bar{F}_{\text{Ca}}^{\text{or}}$ at equilibrium.)

In the case of mixed crystals or liquid solutions, the concept of partial molal free energy is very important because it can be used to evaluate the particular composition of a solution (solid or liquid) which is in chemical equilibrium with other phases (p. 22).

CHANGE OF FREE ENERGY WITH PRESSURE

It is impossible to determine the free energy of a phase in absolute values. Fortunately, that is of little consequence, since we are solely interested in differences in free energy between different phases or states. Let us suppose we have fixed the situation at which two phases (minerals) are in equilibrium; then we know that the two phases have the same molal free energies. It is evident, now, that we would be able to follow the equilibrium situation in a P,T,X-coordinate system if the change of free energy with P, T, and X was known for the two phases (X denotes bulk chemical composition of the assemblage).

For the benefit of such calculations for rock-making minerals, let us consider briefly the relation between F and pressure, temperature, and concentration. The total molal free energy of a phase changes with pressure as follows:

$$\left(\frac{\partial F}{\partial P}\right)_{T,X} = V\,, \tag{8}$$

where V is the molal volume of the phase considered and the subscript T,X indicates that T and X are kept constant.

The partial molal free energy of a particular type of atom, ion, or molecule in the phase changes with pressure as follows:

$$\left(\frac{\partial \bar{F}_i}{\partial P}\right)_{T,X} = \bar{V}_i\,, \tag{9}$$

where $\bar{V}_i$ is defined as the partial molal volume of component i in the phase. The partial molal volume of a component i in a phase is definable as the volume change which an infinitely large volume of the phase undergoes with loss or gain of 1 mol of component i. Under certain circumstances, $\bar{V}_i$ may be negative, as, for example, when hydrogen is dissolved in some metals. The term "infinitely large volume" enters the definition to secure that no change of composition takes place by the loss or gain of 1 mol of the *i*-component.

The importance of the relationship between free energy and pressure may be seen by considering the effect of pressure upon the stability of two crystalline modifications of a compound. Let us consider a certain temperature and a low pressure P_0, under which conditions modification A' is stable. The unstable modification is A''. The free energy of A', $F^{A'}$, must be less than the free energy, $F^{A''}$, of the unstable A''. If we increase the pressure on the stable as well as the unstable modifications, then $F^{A'}$ and $F^{A''}$ will rise according to equation (8) (p. 13), the slope of the F–P curves being determined by the molal volumes. Supposing that the molal

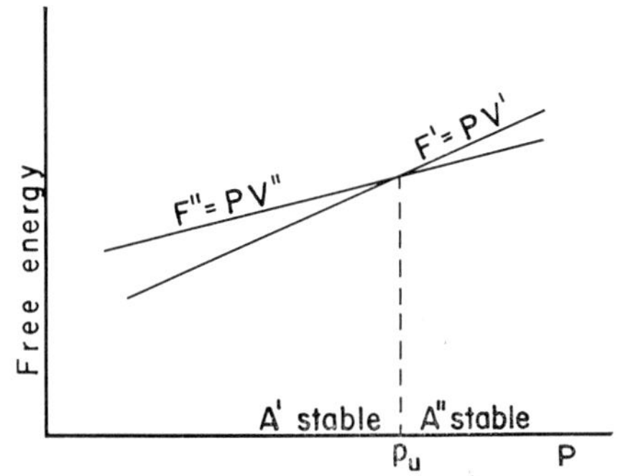

FIG. 5.—Change of free energy with pressure at constant temperature for two modifications with different molal volumes. P_u is the transition pressure at the given temperature.

volume of A'' is less than that of A', then the free energies will change with pressure, as shown in Figure 5. At a certain high pressure, A'' will be stabilized. From these considerations, it is seen that the well-known fact that dense minerals tend to form at the expense of less dense ones at high pressure is a direct consequence of the relationship between free energy, molal volume, and pressure.

THE FORCE OF CRYSTALLIZATION AS DEFINED BY THERMODYNAMICS

The relation between free energy and pressure enables us to define the so-called "force of crystallization" of minerals (Seng, 1937; Ramberg, 1947). A mineral in chemical equilibrium with its environment (at constant P and T) cannot develop a so-called "force of crystallization" or pressure of growth. The original equilibrium conditions are characterized by equal molal free energy of crystal, F_c, and of corresponding components in the surroundings, F_s, i.e., $F_c = F_s$.

According to equation (8), any increase of pressure affecting *only the crystal* will increase F_c correspondingly, so that the crystal will "dissolve" in the surroundings (which may be gases, liquids, or other crystals). If, on the other hand, the surroundings are supersaturated with respect to the mineral, i.e., when $F_s > F_c$, the mineral can grow even if the pressure acting on it is greater than that existing in the surroundings. Accordingly, the force or pressure of crystallization can be defined as the excess pressure which must be applied on the crystal in order that the crystal may be in chemical equilibrium with its surroundings, which are supersaturated to a given degree. The degree of supersaturation may most logically be measured by the difference in free energy between the saturated and the supersaturated conditions.

The force of crystallization of a mineral, therefore, can be defined as follows:

$$f_c = \left(\frac{\partial P}{\partial F}\right)_{T,X} = \frac{1}{V} = \frac{d}{M}, \qquad (10)$$

where V is the molal volume of the crystal, d its density, and M its molecular weight. We shall see that it is in harmony with the empirically established "porphyroblastic" series of minerals (p. 131). The less the molal volume or the higher the packing index (Fairbairn, 1943), the greater the force of crystallization. The force of crystallization here defined should not be identified with the ability of some minerals to form euhedral grains when growing in the solid matrix of metamorphic rocks. This latter ability depends upon a rather vaguely understood property termed "shape-" or "form-energy" of crystals.

CHANGE OF FREE ENERGY WITH TEMPERATURE

The change of free energy with temperature at constant pressure and composition is:

$$\left(\frac{\partial F}{\partial T}\right)_{P,X} = -S, \qquad (11)$$

where S is the molal entropy of the phase in question. The entropy of a phase is equal to all the heat introduced in this phase divided by the absolute temperature at which the heat has been introduced (see p. 11).

The partial molal free energy of component i in a phase changes as follows:

$$\left(\frac{\partial \bar{F}_i}{\partial T}\right)_{P,X} = -\bar{S}_i, \qquad (12)$$

where $\bar{S}_i$ is the partial molal entropy as assigned to 1 mol of i in the phase.

The fact that the slope of the free-energy–pressure curves is given by the molal volume explains why dense phases or phases with low mol volume tend to be stabilized at high pressure. In a strictly analogous way, the slope of the free-energy–temperature curves is determined by the entropy, thereby explaining why phases with high entropy are stabilized

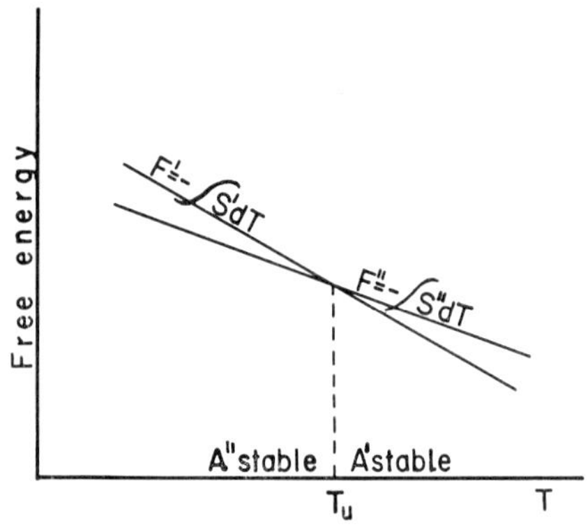

FIG. 6.—Change of free energy with temperature for two modifications at constant pressure. T_u is the transition temperature at the given pressure.

at high temperature (Fig. 6). That is, endothermic reactions take place with rising temperature.

CLAUSIUS CLAPEYRON'S AND POYNTING'S LAWS

From the equation $F = H - TS$, the following important relation can be obtained:

$$dF = V\,dP - S\,dT\ , \tag{13}$$

which combines the change of free energy with pressure and temperature changes.

This equation can be used to compute Clausius Clapeyron's law, which relates equilibrium temperature of chemical reactions to pressure and is important for petrology. Let subscript 1 denote phase 1, and subscript 2 denote phase 2. The conditions for maintenance of equilibrium are now that $dF_1 = dF_2$; consequently, the following relationship must be valid:

$$V_1\ dP - S_1 dT = V_2 dP - S_2 dT\ ,$$

$$\frac{dT}{dP} = \frac{V_1 - V_2}{S_1 - S_2} = \frac{\Delta V}{\Delta H}\,T\ , \tag{14}$$

where ΔV is volume change during the transition and $\Delta H/T = \Delta S$ is the entropy of transition. This equation is of significance when the equilibrium temperature as a function of pressure is considered (Fig. 7).

It is believed by several authorities on metamorphism that a non-homogeneous pressure may exist in solid rocks in such a way that the minerals take up most of the pressure, whereas the fluids (pore solutions),

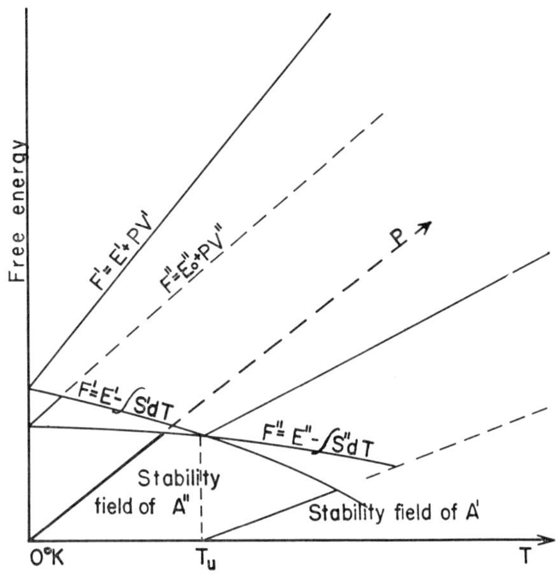

FIG. 7.—Free energy–temperature–pressure surfaces for two modifications, showing how the transition temperature is related to pressure after Clausius Clapeyron's law.

which are mobile in rocks, escape from places of high pressure. As a consequence, rising pressure should always act in such a way that the melting point decreases and the solubility in a pore fluid increases. The relation between melting point and pressure is then embodied in Poynting's law. Let V_s be the mol volume of the crystal, V_l of the melt, S_s the entropy of the crystal, and S_l the entropy of the melt. At equilibrium, we have: $V_s dP - S_s dT = V_l dP - S_l dT$. The pressure acting on the liquid is constant, hence: $V_l dP = 0$, and

$$\frac{dT}{dP} = \frac{V_s}{S_s - S_l} \tag{15}$$

or

$$\frac{dT}{dP} = \frac{V_s T}{\Delta H}, \tag{16}$$

which shows that the melting point decreases rapidly with rising pressure (Fig. 8). (Whether or not such a differential pressure can be realized, however, invites further discussion, p. 215.)

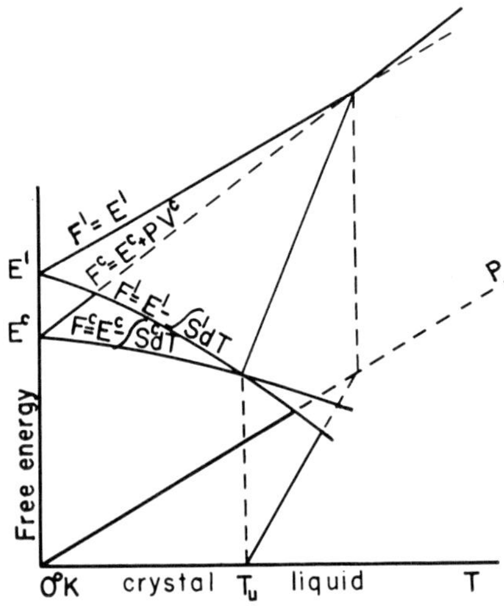

FIG. 8.—Free energy–temperature–pressure surfaces for solid and liquid in case pressure does not act upon the liquid, i.e., $PV^l = 0$. The decrease of melting temperature with increasing pressure is visualized (Poynting's law).

CHANGE OF FREE ENERGY WITH COMPOSITION OF PHASES

In an ideal[1] crystalline or liquid solution, the partial molal free energy of a constituent i changes with concentration as follows:

$$\partial \bar{F}_i = RT\partial\,(\ln X_1)\,, \tag{17}$$

where X_1 denotes the mol fraction of species i in the ideal mixture. Consider, for example, rhombic pyroxene $(Mg,Fe)SiO_3$ as an ideal mixed crystal, and suppose that at a given P,T-combination the partial molal free energy of the pure enstatite $MgSiO_3$ is F_{en}. The partial free energy of enstatite in any mixture of enstatite and ferrosilite ($FeSiO_3$) is given by the following equation:

$$\bar{F}_x = RT \ln X + F_{en}\,, \tag{18}$$

where X denotes the mol fraction $Mg/(Mg + Fe)$ in the pyroxene. Figure 9 shows the conditions graphically.

It is important to note that $\bar{F}_x \rightarrow -\infty$ when $x \rightarrow 0$ (provided that F_{en}

[1] An ideal solution exhibits no heat of mixing, and the partial molal volumes are independent of concentration.

is a finite quantity). A large negative partial molal free energy indicates a strong "affinity" for the constituent in question; hence we see that pure ferrosilite has a very strong tendency to absorb some magnesium silicate from the surrounding minerals and incorporate it as enstatite in the pyroxene lattice. This certainly does not hold only in the special case here considered; generally it is true that the end-members of mixed crystal series have a very strong tendency to incorporate some amount of the opposite "end-members" in their lattices. This tendency is clearly visualized by the relationship between free energy and mol fraction, as shown in

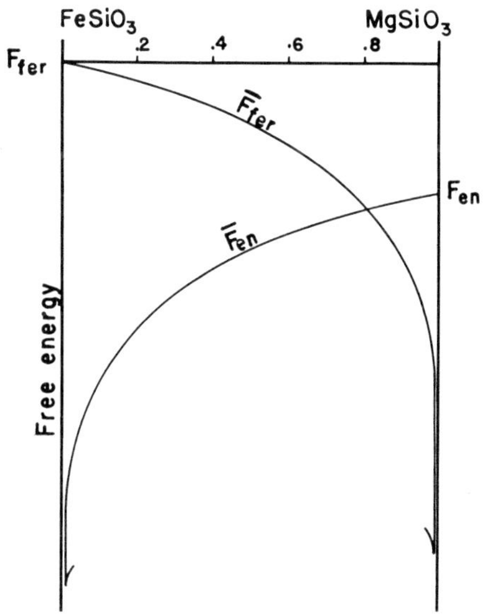

FIG. 9.—Change of partial molal free energy, $\bar{F}$, with mol fraction in the mixed crystal enstatite-ferrosilite: $(Mg,Fe)SiO_3$. The mixed crystal is supposed to be ideal.

equation (17). For these reasons the pure end-members of mixed crystal series are rare in natural rocks.

When the mixed crystals have a more complicated composition than the rhombic pyroxenes considered above, the relationship between partial molal free energy and mol fraction may be somewhat different. Let us consider olivine as another example. The chemical formula is $(Mg,Fe)_2SiO_4$. Provided that the olivine is an ideal mixture and that the Mg and Fe exist as discrete individual ions in the lattice, as we know they do, the relationship between the partial molal free energies of, say, forsterite, Mg_2SiO_4 in an olivine of intermediate composition is, to a good approximation,

$$(\bar{F}^{\text{ol}}_{\text{for}})_{T,P} = 2RT \ln X + (F_{\text{for}})_{T,P} \tag{19}$$

where $(F_{\text{for}})_{T,P}$ is the molal free energy of pure forsterite at T and P, and X is the mol fraction of magnesium silicate in the mixture.

Consider as a third example the almandine-spessartite garnet: $(Fe,Mn)_3Al_2Si_3O_{12}$. Provided that the garnet is an ideal mixed crystal, the partial molal free energy of the almandine component is:

$$(\bar{F}^{\text{gar}}_{\text{alm}})_{T,P} = 3RT \ln X + (F_{\text{alm}})_{T,P}\,, \tag{20}$$

where X denotes the mol fraction of almandine in the mixture. We see that the general rule is that the term $RT \ln X$ is multiplied by the number of exchangeable ions which constitute the chemical formula: 1 in pyroxene: $(Mg,Fe)SiO_3$; 2 in olivine: $(Mg,Fe)_2SiO_4$; and 3 in garnet: $(Fe,Mn)_3Al_2Si_3O_{12}$; etc. This rule holds only if the mixed crystals are ideal solutions and the molal free energy of SiO_2 in the crystals is independent of the Mg/Fe ratio.

Very few mixed crystals are strictly ideal solutions. Most of them show heat of mixing, so that the equations above do not hold except as an approximation when X is small. It is important to note, however, that in any thermodynamically stable mixed crystal or liquid solution the partial free energy will always *increase* when the concentration of the corresponding component increases.

In the course of the evolution of modern crystal chemistry, the rigid distinction between mixed crystals and stoichiometric crystals has almost vanished. In order to understand chemical reactions and diffusion processes in the crystalline state, one is forced to assume that most crystals can dissolve some amounts—although often only very minor percentages—of almost any element in its ionic or atomic form. Likewise, we are forced to accept the fact that the stoichiometric ratio can usually be displaced to a small extent. These apparent negligible changes of concentrations in so-called "stoichiometric" crystals are important enough to create rather drastic changes of the corresponding partial free energies.

The relation between the partial molal free energy of a certain constituent and its concentration in a mixed crystal is a measure of the tendency of the crystal to incorporate in its lattice the particular type of particle. If the partial free energy of a constituent increases rapidly with its concentration, the tendency to incorporate the element is small, and vice versa. As an example, let us consider magnetite in relation to pyroxene. In magnetite, hardly 1–2 per cent of the iron can be replaced by magnesium, whereas the Mg/Fe ratio may have any value in the pyroxene series. We shall assume now that almost pure enstatite exists in equilibrium with magnetite:

$$\underset{\text{pyroxene}}{(Mg)\,SiO_3}\,; \qquad \underset{\text{magnetite}}{(Fe,Mg)\,OFe_2O_3}\,.$$

This magnetite must have the same partial free energy of Mg as has the practically pure enstatite (pyroxene). Consequently, a change of Mg/(Mg + Fe) from 0 to some $\frac{2}{99}$ in magnetite corresponds to exactly the same change in partial free energy of Mg as does a change of the Mg/(Mg + Fe) ratio from 0 to 1.00 in the pyroxene (Fig. 10). Similar conditions must exist in several other mineral combinations (see Fig. 11).

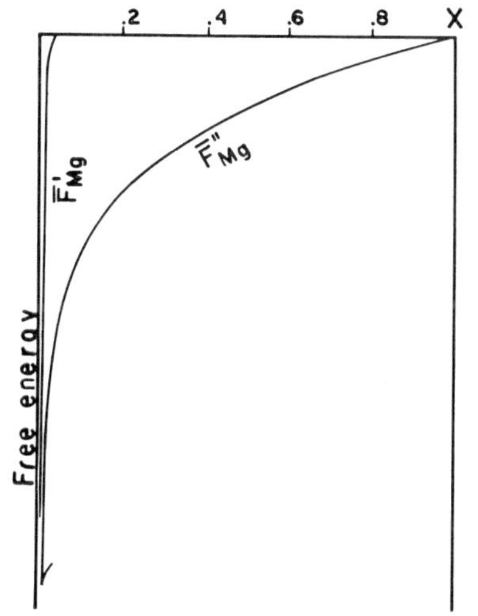

FIG. 10.—The partial molal free energies of Mg in pyroxene and in magnetite plotted against mol fraction X = Mg/(Mg + Fe). $\bar{F}'_{Mg}$ refers to magnetite, $\bar{F}''_{Mg}$ to pyroxene.

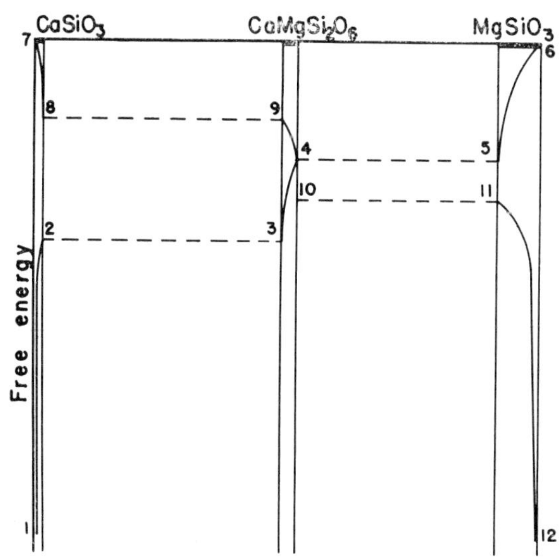

FIG. 11.—The partial molal free energies of Mg and Ca in wollastonite, diopside, and enstatite at constant P,T. The curves *1–2*, *3–4*, and *5–6* show the free energy of Mg or Mg-silicate; the curves *7–8*, *9–10*, and *11–12* show the free energy of Ca or Ca-silicate.

We have, so far, seen that the partial molal free energy changes with temperature, with pressure, and with composition of mixtures. We have already seen that a combination of the variation of free energy with pressure and temperature can give rise to the important Clausius Clapeyron law. Equally important relations can be derived if we combine the effect of composition (concentration) and temperature or pressure. Thus will be found the change of composition of a mixed crystal which is in equilibrium with other phases at changing temperature or pressure. These relationships are very important for the evolution of metamorphic mineral assemblages; we shall therefore devote the next section to these general rules.

SOME USEFUL RELATIONSHIPS BETWEEN TEMPERATURE AND PRESSURE AND COMPOSITION OF MIXED CRYSTALS IN MINERAL ASSEMBLAGES

Let us first consider the simple case of a compound, AX, which can exist in two crystalline modifications, AX' and AX'', where A is supposed to be a certain metallic ion, whereas X may be composed of any group of elements building up the rest of the crystalline structure. So, for example, in potash feldspar, $KAlSi_3O_8$, K^+ may be symbolized by A and $AlSi_3O_8^-$ by X. We shall now assume that AX' is so characterized that it is able completely to exchange A for another ion B, forming an ideal mixed crystal $(A,B)X'$. The character of the lattice of AX'' does not allow any appreciable substitution of A by B. (There are a number of mineral systems having just these characteristics; see pp. 42–75.)

Assuming constant pressure, we shall seek the relation between the temperature of reversible transition between AX'' and $(A,B)X'$ and the composition (the $A/[A + B]$ ratio) of the latter. Let the transition temperature of the inversion $AX' \rightleftarrows AX''$ be T_u° K. At this temperature $F' = F''$, where F' and F'' denote the molal free energies of AX' and AX'', respectively. At constant pressure, F'' is dependent only upon temperature, because the composition of the modification AX'' is not changeable. The free energy of 1 mol of AX' in the mixed crystal $(A,B)X'$, however, will also depend upon the $A/(A + B)$ ratio, (X), in the mixed crystal, as we have seen previously.

At the equilibrium temperature T_u, the free energy of AX' in its pure state equals the free energy of AX''. If temperature is changed to T_1, then the thermodynamic requirement for maintenance of equilibrium must be that the change of molal free energy of AX'' equals the change of the partial molal free energy of AX in the mixture $(A,B)X'$:

$$F''_{T_u} - F''_{T_1} = F'_{T_u} - \bar{F}'_{T_1}, \tag{21}$$

where F''_{T_u} and F'_{T_u} are the free energies of the two modifications in their pure state at the temperature T_u; F''_{T_1} is the free energy of modification AX'' at T_1; and $\bar{F}'_{T_1}$ is the partial free energy of AX' in the mixed crystal at T_1. Provided that the entropies are constant over the temperature range T_u–T_1, we have (see p. 15):

$$-S''(T_u - T_1) = -S'(T_u - T_1) - RT_1 \ln X , \tag{22}$$

where X is the $A/(A + B)$ fraction in $(A,B)X'$. Hence:

$$\ln X = \frac{(S'' - S')(T_u - T_1)}{RT_1} \sim \frac{\Delta H\,(T_u - T_1)}{RT_1^2} , \tag{23}$$

where $S'' - S'$ is the molal entropy difference between AX'' and AX' at the temperature T_1, and ΔH is the heat of reaction.

If $S' > S''$, the mixed crystal is the high-temperature phase (see p. 16), and the general equilibrium diagram is shown in Figure 12. If $S'' > S'$, the low-temperature modification is the mixed crystal, and the diagram is inverted (Fig. 13).

It may happen that the compound BX also exists in two modifications, BX' and BX'', and that BX' is able to exchange B with A completely, but BX'' cannot do so. Then we find that the general equilibrium diagram is entirely analogous to the common eutectic diagrams of equilibrium between liquids and crystals (Fig. 14).

The transition temperature of the end-members $AX' \rightleftarrows AX''$ and $BX' \rightleftarrows BX''$ will usually be effected by changing pressure in accordance with Clausius Clapeyron's law (p. 16), so that, in by far the most cases (Bridgman, 1946), the transition temperature will increase with pressure. (According to Slater [1939, p. 181], there is a tendency for polymorphic transition temperatures of many substances to increase about 200° with some 12,000 atm. increase of pressure.)

Figure 15 shows the general effect of pressure upon the equilibrium between the mixed crystal $(AB)X'$ and the modifications BX'' and AX''. An isothermic plane in this diagram is shown in Figure 16, where the equilibrium curves are given by the following equations:

$$\frac{d\,(\ln X)}{dP} = \frac{V_{AX''} - V_{AX'}}{RT} \tag{24}$$

and

$$\frac{d\,[\ln(1 - X)]}{dP} = \frac{V_{BX''} - V_{BX'}}{RT} , \tag{25}$$

provided that the mixed crystal is ideal to the extent that it does not exhibit heat of mixing. Here $V_{AX'}$, . . . , etc., are the molal volumes of the corresponding components. It is supposed that the partial molal volumes

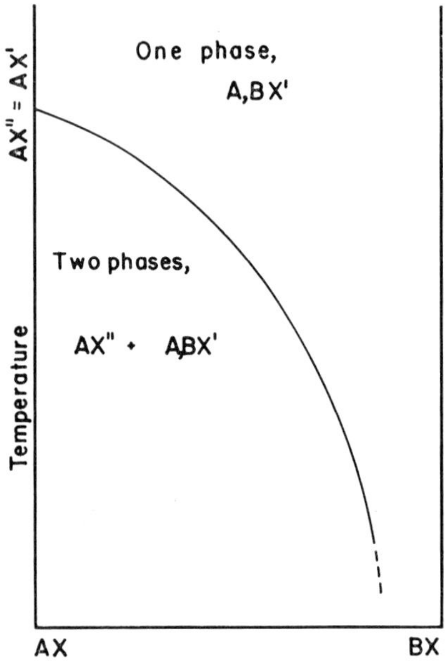

FIG. 12.—Equilibrium diagram of the crystalline compounds AX and BX, of which AX occurs in two modifications AX' and AX'', the high-temperature form of which is completely miscible with BX, the other not. The curve gives the composition of $(A,B)X'$, which is in equilibrium with AX''.

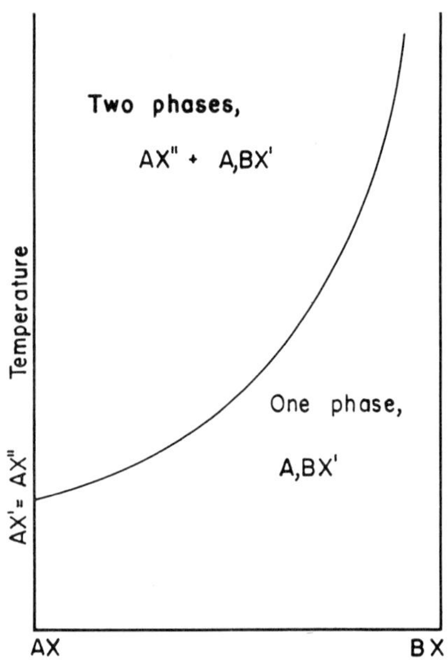

FIG. 13.—Similar situation as shown in Fig. 12, except that here the low-temperature modification of AX is completely miscible with BX, and the high-temperature form of AX is not able to accept B in its structure.

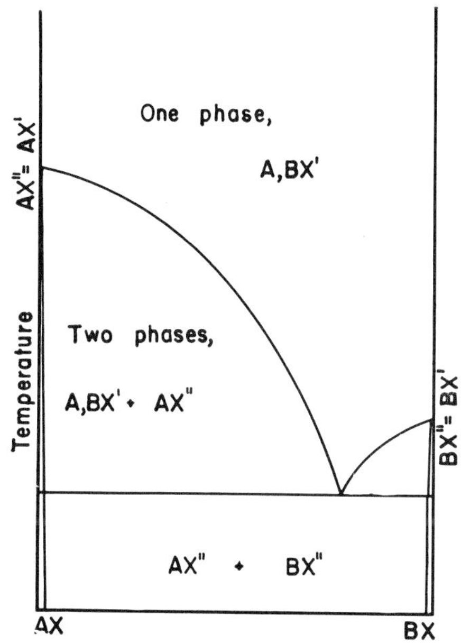

FIG. 14.—Equilibrium diagram of AX and BX when both substances occur in two modifications, the high-temperature forms of which are miscible in all proportions, and the low-temperature forms are unable to intermix.

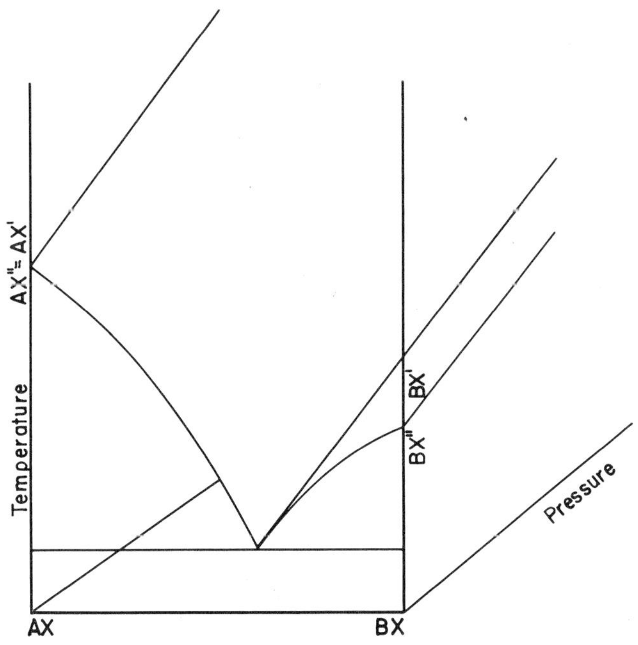

FIG. 15.—The effect of pressure on equilibrium diagram of type shown in Fig. 14

of AX' and BX' in the mixed crystal do not change with composition, so that they are equal to the molal volume of the end-members in their pure state. The compressibility is disregarded.

Among natural minerals, it is not usual to find simple modification changes of such a character that the one modification is able to substitute extensively or completely one ion, A, for another, B, at the same time as

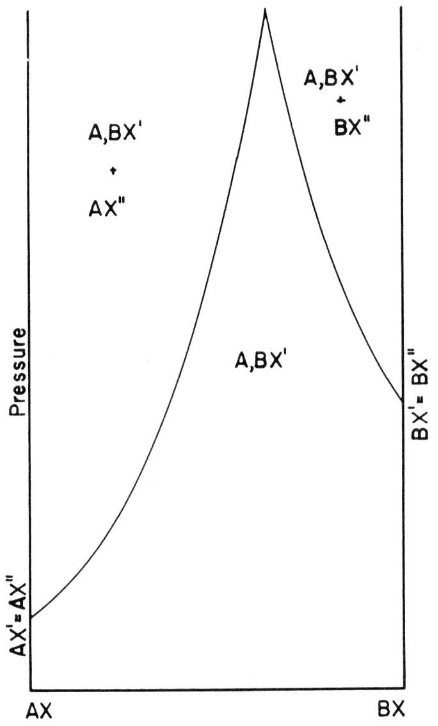

FIG. 16.—An isothermal section through Fig. 15, showing the effect of pressure upon the equilibria $(A,B)X' \rightleftarrows AX''$ and $(A,B)X' \rightleftarrows BX''$.

the other modification is practically unable to execute the same ionic substitution. One of the very few examples of this type may be provided by parts of the series of rhombic and monoclinic pyroxenes. Clino-enstatite and rhombic enstatite (or ferrosilite) are two modifications of the compound $Mg_2Si_2O_6$ of which the monoclinic modification can exchange extensively, at high temperature, the Mg^{2+} ions for the Ca^{2+} ions, whereas practically no Ca^{2+} can enter the rhombic pyroxene.

Among the rock-making minerals, it is much more common for mineral interactions of the following general types to take place:

$$kA_aX + lB_bY \ldots \rightleftarrows mA_cZ + nB_dQ \ldots, \tag{26}$$

where A and B are individual ions, whereas $X, Y, Z, Q, \ldots$, are the rest of the chemical formulae corresponding to the minerals. Very often in such a mineral reaction assembly, one of the mineral species is able to replace one type of ion, symbolized here by A (or B), with another ion, say C, which is unable to enter the lattices of any other of the minerals constituting the assembly. (Incidentally, this is an interesting fact, which shows that it is by no means only the size of the ions and their number of charge units which determine the interchanges of ions in the mineral lattices. There are less simple interatomic bonding forces which play a significant role; see p. 282.)

As an example, let us consider the following mineral reaction:

$$\underset{\text{calcite}}{6CaCO_3} + \underset{\text{talc}}{5Mg_3Si_4O_{10}(OH)_2} + 4SiO_2 \rightleftarrows \underset{\text{tremolite}}{3Ca_2Mg_5Si_8O_{22}(OH)_2} + 2H_2O + 6CO_2\,. \tag{27}$$

Among these minerals, tremolite is the only one able to substitute Mg^{2+} for Fe^{2+} to any appreciable extent.

Under these circumstances, there will be a relationship between temperature and pressure of equilibrium and the composition of the mixed crystal which is analogous to the case of simple modification transition just considered. Instead of the general reaction equation (26), let us treat the talc-tremolite example above, in order to find the relation between composition of the mixed crystal and the temperature of equilibrium.

At given partial water vapor and carbon dioxide pressures, the reaction as shown in equation (27) has a certain equilibrium temperature, T_u, at a given rock pressure. At this temperature, the sum of the free energies on either side of the equation is the same:

$$6F_{CaCO_3} + 5F_{talc} + 4F_{SiO_2} = 3F_{trem} + 2F_{H_2O} + 6F_{CO_2}\,. \tag{28}$$

Let us now change the temperature to T_1 under constant pressure and, at the same time, introduce just so much $Ca_2Fe_5Si_8O_{22}(OH)_2$ into the tremolite that equilibrium is maintained. The requirement for maintenance of equilibrium is that the free-energy changes on either side of the equation be identical:

$$-(6S_{CaCO_3} + 5S_{talc} + 4S_{SiO_2})(T_u - T_1) = -(3S_{trem} + 2S_{H_2O} + 6S_{CO_2}) \times (T_u - T_1) - 15RT \ln X \tag{29}$$

or

$$\ln X = \frac{\Delta S \cdot \Delta T}{15RT}\,, \tag{30}$$

where X is the mol fraction $Mg/(Mg + Fe)$ in tremolite, $\Delta T =$

$(T_u - T)$, and

$$\Delta S = 6S_{CaCO_3} + 5S_{talc} + 4S_{SiO_2} - (3S_{trem} + 2S_{H_2O} + 6S_{CO_2})\,.^{2}$$

Knowing from field experience that the tremolite side of the equation is the high-temperature side, we also know that ΔS is negative and that X is 1 when $\Delta T = 0$ and decreases when T decreases. Figure 33 (p. 64) shows the general equilibrium diagram in such cases.

In our particular example the effect of pressure is especially important because of the participation of the gases CO_2 and H_2O in the reaction. In a P,T,X-diagram, the equilibrium will be basically similar to Figure 15, provided, of course, that pressure acts with the same magnitude on all phases in the system. In an isothermal section through the equilibrium temperature for reaction (27) at 1 atm. pressure, one may calculate the effect of pressure upon the equilibrium at constant temperature, which we shall assume is 650° K. or just above the critical temperature of H_2O. In strict analogy to equation (24) (p. 23) we have:

$$\frac{\partial\,(\ln X)}{\partial P} = \frac{V'' - V'}{15RT}\,, \qquad (31)$$

where V'' is the sum of the molal volumes in the calcite-talc side of equation (27) and V' is the sum of the molal volumes of the tremolite-CO_2 side of the equation. Disregarding the volume of the solids on either side of the equation, $V'' - V'$ is equal to the volume of 2 mol of H_2O vapor and 6 mol of CO_2 gas. Since the water vapor is supercritical and dense, we shall disregard its volume and assume that $V'' - V' = -6$ mol volumes of gas. Hence $V'' - V' = -6\ RT/P$, according to the gas law $PV = RT$. Equation (31) then becomes:

$$\ln X = -\tfrac{6}{15}\ln P + C\,, \qquad (32)$$

where C is determined by the conditions that $X = 1$ when $P = 1$, i.e., $C = 0$. Hence:

$$X = P^{-2/5}\,. \qquad (33)$$

Figure (17) shows the equilibrium curve calculated thus in the isothermal section plane at 650° K., provided that this temperature corresponds to the reversible reaction (27) at 1 atm. pressure.

Very commonly, in the world of natural minerals, it is true that two or more end-members of a mixed crystal series undergo strictly analogous polymorphic transitions or chemical reactions. A good example is provided by the so-called "isodimorphic" series, as, for example, the series of clino-

[2] The tremolite-ferrotremolite mixture is treated as an ideal solution in order to simplify the calculations.

enstatite–clino-ferrosilite and its orthorhombic modifications. In the case of isodimorphism, we generally have the following conditions:

$$A' \rightleftarrows A'' \qquad \text{and} \qquad B' \rightleftarrows B'' ,$$

where A' and B' are isomorphous and can intermix in any ratio, and so can A'' and B''. If both series approach ideal mixed crystals, the phase diagram will be as shown in Figure 18, where the equilibrium curves can be calculated by using a method similar to that used previously (Marsh, 1935). The equilibrium curves depend upon the difference $T_A - T_B =$

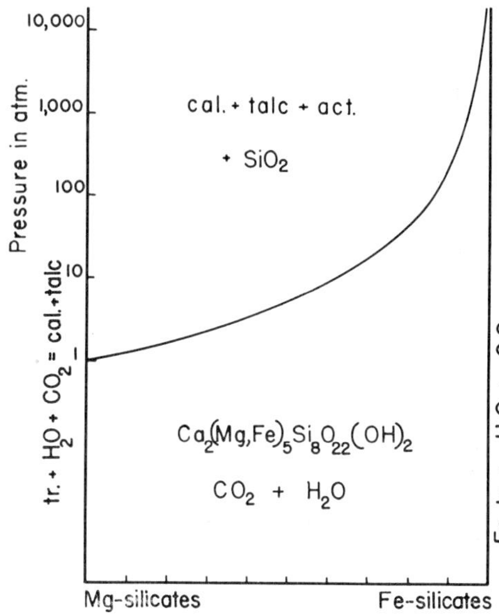

FIG. 17.—The theoretical and simplified effect of pressure upon the composition of actinolite in equilibrium with calcite, talc, H_2O, and CO_2. The curve gives the Mg/(Fe + Mg) ratio of actinolite in equilibrium with calcite, talc, and silica.

ΔT and upon the entropy differences, $\Delta S_A = S_{A'} - S_{A''}$ and $\Delta S_B = S_{B'} - S_{B''}$ in such a way that the gap between the curves will become wider, the greater the differences ΔT, ΔS_A, and/or ΔS_B. (It should be noticed that the curves do not approach a straight line between T_A and T_B when the crystals approach ideal solid solutions, as it may be tempting to assume.)

Isodimorphic series are not too common among minerals, the usual case is that analogous reactions of the following type take place:

$$\begin{aligned} &k A_a X + 1 B_b Y \ldots \overset{T_{u'}}{\rightleftarrows} m A_c Z + \ldots + n B_d q \ldots , \\ &k C_a X + 1 D_b Y \ldots \overset{T_{u''}}{\rightleftarrows} m C_c Z + \ldots + n D_d q \ldots , \end{aligned} \qquad (34)$$

where, for example, A and C are mutually substitutable in $(A,C)_aX$ and in $(A,C)_cZ$.

The following mineral reactions may provide an illustrative example:

$$\begin{aligned} &H_2O + \underset{\text{anorthite}}{2CaAl_2Si_2O_8} + \underset{\text{enstatite}}{3MgSiO_3} \rightleftarrows \underset{\text{Mg-hornblende}}{Ca_2Mg_3Al_4Si_6O_{22}(OH)_2} + SiO_2 , \\ &H_2O + \underset{\text{anorthite}}{2CaAl_2Si_2O_8} + \underset{\text{ferrosilite}}{3FeSiO_3} \rightleftarrows \underset{\text{Fe-hornblende}}{Ca_2Fe_3Al_4Si_6O_{22}(OH)_2} + SiO_2 . \end{aligned} \quad (35)$$

Figure 19 indicates the general equilibrium diagram of such a set of minerals.

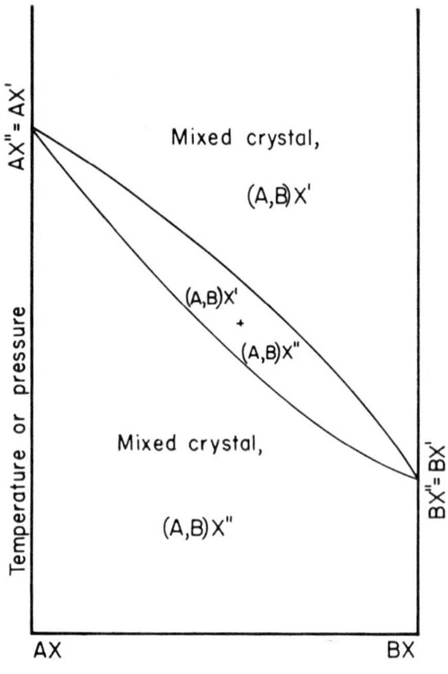

FIG. 18.—Generalized equilibrium diagram of the case $(A,B)X' \rightleftarrows (A,B)X''$, where $(A,B)X'$ is the high-, and $(A,B)X''$ the low-temperature modifications.

Usually, pressure will be of some consequence for these equilibrium conditions, inasmuch as the temperatures of equilibrium of the end-members ($T_{u'}$ and $T_{u''}$) will change (usually increase) with rising pressure. Figure 20 illustrates these conditions.

It is of particular interest to recognize that the equilibrium conditions considered so far may provide a clue to pressure and temperature determinations in rocks in which the mineral associations have reached approximate mutual equilibrium. If pressure is known, the temperature at which equilibrium was obtained in a certain rock is directly given by the composition of one or more mixed crystals existing in a set of minerals corresponding to equations (26) and/or (34). If the pressure is not known, it is evident that this particular composition may correspond to any temperature within certain limits. Suppose now, however, that in the same rock there exists another set of minerals of the general type given by equation (26) or equation (34); then these two sets of minerals may, in combination, fix pressure as well as temperature at the same time. The reason for this is that a given composition of a mixed crystal in the above-mentioned

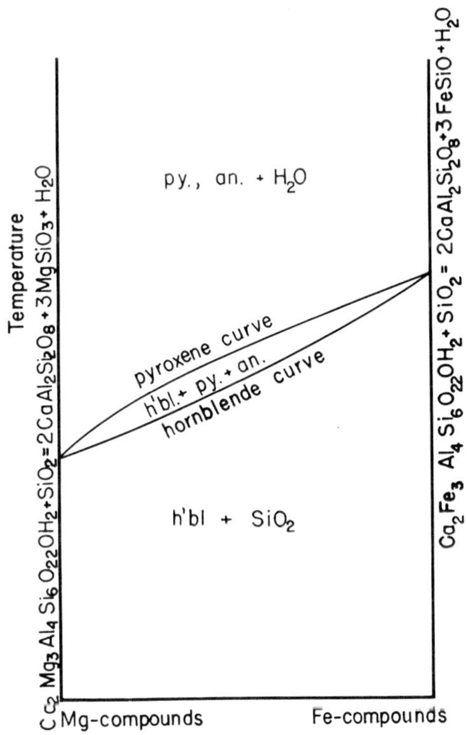

FIG. 19.—Simplified equilibrium diagram, showing the Mg/(Mg + Fe) ratio in aluminous hornblendes and in coexisting orthopyroxenes when "transition" takes place according to eq. (35) (p. 30). Based upon field data.

set of minerals is determined by P,T-combinations following a certain curve in the P,T-diagram. Two mineral sets give two curves which may intersect, thus determining pressure as well as temperature. A knowledge of a sufficient number of thermodynamic data of rock-making minerals will indeed enable us to read temperature and pressure directly out of the mineral assemblages of rocks.

Up to now, we have considered only the equilibrium conditions of mixed crystals within the pressure and temperature interval in which phase transitions take place, i.e., within the temperature interval $T_{u'} - T_{u''}$ (Figs. 12–14, 18, and 19).

It is also of importance to consider the equilibrium relations between mixed crystals outside the field of true phase transitions. Let us, to this end, consider stable coexisting olivine and hypersthene (Ramberg and DeVore, 1951). Provided that tridymite (or α-quartz) is also a member of this stable assembly, the distribution of Mg^{2+} and Fe^{2+} in the coexisting olivine and hypersthene is determined by the equilibrium curves in the system: $(Mg,Fe)_2Si_2O_6 \rightleftarrows (Mg,Fe)_2SiO_4 + SiO_2$ (Fig. 21). At a giv-

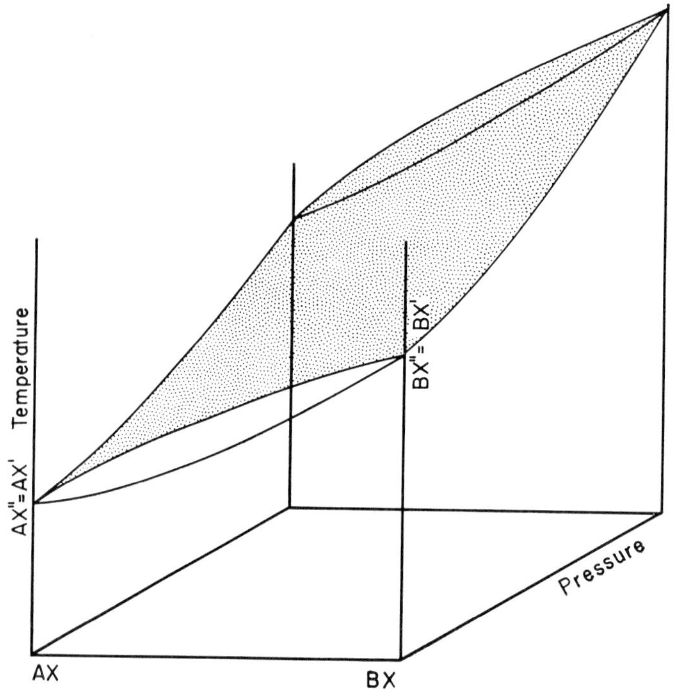

FIG. 20.—The effect of pressure upon the equilibrium of type $(A,B)X' \rightleftarrows (A,B)X''$

en pressure, the only possible area of stable coexistence of olivine, hypersthene, and SiO_2 is the gap between the two equilibrium curves. Below the hypersthene curve, olivine cannot coexist with SiO_2 (tridymite or quartz) in a stable situation. On the other hand, if free silica is not present, there is nothing unstable about the presence of olivine below the hypersthene curve in Figure 21. The question is how magnesium and iron will distribute themselves between coexisting olivine and pyroxene. Let us assume that equilibrium is attained between olivine, hypersthene, and silica at a temperature T, at which olivine of composition o exists, together with pyroxene of composition p. At this point we shall remove every trace of free silica and let the temperature drop a small interval, ΔT.

The only possible process which may now take place is that some Fe^{2+} ions go from olivine to hypersthene or vice versa, at the same time as the same quantity of Mg^{2+} ions goes in the opposite direction. For the sake of continuity, one must believe that this possible change of the Mg/Fe ratio in olivine and pyroxene takes place along smooth curves down from points o and p in Figure 21.

The exchange of Mg^{2+} and Fe^{2+} ions may be identified with a sort of circular reaction in such a way that enstatite is formed from forsterite and silica at the same time as ferrosilite splits up into fayalite and silica,

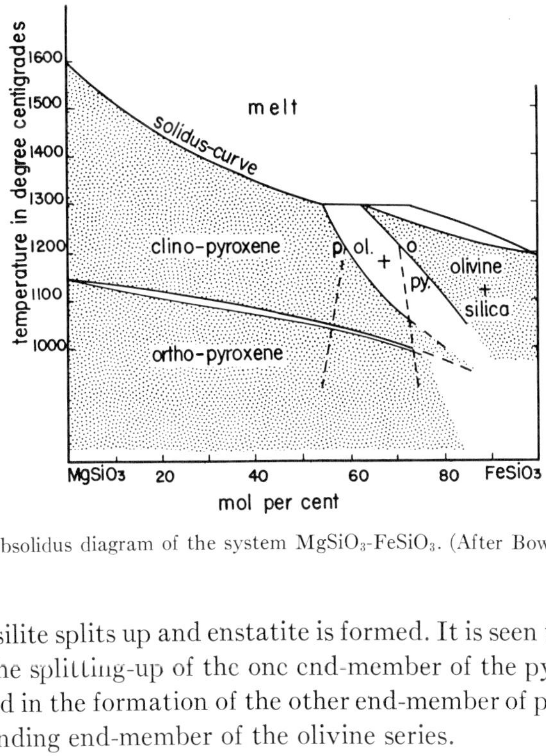

FIG. 21.—Subsolidus diagram of the system $MgSiO_3$-$FeSiO_3$. (After Bowen and Schairer, 1935.)

or that ferrosilite splits up and enstatite is formed. It is seen that the silica released in the splitting-up of the one end-member of the pyroxene series is totally used in the formation of the other end-member of pyroxene from the corresponding end-member of the olivine series.

$$\underset{\text{olivine}'}{(Mg_{X'_o}Fe_{1-X'_o})_2SiO_4} + 2\,\underset{\text{pyroxene}'}{(Mg_{X'_p}Fe_{1-X'_p})SiO_3} \overset{T'}{\underset{T''}{\rightleftarrows}} \underset{\text{olivine}''}{(Mg_{X''_o}Fe_{1-X''_o})_2SiO_4} + 2\,\underset{\text{pyroxene}''}{(Mg_{X''_p}Fe_{1-X''_p})SiO_3}\,. \tag{36}$$

At T', olivine′ with a mol fraction of Mg_2SiO_4 equal to X'_o is in equilibrium with pyroxene with a mol fraction of $MgSiO_3$ equal to X'_p. Certainly, at this temperature, the mol fraction X'_o may change within wide limits, accompanied by definite changes of the mol fraction X'_p of the coexisting pyroxene. At T'', an olivine of composition X_o'' is in equilibrium with a pyroxene of composition X_p''. Also, at this temperature, X_o'' may change within wide limits, and now the composition X_p'' of the coexisting pyroxene also follows the changing X_o'' according to certain rules.

With the aid of the general thermodynamic equations, it can be demonstrated that the following relationship holds between temperature and composition of coexisting olivine and pyroxene, provided that both series are ideal:

$$\frac{X_p}{1-X_p}\cdot\frac{1-X_o}{X_o}=\left(\frac{X_p'}{1-X_p'}\cdot\frac{1-X_o'}{X_o'}\right)^{T'/T}\cdot e^{\Delta S(T-T')/RT}, \qquad (37)$$

where X_p and X_o are the mol fractions of Mg-silicate in pyroxene and olivine, respectively, at $T°$ K.; X_p' and X_o' are the same at $T'°$ K.; and $\Delta S = \frac{1}{2}(S_{\text{for}} - S_{\text{fay}}) - (S_{\text{en}} - S_{\text{fer}})$ where S_{for}, S_{fay}, S_{en}, and S_{fer} are the molal entropies of forsterite, fayalite, enstatite, and ferrosilite between $T°$ and $T'°$ K.

Equation (37) gives the effect of temperature upon the distribution of Mg and Fe in coexisting pyroxene and olivine in quartz-free rocks. If temperature is kept constant and pressure changes, the Mg-Fe distribution will also change. Equations (8) and (17) (pp. 13 and 18) enable us to calculate the following relationship:

$$\frac{X_p}{1-X_p}\cdot\frac{1-X_o}{X_o}=\frac{X_p'}{1-X_p'}\cdot\frac{1-X_o'}{X_o'}\cdot e^{\Delta P\Delta V/RT'}, \qquad (38)$$

where $\Delta V = V_{\text{for}} - 2V_{\text{en}} - (V_{\text{fay}} - 2V_{\text{fer}})$ and $\Delta P = P - P'$, where P is the pressure acting on the minerals and P' is the pressure at which the distribution function is determined to be $X_p'/(1 - X_p') \cdot (1 - X_o')/X_o'$.

The effect of pressure (at constant temperature) upon the Mg/(Fe + Mg) distribution between olivine and hypersthene in the absence of free silica is governed by the ΔV as defined above. The ΔV is a measure of the difference in tightness of packing of the iron- and the magnesium-olivine as compared with the iron- and magnesium-pyroxene.

The implication of these distributions of exchangeable elements in minerals and their relations to pressure and temperature is of great interest for metamorphic petrology. Actually, any pair of coexisting mixed crystals containing the same exchangeable ions is subjected to mutual ionic interchanges similar to those expressed by the equations above.

A subsolidus equilibrium diagram considerably different from the previous ones will now be discussed briefly. We shall consider the assemblage pyroxene, geikielite ($MgTiO_3$), and ilmenite ($FeTiO_3$) which was studied by Sahama and Torgeson (1949). In pyroxene, iron and magnesium are completely miscible in all proportions. Geikielite, however, is unable to take appreciable amounts of iron in place of magnesium; and ilmenite is not likely to take much magnesium in its structure.

It has been established by free-energy calculations that geikielite and ferrosilite represent an unstable assemblage at low temperature (below some 500° C.), so that the following reaction takes place:

$$\underset{\text{ferrosilite}}{FeSiO_3} + \underset{\text{geikielite}}{MgTiO_3} \rightarrow \underset{\text{enstatite}}{MgSiO_3} + \underset{\text{ilmenite}}{FeTiO_3} \,. \tag{39}$$

As mentioned above, geikielite and ilmenite do not form mixed crystals. In an attempt to evaluate the equilibrium conditions in this system, one may reason as follows:

The enstatite formed by the reaction between geikielite and ferrosilite will not occur in its pure state but must mix with ferrosilite, forming $(Mg,Fe)SiO_3$. Thereby, the molal free energy of ferrosilite in the pyroxene must drop a certain amount at the same time as the molal free energy of $MgSiO_3$ in the pyroxene phase will increase somewhat. If there is an excess of geikielite present, a point will be reached at which the partial molal free energy of $FeSiO_3$ in pyroxene is so depressed that no reaction between geikielite and ferrosilite takes place. At the same time, the partial molal free energy of $MgSiO_3$ in the pyroxene has risen to such a value that ilmenite and $MgSiO_3$ are on the threshold of interacting to form geikielite and ferrosilite. Then there is chemical equilibrium among geikielite, ilmenite, and pyroxene of a certain composition. The calculation of the composition of the pyroxene in equilibrium with ilmenite and geikielite at different temperatures may be undertaken as follows:

Suppose that we start with a temperature T' at which pyroxene of a certain mol fraction, X', is in equilibrium with ilmenite and geikielite. That is, reaction (39) has passed to the right until equilibrium is reached. Then the following relations must be fulfilled:

$$F'_f + RT' \ln(1 - X') + F'_g = F'_e + RT' \ln X' + F_i \,, \tag{40}$$

where F_f, F_g, F_e, and F_i are the molal free energies of ferrosilite, geikielite, enstatite, and ilmenite at T' degrees and X' is the mol fraction of enstatite in the pyroxene. At another temperature, T, equilibrium is attained if X' has changed to X, and the following conditions must be fulfilled:

$$F'_f - S_f\Delta T + RT \ln(1 - X) + F'_g - S_g\Delta T = F'_e - S_e\Delta T + RT \ln X + F'_i - S_i\Delta T \,, \tag{41}$$

where S_f, S_g, etc., are the entropies of ferrosilite, geikielite, etc., and

ΔT is $T - T'$. Equations (40) and (41) give:

$$\ln \frac{1-X}{X} = \frac{T'}{T} \ln \frac{1-X'}{X'} \cdot \frac{(T-T')(S_f+S_g-S_e-S_i)}{RT} . \qquad (42)$$

If one point on the equilibrium curve, X' and T', is fixed experimentally, it is possible to determine any other point on the equilibrium curve or the corresponding X and T at any temperature at which the entropies are

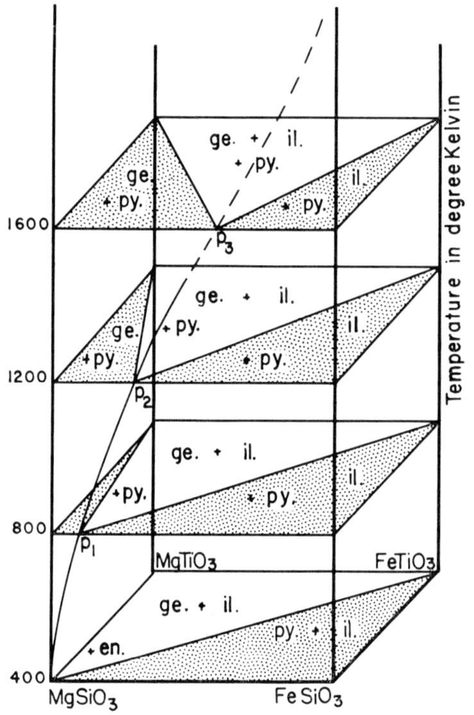

FIG. 22.—The equilibrium relationships of geikielite, ilmenite, and pyroxene. In the triangles *il.* + *py.*, ilmenite and pyroxene of any composition from $FeSiO_3$ to P_1, P_2, or P_3 can exist. In the triangles *ge.* + *py.*, the pyroxene can have any composition varying from $MgSiO_3$ to P_1, P_2, or P_3. In the triangles *ge.* + *il.* + *py.*, however, the composition of the pyroxene is fixed at any temperature to the curve through P_1, P_2, or P_3. Temperature in degrees Kelvin. (Data after Sahama and Torgeson, 1949.)

known. The principal trend of the equilibrium curve is shown in Figure 22.

We have so far studied equilibrium conditions among ideal mixed crystals under changing pressure and temperature. Because of heat of mixing, real crystals deviate more or less from the ideal mixed crystals. The real crystals may deviate so severely from ideality that unmixing occurs in the intermediate compositions. The thermodynamic reason for such an unmixing is found in the fact that the partial molal free energies of the components of the crystal do not increase with increasing concentration of the respective components, but, on the contrary, increasing concentration causes a decrease in the partial free energies. Figure 23 shows the situation of unmixing in view of the partial free energies.

Petrological applications of the matter discussed above are found in later chapters.

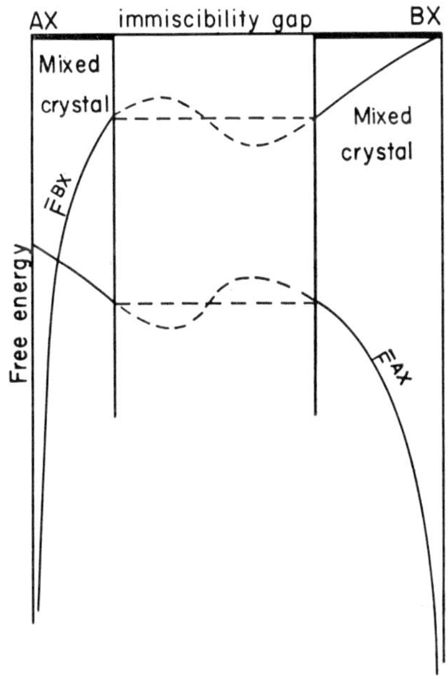

FIG. 23.—The partial molal free energies of a mixed crystal series with immiscibility gap. The dashed curves are unstable.

THE RELATIONSHIP BETWEEN FREE ENERGY AND POSITION OF A PHASE IN THE EARTH'S GRAVITATIONAL FIELD

It is of particular significance for some aspects of petrology to note that the free energy of a phase depends upon the level at which the phase is situated in the gravitational field, if pressure, temperature, and composition of the phase are constant. The higher up in the gravitational field a phase is situated, the higher its free energy. Since transportation of matter in the earth tends to equalize the partial free energies (see the section on granitization, p. 238), the said relationship plays some role in the evolution of the metasomatic rocks. Denoting the height in the gravitational field, h, and the molecular weight of the phase, M, we have:

$$\left(\frac{\partial F}{\partial h}\right)_{T,\,P,\,X} = Mg\,, \qquad (43)$$

where g is the acceleration due to gravity and h is positive upward. That is: if two bodies of the phase in question (liquid or solid) are separated by a vertical distance Δh, then the molal free-energy difference between them is $Mg\Delta h$, provided that T, P, and X are constant. This means again that if the bodies were inclosed in some solvent which acts as a leader for the dissolved constituents of the phase, then the uppermost body will dissolve, the solution will migrate downward, and the lowermost body will grow (one should note here that the conveying medium, solvent, is supposed to have a smaller density than either of the two bodies in question).

Let us now assume that the condensed phase, which is of a stoichiometrically fixed composition, is situated as a continuous concentric layer in the gravitational field. In this situation the pressure cannot be the same from top to bottom if hydrostatic equilibrium be maintained and no bodily movement occurs. However, temperature can be kept constant, and the composition is also constant by definition of a stoichiometrically composed substance (see above). Hence the change in free energy through the vertical direction of the layer is given only by the change in heights and pressure:

$$dF = \left(\frac{\partial F}{\partial h}\right)_{T,P,X} dh + \left(\frac{\partial F}{\partial P}\right)_{T,X,h} dP . \tag{44}$$

Furthermore, we have:

$$dP = -d\,g\,dh\,; \qquad d = \frac{M}{V}\,; \qquad \left(\frac{\partial F}{\partial P}\right)_{T,X,h} = V\,,$$

and

$$\left(\frac{\partial F}{\partial h}\right)_{T,P,X} = M g\,,$$

where d is the density, V the mol volume, M the molecular weight, and g the acceleration due to gravity. Hence equation (44) becomes:

$$dF = M g dh - M g dh = 0\,. \tag{45}$$

One finds, therefore, that, in a *freely resting homogeneous stable layer in the gravitational field, the total molal free energy is the same from top to bottom.*

The situation will be different, though, if the composition of the layer is changeable as it is in a magma or in a mixed crystal layer. In this case, we need to consider the partial molal free energies of the several exchangeable constituents of the mixture. The requirement for thermodynamic equilibrium from top to bottom will now be that the partial free energies do not change along the vertical direction. Usually this requisite is followed by a consequent change in relative amount of exchangeable constituents. Let us consider the mixture (A,B) as an example.

$$\left(\frac{\partial \bar{F}_A}{\partial h}\right)_{T,P,X} = M_A g \quad \text{and} \quad \left(\frac{\partial \bar{F}_B}{\partial h}\right)_{T,P,X} = M_B g\,, \tag{46}$$

give the change of $\bar{F}_A$ and $\bar{F}_B$ with heights at constant T, P, and X. The requirements for chemical equilibrium along the vertical axis are

$$\frac{d\bar{F}_A}{dh} = 0, \quad \text{and} \quad \frac{d\bar{F}_B}{dh} = 0. \tag{47}$$

Furthermore, temperature must be constant throughout, and hydrostatic pressure must prevail. But the composition, X, is changeable; hence we have:

$$\begin{aligned} a)\ \ d\bar{F}_A &= \left(\frac{\partial \bar{F}_A}{\partial h}\right)_{T,P,X} dh + \left(\frac{\partial \bar{F}_A}{\partial P}\right)_{T,h,X} dP + \left(\frac{\partial \bar{F}_A}{\partial X}\right)_{T,P,h} dX = 0, \\ b)\ \ d\bar{F}_B &= \left(\frac{\partial \bar{F}_B}{\partial h}\right)_{T,P,X} dh + \left(\frac{\partial \bar{F}_B}{\partial P}\right)_{T,h,X} dP + \left(\frac{\partial \bar{F}_B}{\partial X}\right)_{T,P,h} dX = 0, \end{aligned} \tag{48}$$

where X denotes the mol fraction of A in (A,B). The following additional conditions are also valid:

$$dP = -d_{AB}\ g dh; \quad d_{AB} = \frac{M_{AB}}{V_{AB}}; \quad \left(\frac{\partial \bar{F}_A}{\partial P}\right)_{T,h,X} = \bar{V}_A;$$

$$\left(\frac{\partial \bar{F}_B}{\partial P}\right)_{T,h,X} = \bar{V}_B; \quad \left(\frac{\partial \bar{F}_A}{\partial h}\right)_{T,P,X} = M_A\ g;$$

$$\left(\frac{\partial \bar{F}_B}{\partial h}\right)_{P,T,X} = M_B\ g; \quad \left(\frac{\partial \bar{F}_A}{\partial X}\right)_{T,P,h} = \frac{RT}{X},$$

where the subscript A refers to the component A, subscript B to component B, and subscript AB to the mixture AB. Introducing these relations into equation (48*a*) gives:

$$d(\ln X) = \frac{(\bar{V}_A[M_{AB}/V_{AB}] - M_A)\,g}{RT}\,dh, \tag{49}$$

showing that the stable X will increase with rising heights if $M_A/V_A < M_{AB}/V_{AB}$ and decrease with rising heights if $M_A/V_A > M_{AB}/V_{AB}$. That is, in a homogeneous mixture layer in thermodynamic stability in the gravitational field, light constituents with large partial mol volume will concentrate toward the top, whereas heavy constituents with small mol volume will concentrate at the bottom.

The conditions here considered may also lead to the development of new phases in the upper and/or lower parts of the original layer, because the concentration changes may surpass those which are permitted by the homogeneity field of the original layer. These new phases, which may be stabilized, will usually have a bulk chemical composition different from the original layer, so that chemical migration up and down in the earth's crust creates a differentiation of the crust in an attempt to approach thermodynamic equilibrium. There are strong indications that parts of the differentiation of the earth's crust actually follow these thermodynamic laws (see p. 261). For further information the reader is referred to Gibbs, 1928; Ramberg 1944*b*, *c*–48; Verhoogen, 1949; Brewer, 1951.

MINERALS IN EQUILIBRIUM WITH HYDROUS SOLUTIONS

Water plays an important role in the formation of metamorphic minerals, acting as a solvent and a catalyst. We shall therefore discuss briefly some general laws of solutions. Only a few natural minerals are appreciably soluble in water, which is natural enough because otherwise the geochemical reservoir of their elements would not have been the rocks but the oceans. Only a few data as to the solubility of common rock-forming minerals are known. Only in rather strong alkaline solutions (water glass) and hydrofluoric acid solutions is the solubility of silicates, aluminates, oxides, and sulfides, which build up the natural rocks and ore bodies, greater than a small fraction of 1 per cent. The hydrous solutions existing in metamorphic rocks, therefore, do not function as a reservoir of mineral-building elements but merely as a conductor of material from one mineral to the other. The thermodynamic parameters (free energies, etc.) of these solutions are not too significant, because the solutions simply reflect the free energies of the minerals with which they almost always are in close equilibrium. Hence, provided that the thermodynamic properties of the minerals are known, it is usually unnecessary to consider the concentrations in the solutions at all, as long as the main interest is not concerned with the velocity of the reactions and their kinetics. A few words will nevertheless be said about the thermodynamics of such solutions.

At a given temperature, the saturated hydrous solution of, say, potash feldspar has a certain concentration of the several constituents of the mineral. The ions K^+, $Al(OH)^{2+}$, Al^{3+}, SiO_4^{4-}, and several other combinations of ions exist in the solution. When equilibrium is attained, the partial molal free energies of K^+, Al^{3+}, Si^{4+}, and O^{2-} in the feldspar are equal to those of the corresponding constituents in the solution.

THE MASS-ACTION RULE

The mass-action law is widely applied when metasomatism of rocks is studied (Goldschmidt, 1921; Eskola, 1939). When a stoichiometric mineral like enstatite $MgSiO_3$ dissolves in a hydrous solution, the product of the concentrations (or, more correctly, the activities) of the several dissolved ions remains constant at constant P and T, provided that the solution is saturated. If we suppose that the enstatite dissociates as follows: $MgSiO_3 \rightarrow Mg^{2+} + SiO_3^{-}$, then the product of the concentrations should be constant at constant P and T:

$$C_{Mg^{2+}}C_{SiO_3^{2-}} = \text{Constant}\,, \tag{50}$$

where C denotes concentration in the solution. For more complex minerals the conditions are similar. Olivine, for example, may dissociate as follows: $Mg_2SiO_4 \rightarrow 2Mg^{2+} + SiO_4^{4-}$, in which case $C^2_{Mg^{2+}}C_{SiO_4^{4-}}$ should be constant at given P and T conditions.

The chief petrologic importance of this law is implied in the possibility of dissolving a mineral in a solution by withdrawing but one of its constituents from the solution, or, vice versa, to precipitate a mineral by adding an excess of only one or a few of the constituents of this mineral.

Instead of stating that the product of the concentrations (or some power of the concentrations, depending upon the formula of the mineral we consider) should remain constant, one may also state that the sum of the partial molal free energy remains constant at constant P and T by reason of the logarithmic relationship between free energy and concentration (p. 18).

Consider again the dissociation of enstatite:

$$MgSiO_3 \rightarrow Mg^{2+} + SiO_3^{2-}\,.$$

We shall call the partial molal free energy of Mg^{2+} in the saturated solution, $\bar{F}_{Mg^{2+}}$, and that of SiO_3^{2-}, $\bar{F}_{SiO_3}$. It can then be shown that equation (51) is valid for saturated solutions:

$$\bar{F}_{Mg^{2+}} + \bar{F}_{SiO_3^{2-}} = F_{enst} = \text{Constant}\,. \tag{51}$$

Hence we see that in a saturated ideal solution the sum of the partial molal free energies is constant at constant P and T. In more complex silicates, like Mg_2SiO_4, we have:

$$F_{ol} = 2\bar{F}_{Mg^{2+}} + \bar{F}_{SiO_4^{4-}} = \text{Constant}\,, \tag{52}$$

or, for $Mg_3Al_2Si_3O_{12}$ (pyrope):

$$F_{pyr} = \bar{F}3_{Mg^{2+}} + 2\bar{F}_{Al^{3+}} + 3\bar{F}_{SiO_4^{4-}} = \text{Constant}\,. \tag{53}$$

That is, if the partial molal free energy of one type of particle increases in the saturated solution, the free energy of the other particles constituting the precipitate must decrease correspondingly, in order to maintain saturation equilibrium. The constants F_{enst}, F_{ol}, and F_{pyr} will usually vary with temperature and pressure, depending upon the heat of solution and upon the difference in mol volumes between the solute in the solution and the precipitate.

EQUILIBRIUM DIAGRAMS OF METAMORPHIC MINERALS

THERE now exists a large number of empirical equilibrium diagrams of silicate systems at melting conditions, thanks chiefly to the work of Bowen and his school. These diagrams have helped greatly to place igneous petrology on a sound and fruitful level. So far, no similar systematic exploration of the subsolidus equilibrium conditions of metamorphic mineral systems has been undertaken. This is doubtless due to the almost impassable difficulties in synthesizing many metamorphic minerals. Very little is known about the temperature of reaction and transition of most metamorphic minerals, and almost nothing is definitely known about the change of composition which mixed crystals undergo during change in pressure and/or temperature.

Complete information about the equilibrium conditions of metamorphic minerals would add greatly to our understanding of metamorphism in general. The best we can do at the present, however, is to organize our somewhat vague experience from natural mineral assemblages and parageneses and compile these results into equilibrium diagrams, which, unfortunately, are fated to be of a rather approximate and qualitative nature. To secure comparatively reliable results, we shall, at each step of our compilation, follow the general laws of thermodynamics discussed previously.

As the following diagrams are based mainly on the study of mineral assemblages in metamorphic rocks, the temperature and pressure scales of the diagrams are of a relative nature. We have but few fixed points on the pressure and/or temperature axis of our diagrams. We do have, however, fairly good information about the relative P,T-conditions involved. This information about the relative P,T-conditions is based on observations of metamorphic rocks and has, in the last few decades, been compiled on the facies principle of Eskola. According to Eskola's idea, one facies of rocks represents rocks formed or recrystallized within a certain P,T-interval confined by some crucial mineral reaction "points," or curves. The higher the facies, the higher the P,T-conditions (see p. 135). The sequence of regional metamorphism is as follows: (1) green schist facies, (2) epidote-amphibolite facies, (3) amphibolite facies, (4) granulite facies (Fig. 73, p. 137). The diagrams in the following section have ordinates denoting increasing facies. The reader should be aware of the necessary simplifications and inaccuracies involved in the several diagrams.

SOME SINGULAR SYSTEMS

In a thermodynamically stable mineral assemblage, each mineral species must also be stable in itself when separated from the original assemblage and considered as an individual system. It may, therefore, be convenient to consider, first, some separate minerals and their stability at changing P and T; such a separate mineral is certainly a one-component system within certain P,T-limits.

We shall find, however, that the stability conditions of isolated minerals are not of much significance for metamorphic petrology, because polymorphic transitions and simple chemical splitting-up of minerals are not too common phenomena among natural silicates. Usually, chemical interactions in which several mineral species are involved take place. As a consequence, polycomponent systems must usually be considered.

SILICA

As silica occurs in several modifications, it offers a simple, but instructive, one-component system. Only the α- and β-quartz and perhaps tridymite are of significance for the study of metamorphism. The volume difference and entropy of transition between α-quartz and tridymite are so related that rising pressure rapidly elevates the transition temperature beyond that of even a high degree of metamorphism (Fig. 24). Only in the case of low-pressure pyrometamorphism connected with inclusions in lavas and shallow magmas would tridymite be able to form under equilibrium conditions by recrystallization. (Cristobalite recorded from low-grade sediments, chert, is always formed metastably.)

The transition α–β-quartz is kinetically of a fast-running displacive type (Buerger, 1948); so α-quartz cannot persist as an unstable relict below the transition curve. All quartz in metamorphic rocks—and magmatic, so far—is consequently β-quartz, which only by special investigations (Wright and Larsen, 1909; Sosman, 1927; Tuttle, 1949*b*) can be identified as having originally formed as high quartz or grown directly as low quartz.

FELDSPARS

Potash feldspar can occur in several modifications which still are only incompletely known regarding structure as well as stability (Laves, 1950, 1952). It seems justifiable to state, however, that the triclinic microcline and the monoclinic orthoclase are actually two different modifications with different unit cells, thus evidently having somewhat different lattice energies and stability ranges. Although the transition P,T-curve between orthoclase and microcline is not known, our experience from rocks indicates that orthoclase is stable at high temperatures and microcline at medium and lower temperature. Studies of rocks formed in granulite facies (Holland, 1900; Groves, 1935; Ramberg, 1949) also show that orthoclase forms at high pressure, provided that the temperature is high enough. (It is unlikely that orthoclase is formed metastably in these thoroughly recrystallized rocks.)

Independent of these uncertainties, it is evident that potash feldspar, in one modification or another, is physically stable over the entire range

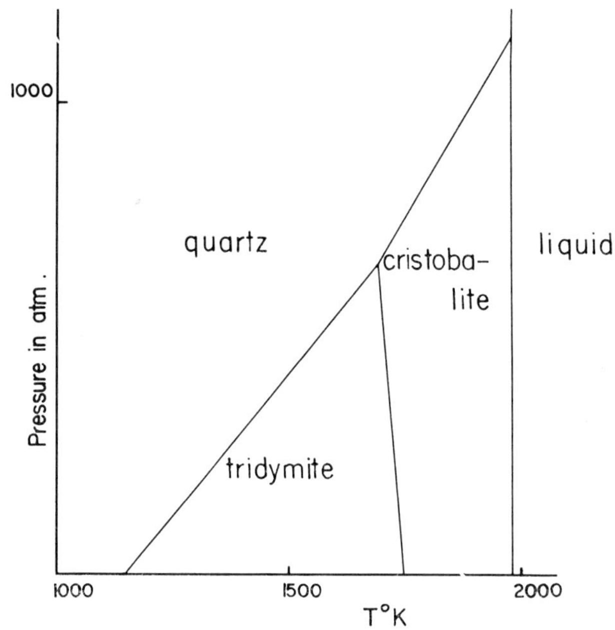

FIG. 24.—Phase diagram of silica based upon thermodynamic data. (After Mosesman *et al.*, 1941.)

of temperature and pressure at which metamorphic rocks form. It has often been demonstrated that potash feldspar can grow, in a suitable chemical environment, authigenetically even at the low temperature of cementation of sediment (Gilbert, 1949; Pettijohn, 1949); and it should not be necessary to state that potash feldspar in the form of microcline and orthoclase readily grows in medium- and high-grade rocks.

Albite is also physically stable and can grow readily (under suitable chemical conditions) within the P,T-range covered by metamorphic rocks. There are at least two modifications of albite, as indicated by field observations (Oftedahl, 1948; Köhler, 1949). Recently, Tuttle and Bowen (1950) have synthesized the high-temperature albite, which appears to be stable above some 700° C.

Anorthite is evidently stable at high and medium degrees of metamorphism, whereas the stability of pure anorthite at lower temperatures is more doubtful. It is clear enough that anorthite is unstable at low temperature in the presence of H_2O, calcite, potash feldspar, etc. (see p. 50). But this fact does not prove that anorthite as an isolated mineral is unstable and would tend to break up into new minerals or undergo a modification change at low temperature. So far, such an "isochemical" splitting-up of anorthite has not been demonstrated, either in laboratories[1] or in natural rocks. On the other hand, the natural rocks usually contain water and some minerals, which will react with anorthite at low temperatures, according to one of equations (56) and (57) (p. 50), showing that, even if pure anorthite is stable in an anhydrous system at low temperature, it is unstable in most natural rocks.

BIOTITE AND MUSCOVITE

Biotite has, as an isolated mineral, a wide temperature-pressure range of stability, inasmuch as it forms even at a rather low temperature (high green schist facies) and remains stable up through granulite facies. Some investigations indicate, however, that the high-temperature biotite needs to contain a rather high percentage of titanium in order to remain stable (Ramberg, 1948, 1949).

Muscovite forms at lower temperatures than biotite and does not persist stably up to a higher degree of metamorphism than the border line between amphibolite and granulite facies (eq. [54], p. 48). Noll (1932) has synthesized muscovite at some 200° C. This synthesis has been reproduced under varying chemical milieus by Norton (1939) and O'Neill (1948). Muscovite decomposed at high temperature has been resynthesized at 650° C. under high water-vapor pressure by Roy (1949). This experiment shows that muscovite is stable up to at least 650° C.

ALUMINUM SILICATES

Andalusite, *kyanite*, and *sillimanite* are the known three modifications of Al_2SiO_5. From petrologic study it seems certain that andalusite is stable at high temperature but low pressure; kyanite is stable at high pressure and intermediate temperature; whereas sillimanite is stable at high temperature as well as the highest pressure known in metamorphic rocks, viz., granulite facies. The limitations of the stability fields of the several modifications are not known. There are absolutely no petrologic reasons to believe, however, that any one of the three modifications would be unstable altogether, as proposed by Neumann (1925). It is interesting

1. Recent experiments by Dr. J. Goldsmith and Mr. E. Ehlers, at the University of Chicago, have shown that an apparently metastable modification of the same composition as anorthite forms at moderately low temperatures.

to note that sillimanite has recently been reported to form at 450°–500° C., and at about 50 atm. water-vapor pressure in the presence of boron and sodium salts (see Michel-Lévy, 1950).

BINARY AND POLYCOMPONENT SYSTEMS

POTASH FELDSPAR—ALBITE

The ability of potash feldspar and albite to form mixed crystals is a well-known fact which is of great interest for metamorphism and the theory of origin of quartzo-feldspathic rocks. It is known that the alkali feldspars are miscible in all proportions above a certain temperature, which may vary with pressure. Recent experiments by Bowen and Tuttle (1950) and by Laves (1952) have established the top of the immiscibility gap to be about 660° C. for the sanidines and about 550° C. for the microclines. Figure 25 shows the rather complicated relationships found in the alkali feldspar system.

The apparently monoclinic *orthoclase* is probably a partly microclinized sanidine (Laves); and the *adularia*, which is typically formed in very low-temperature alpine veins, seems to have formed metastably as some sort of sanidine much below the stability field of sanidine (Laves).

PLAGIOCLASE

It has generally been assumed by petrologists that the plagioclases constitute a complete mixed crystal series, because unmixing structures, like the perthites of the alkali feldspars, have not been observed. The X-ray investigations by Chao and Tayler (1940), however, indicate that the plagioclases may not be such homogeneous mixed crystals as has been assumed. At first glance, such an assumption seems entirely in contradiction to all petrological experience, because two or more distinct plagioclases do not occur in the same mineral assemblage unless, at the same time, it is justifiable to assume that stable equilibrium was not attained. On the other hand, it is well known that only acid plagioclase is stable at a low degree of metamorphism, simply because the anorthite component is unstable in association with H_2O (p. 50). Therefore, a great field in the plagioclase series is replaced by zoisite, epidote, etc. (p. 51, Fig. 27) in metamorphic rocks. The plagioclases within this field are unstable in a hydrous milieu; consequently, if the unmixing gap in the plagioclase series is situated inside this unstable field of the plagioclases, we would have but a negligible opportunity to detect the phenomenon of unmixing in plagioclases in natural rocks.

Incidentally, it is in accordance with crystal chemistry that all mixed crystals are unstable at temperatures approaching 0° K. (Slater, 1939), because the thermodynamic properties of the exchangeable particles are

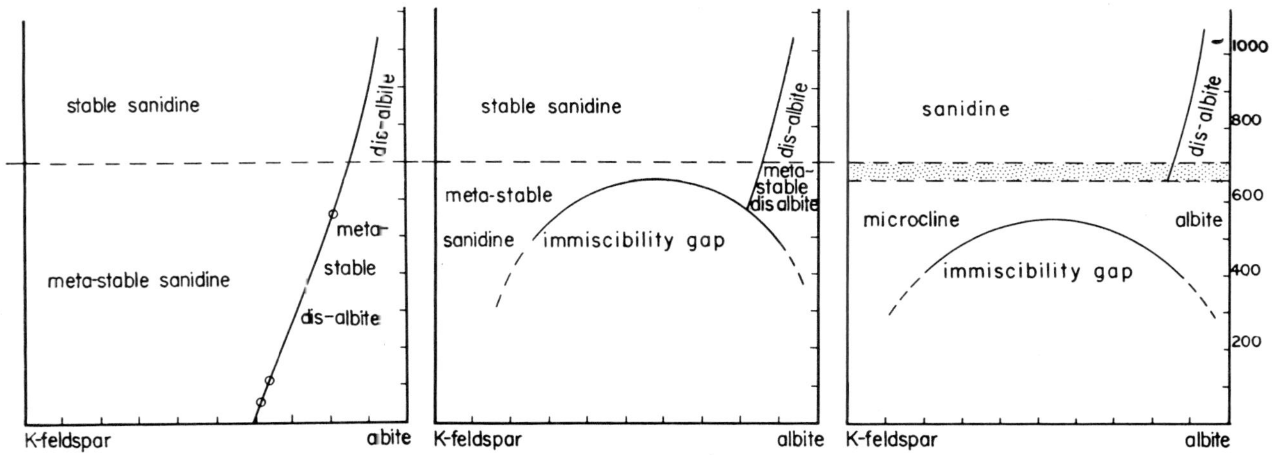

a *b* *c*

FIG. 25.—Subsolidus diagram of the alkali feldspars after Laves (1952, and oral communication), and Bowen and Tuttle (1950). *a* shows the α-β transition (displacive transition) between sanidine and dis-albite, which have different structures (Laves). The lowest part of the diagram is metastable, but it can be investigated because of the exceedingly sluggish transitions sanidine $\rightleftharpoons$ microcline and dis-albite $\rightleftharpoons$ albite and because the unmixing is also slow at low temperatures. *b* shows the metastable unmixing curve in the system metastable sanidine–metastable dis-albite (after Bowen and Tuttle). *c* shows the unmixing in the stable microcline-albite system, as indicated by natural alkali feldspars investigated by Laves. Note that the name "dis-albite" originally suggested by Laves has been changed to "an-albite" in his 1952 publication.

never identical. The temperature at which this splitting-up of a mixed crystal series starts and the detailed trend of this process depend upon the particular character of the mixed crystal in question.

MUSCOVITE, POTASH FELDSPAR, AND SILLIMANITE OR ANDALUSITE

The assemblage muscovite and quartz is stable over a large part of the P,T-range of metamorphism. At high temperature, however, the following reaction takes place:

$$\underset{\substack{\text{muscovite}\\140\text{ cc.}}}{KAl_3Si_3O_{10}(OH)_2} + \underset{22\text{ cc.}}{SiO_2} \rightleftarrows \underset{\substack{\text{orthoclase}\\\text{(or microcline)}\\110\text{ cc.}}}{KAlSi_3O_8} + \underset{\substack{\text{sillimanite}\\50\text{ cc.}}}{Al_2SiO_5} + H_2O\uparrow . \qquad (54)$$

Mol volumes: muscovite 140 cc.; SiO_2 22 cc.; orthoclase (or microcline) 110 cc.; sillimanite 50 cc.

According to numerous observations, this reaction takes place at the border between amphibolite facies and granulite facies, where temperature as well as pressure is high. If pressure is low, andalusite takes the place of sillimanite (pyroxene-hornfels facies; p. 162) in reaction (54).

The equilibrium temperature of this reaction is not determined by the rock pressure alone but also by the partial vapor pressure of water. (We assume here that the composition of the minerals is constant.) The partial vapor pressure of water is commonly not so high as the total rock pressure (p. 180). If the water-vapor pressure was the same as the total rock pressure corresponding to the load of the superincumbent column of rocks, then indeed the rock would be at the boiling point and a small rise in temperature would create an explosion. We have no evidence that such an explosion takes place when muscovite is altered into potash feldspar and sillimanite. On the contrary, the process seems to be quiet and slow, without disturbing the structure of the rocks, as one would expect if the water molecules migrated slowly away from the place of reaction. There will be dynamic equilibrium with reference to the migration of water vapor when the partial vapor pressure is the same over a certain volume of the rock complex.

The higher the water-vapor pressure in the intergranular of the rock, the higher the temperature of reaction of equation (54). At low vapor pressure, potash feldspar and some modification of Al_2SiO_5 should be stable at considerably low temperature (the rock pressure is here considered constant).

The effect of the rock pressure upon the temperature of reaction can be found qualitatively if the mol volume of H_2O is disregarded in equation (54). It appears that high pressure slightly favors the formation of sillimanite and feldspar, if pressure does not act upon the water vapor; but muscovite if the water is under high pressure also.

The resynthesis of muscovite (without fluorine) at a maximum temperature of 650° C. under a water-vapor pressure of from 1,000 to 10,000 lb/sq in (Roy, 1949) certainly does not define the equilibrium temperature of reaction (54). It is interesting, though, that various independent circumstances make us believe that a temperature of some 600° C. is the most likely one for the boundary between amphibolite facies and granulite facies (see pp. 156 and 161).

SILICA AND CALCITE, THE "WOLLASTONITE EQUILIBRIUM"

At low pressure (about 1 atm.) the following reaction takes place at some 400° C. (A. Danielsson, 1950):

$$SiO_2 + CaCO_3 \rightleftarrows CaSiO_3 + CO_2 \,. \qquad (55)$$

Mol volumes: 22 cc. + 37 cc. 41 cc.

Increasing pressure increases the equilibrium temperature, provided that the pressure also acts on the CO_2 gas. The effect of pressure has been calculated by Goldschmidt (1912); but Danielsson (1950) has recalculated the effect on the basis of new data (Fig. 26).

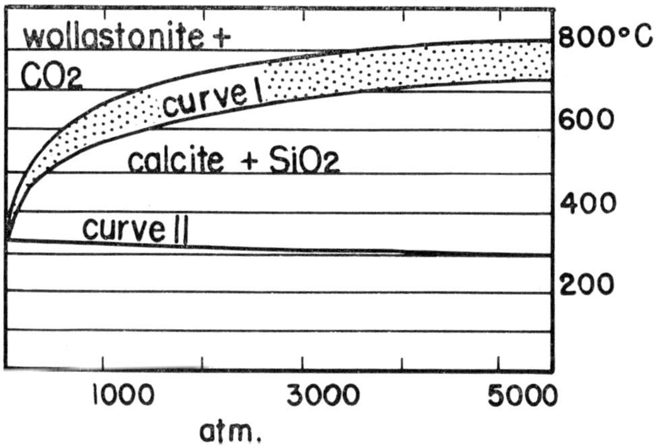

FIG. 26.—The effect of pressure upon the equilibrium temperature of the reaction wollastonite + $CO_2 \rightleftarrows$ calcite + SiO_2. Curve *I* after Goldschmidt (1912); and Danielsson (1950); curve *II* after equation on p. 50.

For the same reasons as those mentioned under the discussion of muscovite above, it is likely that the partial CO_2 pressure is commonly less than the rock pressure because the gas can migrate through the rocks with great ease. Under such circumstances, the effect of rock pressure upon the equilibrium curve of equation (55) can be calculated by disregarding the volume of the CO_2 gas, as suggested by the writer in several papers (see, e.g., Ramberg, 1946).

Introducing ΔV = 18 cc. and the molal heat of reaction (24, 300 cal.)

into Clausius Clapeyron's law (p. 16) gives approximately:

$$\frac{dT}{dP} = \frac{-18 \text{ cc.} \cdot 673^\circ \text{K.}}{10^6 \text{ cc. atm.}} = -0.012 \text{ degree/atm.}$$

This result is plotted in Figure 26, together with the curve of Goldschmidt and Danielsson. This equation is valid only if the partial CO_2 pressure remains unchanged and *equal to* the CO_2 pressure at which the reaction has an equilibrium temperature of about 400° C. There is certainly no reason to believe that precisely this CO_2 pressure exists throughout the earth's crust. If the CO_2 pressure is lower, wollastonite will form at lower temperatures than indicated by the equation above; if it is higher, the assemblage calcite-quartz is stable at correspondingly higher temperatures, but the upper limit will be that given by the Goldschmidt curve.

SOME EQUILIBRIA IN WHICH PLAGIOCLASE IS INVOLVED

Plagioclase and epidote.—It has long been recognized that the anorthite content of plagioclase in metamorphic rocks often increases with increasing degree of metamorphism (Turner, Becke). The reason is that calcium is an element constituting several metamorphic minerals, whereas sodium in rocks is mainly found in the albite "molecule" of the plagioclase. Therefore, there take place chemical interactions between the anorthite of plagioclase and other Ca-carrying silicates. These interactions are such that, most commonly, the anorthite percentage increases with rising temperature. Most simple is the interaction between plagioclase and epidote. Equations (56) and (57) should represent the most common mineral reactions in which anorthite as well as epidote are involved:

$$\underset{\text{epidote}}{2Ca_3Al_3Si_3O_{12}OH} + \underset{\text{kyanite}}{3Al_2SiO_5} + 3SiO_2 \rightleftarrows \underset{\text{anorthite}}{6CaAl_2Si_2O_8} + H_2O\,, \tag{56}$$

$$\underset{\text{epidote}}{2Ca_2Al_3Si_3O_{12}OH} + \underset{\text{muscovite}}{KAl_3Si_3O_{10}(OH)_2} + 2SiO_2 \overset{T_u}{\rightleftarrows} \underset{\text{anorthite}}{4CaAl_2Si_3O_8} + \underset{\text{orthoclase}}{KAlSi_3O_8} + H_2O\,. \tag{57}$$

We know empirically that anorthite and albite are miscible in all proportions, whereas epidote (zoisite) is practically unable to dissolve any sodium compound. Furthermore, we know that the epidote sides of equations (56) and (57) correspond to the low-temperature association. As a consequence, we can construct an equilibrium diagram such as that shown in Figure 27. The diagram is of particular importance as a relative geological thermometer. It also informs us about the process of "saussuritization" of plagioclase: below the temperature T_u, anorthite (or a plagioclase with a minimum percentage of anorthite, depending on temperature and pressure) is unstable, together with potash feldspar and water. These minerals will react and produce epidote (zoisite), muscovite, and a certain acid plagioclase. This reaction does not necessarily require a material contact among the involved minerals, as the active particles can be transported over certain distances. In view of the equilibrium diagram (Fig. 27), therefore, the "saussuritization" of plagioclase in retrogressively metamorphosed rocks becomes evident. The diagram also has some bearing upon the common replacement of perthite lamella by muscovite.

The equilibrium between epidote and plagioclase as represented by

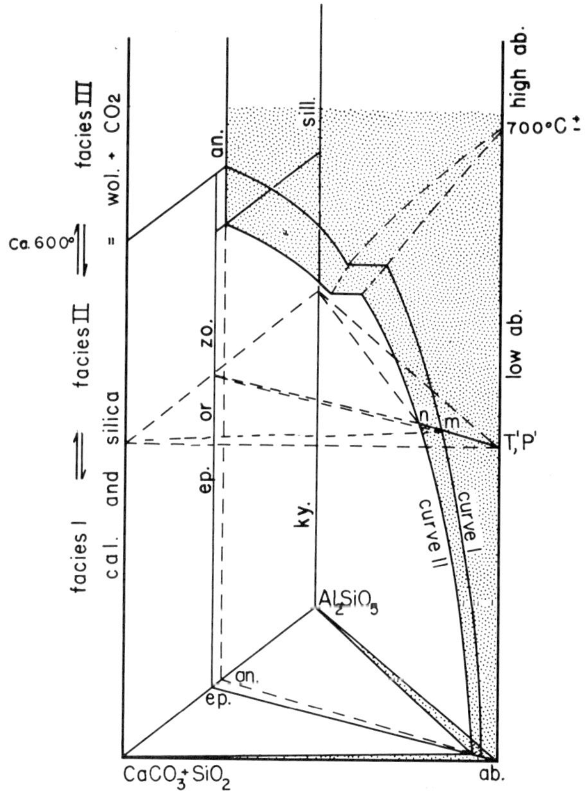

FIG. 27*a*.—Tentative and simplified diagram, showing the change of composition of plagioclase in equilibrium with epidote. Based on field data and the general laws of equilibrium. In the space confined to *ep.* Al_2SiO_5 and curve *II*, epidote, kyanite (or sillimanite), and plagioclase of composition corresponding to curve *II* are stable together. In the space confined by epidote, calcite, and silica and curve *I*, epidote, calcite, and silica exist together with plagioclases of composition given by curve *I*. In the space confined by calcite and silica, albite, and curve *I*, the composition of the plagioclase may vary from pure albite to that given by curve *I*. In the space confined by Al_2SiO_5, albite, and curve *II*, the plagioclase may have any composition from albite to curve *II*. The dotted curves are tentatively based upon the transition high-low albite at 700° C. (Bowen and Tuttle). Facies *I* is the epidote amphibolite facies; facies *II* is the amphibolite facies, and facies *III* is the granulite facies.

Figure 27 is not valid under all circumstances. Several conditions may displace the equilibrium. In particular, it is evident that the partial water-vapor pressure is of importance, as is also the Fe^{3+}/Al^{3+} ratio. As to the directions of the influence of H_2O and Fe^{3+}, the following may be stated. Increasing water-vapor pressure will favor the formation of epidote and muscovite, so that the reaction points of equations (56) and (57) will be elevated gradually as the H_2O pressure rises, and the equilibrium curves

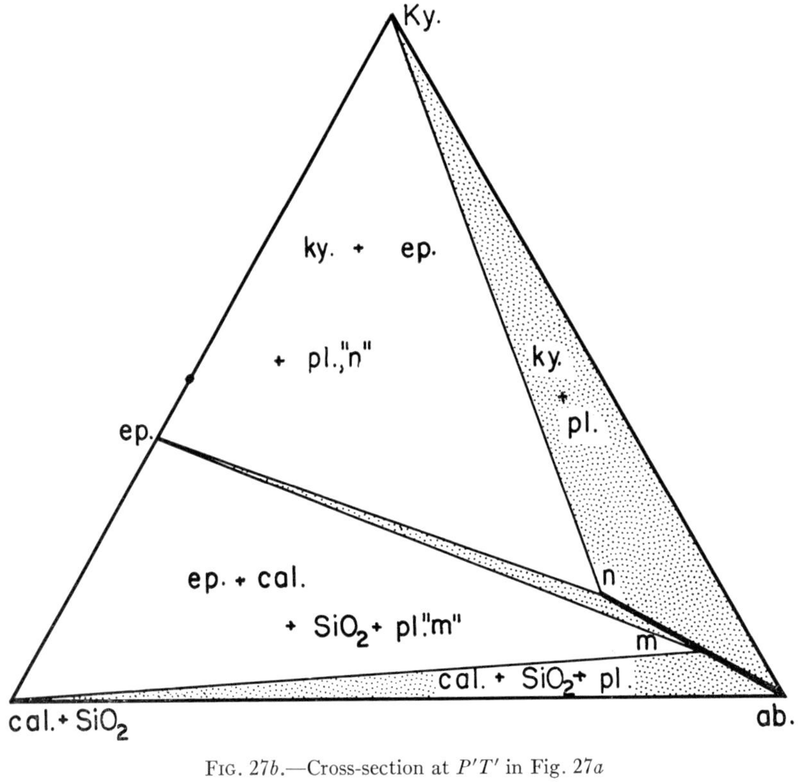

FIG. 27*b*.—Cross-section at $P'T'$ in Fig. 27*a*

of Figure 27 will be displaced toward higher temperatures. (That does not mean that the reaction temperatures are increased by increasing rock pressure, because, as a common rule, the rock pressure and the partial water-vapor pressure are not equal. The rock pressure is usually considerably greater than the partial water-vapor pressure.) The petrological consequence will be that two rocks with the same type of plagioclase in equilibrium with zoisite may have recrystallized (or formed) at different temperatures, depending on the partial water-vapor pressure in the intergranular space of the rocks. At first glance, this may be considered a disadvantage because it shows the equilibrium system to be ambiguous. On the other hand, however, these conditions may enable us to determine the water-vapor pressure which existed in the rocks during their recrystallization.

Considering the effect of Fe^{3+}, it is important to note that epidote is able to replace a rather great amount of Al with Fe^{3+}, whereas plagioclase can hardly take more than 0.5–1.0 per cent of Fe_2O_3 in solid solution. These facts, in connection with the general relationship between partial molal free energy and concentration (p. 18), enable us to conclude that a rising Fe^{3+}/Al^{3+} ratio in a rock favors the formation of epidote. As a consequence, increments in the Fe^{3+}/Al^{3+} ratio stabilize epidote at higher temperatures than usual and displace the equilibrium curve upward.

Attempts to synthesize zoisite hydrothermally have only resulted in formations of anorthite, together with different minerals, none of which is epidote or zoisite. The temperature of these experiments has been 470° C. (Schlaepfer and Niggli, 1914) and as low as 370° C. (E. Ehlers, oral communication). Taking these experiments as indicating that anorthite forms stably down to at least 400° ± under a water-vapor pressure of some 1,000 atm., one should assume that the stability temperature range of epidote is below some 400° C. at pressures ranging up to 1,000 atm. Under the assumption that during regional metamorphism the water-vapor pressure in the pore network of rocks is lower than the rock pressure, we find that the pressure does not significantly increase the equilibrium temperature of reaction (56), even at a depth of some 15 km. The equilibrium temperature of reaction (56) may rise some 100° C., being about 500° C. or less. It is interesting now to note that this reaction (56), according to field experience, takes place just a little below the splitting-up of muscovite (reaction [54], p. 48) (about 650° C.) in the sequence of regional metamorphism.

It should be noticed that reaction (56) is invalid for parageneses rich in calcite. In such cases, the following reactions may take place:

$$\underset{\text{anorthite}}{CaAl_2Si_2O_8} + 2CaCO_3 + SiO_3 \rightarrow \underset{\text{grossularite}}{Ca_3Al_2Si_3O_{12}} + 2CO_2\,, \tag{58}$$

and

$$\overset{\text{high temp.}}{\underset{\text{anorthite}}{3CaAl_2Si_2O_8}} + 2CaCO_3 + H_2O \rightarrow \overset{\text{low temp.}}{\underset{\text{epidote}}{2Ca_2Al_3Si_3O_{12}(OH)}} + 2CO_2\,; \tag{59}$$

and evidently there is no reason to believe that the equilibrium temperatures in this case would be the same as in equations (56) and (57). Observations indicate, in particular, that the equilibrium temperatures of equations (58) and (59) are considerably higher than those of equations (56) and (57).

There are other minerals which constitute an equilibrium system with plagioclase similar to that of epidote (zoisite). These systems, however, are more complicated and much more difficult to evaluate when based on mere petrographic observations without experimental data. All we can do is to list and discuss qualitatively some possibilities. It is hoped that this may stimulate petrologists to look for details in rock parageneses which may help construct more accurately some of the following systems.

Plagioclase and scapolite.—Either of these minerals are mixed crystals where Ca^{+2} and Na^{+} are completely interchangeable. As the simplest reaction between scapolite and plagioclase, let us assume the following:

$$\begin{aligned} &a) \quad \underset{\text{anorthite}}{3CaAl_2Si_2O_8} + CaCO_3 \rightleftarrows \underset{\text{mejonite}}{Ca_4Al_6Si_6O_{24}(CO_3,SO_4,(OH)_2)}\ , \\ &b) \quad \underset{\text{albite}}{6NaAlSi_3O_8} + Na_2Cl_2 \rightleftarrows \underset{\text{marialite}}{Na_8Al_6Si_{18}O_{48}(Cl_2,CO_3,(OH)_2)}\ . \end{aligned} \tag{60}$$

At a certain pressure, reaction *a* goes on at a certain temperature, T_{u1}, and reaction *b* at a temperature T_{u2}, which probably is different from T_{u1}. Unfortunately, very little is known about these reactions; as a matter of fact, we do not know whether T_{u1} is higher than T_{u2} or vice versa, or whether the right or left sides of the equations correspond to the high-temperature associations. The only thing we can state about the scapolite-plagioclase system is that it must correspond formally to the isodimorphic type of equilibrium between mixed crystals (p. 30) and that the distribution of Na and Ca between the two phases is a function of temperature and pressure. On the other hand, this system is complicated by the fact that CO_3, Cl, and SO_4, as well as $(OH)_2$, are interchangeable in the structure.

Plagioclase and hornblende.—Common metamorphic hornblende does not tend to exchange appreciable amounts of Ca with Na in the $(Ca,Na,K)_{2\text{-}3}$ lattice positions; 1–1.5 per cent of Na_2O and 10 per cent of CaO are very common values in hornblende analyses. In many cases hornblende can, therefore, form by reactions between the anorthite component in plagioclase and some ferromagnesian minerals. As a consequence, the composition of plagioclase in equilibrium with hornblende will change gradually with temperature and pressure.

As an example, let us consider the reaction between orthorhombic pyroxene and anorthite at which an aluminum-rich "common" hornblende is formed:

$$\underset{\text{enstatite}}{3MgSiO_3} + \underset{\text{anorthite}}{2CaAl_2Si_2O_8} + H_2O \overset{\text{high temp.}\qquad\qquad\text{low temp.}}{\rightleftarrows} \underset{\text{hornblende}}{Ca_2(Mg_3Al_2)(Si_6Al_2)O_{22}(OH)_2} + SiO_2\ . \tag{61}$$

We shall first disregard the iron members of the pyroxene and hornblende series. Our petrologic experience gives us the following pertinent information:

a) The right-hand side of equation (61) is the low-temperature (high-pressure?) side of the reaction.

b) Orthorhombic pyroxene is not able to dissolve more than some 1–2 per cent of Na_2O and CaO (Poldervaart, 1947).

c) The common hornblende found in the normal facies of regional metamorphic rocks does not contain more than 1–1.5 per cent of Na_2O.

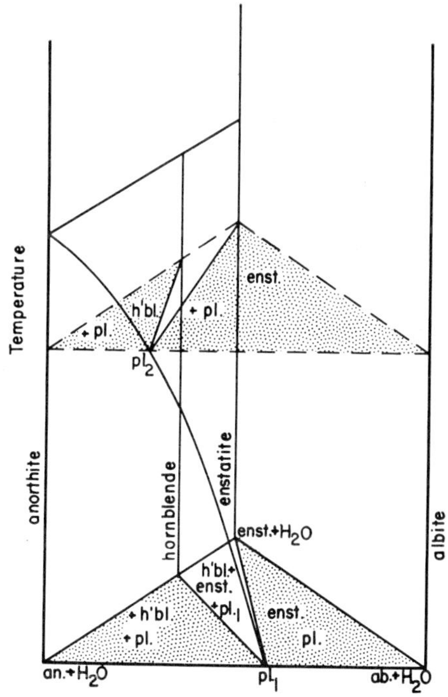

FIG. 28.—Tentative and simplified diagram, showing the effect of temperature upon the composition of plagioclase in equilibrium with hornblende and pyroxene. In the space limited by enstatite ($+H_2O$), *h'bl.*, and the curve through Pl_1 and Pl_2, hornblende ($+SiO_2$) and enstatite ($+H_2O$) are stable together with plagioclase of composition given by the curve through Pl_1 and Pl_2. In the space limited by albite, enstatite, and the plagioclase curve, enstatite and plagioclase of any composition between albite and the curve through Pl_1 and Pl_2 are stable. In the space anorthite, hornblende, and the plagioclase curve, plagioclase of any composition varying from anorthite to the curve through Pl_1 and Pl_2 can be stable.

d) The reaction temperature of equation (61) under regional metamorphic conditions corresponds to something near the boundary between amphibolite and granulite facies (see p. 159).

On these grounds, a simplified and qualitative equilibrium diagram, as shown in Figure 28, can be constructed. The main purpose of this diagram is simply to show how the changing composition of the plagioclase would influence the equilibrium temperature. Several conditions will modify this diagram:

a) The reaction temperature of equation (61) depends upon the partial water-vapor tension(the molal free energy of water) in the rock—a quantity which is very difficult to determine, although it must be a function of the content of water in the hydrous minerals.

b) Enstatite is not stable over the whole temperature range represented by the diagram but will react with water and form anthophyllite, talc, and serpentine at low temperatures. Each of these reaction points will give rise to a discontinuous bend on the equilibrium curve.

c) The aluminum-rich hornblende itself is rendered unstable at lower temperature (see discussion, pp. 65 ff.).

d) In natural rocks we cannot disregard the iron silicates. It is well known that very Fe-rich rhombic pyroxenes are rare. Nevertheless, one may operate with hypothetically pure Fe members. In so doing, it can be shown that the pure Fe-silicates should react according to equation (61) at a somewhat higher temperature than the Mg-silicates (p. 68). In other words, Fe-rich hornblende is stable at higher temperatures than Mg-rich hornblende in the paragenesis now considered.

Plagioclase in relation to grossularite, lawsonite, jadeite, and glaucophane.—Although little is known quantitatively about the equilibrium relationships between plagioclase and grossularite, lawsonite, jadeite, and glaucophane, some principal trends will be discussed. The anorthite component of plagioclase can change into grossularite and lawsonite, both of which have a composition similar to anorthite. Because neither grossularite nor lawsonite is able to take appreciable amounts of sodium in its structure, the equilibrium diagrams of plagioclase and grossularite, or lawsonite, respectively, should be rather similar to the equilibrium between plagioclase and zoisite. These equilibria, when once determined quantitatively, may serve as valuable geological barometers, because of the rather large density difference between the minerals involved. That is, the composition of plagioclase, which at a certain temperature is in equilibrium with either lawsonite or grossularite, must be rather sensitive to changing pressure.

The relationship between anorthite and lawsonite may be written as follows:

$$\underset{\text{anorthite}}{CaAl_2Si_2O_8} + H_2O = \underset{\text{lawsonite}}{CaAl_2Si_2O_8 \cdot H_2O}\,. \tag{62}$$

Similar equilibria must exist between plagioclase and jadeite, $NaAlSi_2O_6$, and the alkali hornblende, glaucophane, both of which can form at the expense of the albite content of plagioclase. As the density of jadeite and glaucophane is relatively high, these two minerals may form at the expense of the albite content in plagioclase, in such a way that increasing pressure gradually transfers the albite component from plagioclase to jadeite or glaucophane. Hence the composition of plagioclase, which is in equilibrium with jadeite or glaucophane, should function as a pressure gauge.

GARNET, MICAS, AND POTASH FELDSPAR

Petrographic experience has demonstrated that the composition of garnets depends greatly on the degree of metamorphism at which the garnets have formed (Eskola, 1921*a;* Wright, 1938). One of these compositional changes can be correlated with an equilibrium system between garnet and biotite plus potash feldspar. The pure Fe-silicate compounds occur in the following reaction:

$$\underset{\text{biotite}}{KFe_3Si_3AlO_{10}(OH)_2} + \underset{\text{muscovite}}{KAl_3Si_3O_{10}(OH)_2} + 3SiO_2 \rightleftharpoons 2\underset{\text{orthoclase}}{KAlSi_3O_8} + \underset{\text{almandine}}{Fe_3Al_2Si_3O_{12}} + 2H_2O\ , \tag{63}$$

which is known to take place, so that the left-hand side is the low-temperature (and low rock-pressure) side. That is, Fe-biotite and muscovite will react to form potash feldspar and garnet at a certain P,T-combination in the sequence of regional metamorphism (low granulite facies). At lower temperature the left-hand side of the equation is stable. If we introduce Mn,Al-silicate into the system, we know that the garnet will absorb the manganese, forming a spessartite-almandine mixture, whereas only negligible amounts of Mn are able to enter the lattice of biotite (Hall, 1941).

In addition, we know that the spessartite garnet is stable down to very low degrees of metamorphism. These observations are sufficient to enable us to construct the equilibrium diagram shown in Figure 29, which has appeared to be in harmony with observations in natural rocks. As the minerals involved also show a great ability to interchange Fe and Mg, the Fe/Mg ratio of the system is of considerable interest. The two equations (63) and (64) should most reasonably have different equilibrium temperatures and/or pressures:

$$\underset{\text{phlogopite}}{KMg_3AlSi_3O_{10}(OH)_2} + \underset{\text{muscovite}}{KAl_3Si_3O_{10}(OH)_2} + 3SiO_2 \rightleftharpoons 2\underset{\text{orthoclase}}{KAlSi_3O_8} + \underset{\text{pyrope}}{Mg_3Al_2Si_3O_{12}} + 2H_2O\ . \tag{64}$$

Available data indicate that the Mg members react at higher P,T-conditions than do the Fe members.[2] These data are not from direct tempera-

2. Pure pyrope seems to be unstable altogether, so that eq. (64) probably represents an imaginary unstable reaction (see Wright's compilation paper, 1938).

ture measurements in laboratories but are merely gathered from the distribution of Fe and Mg in biotite and garnets in the same rocks. Such investigations show that garnet usually has a greater Fe/Mg ratio than does coexisting biotite.

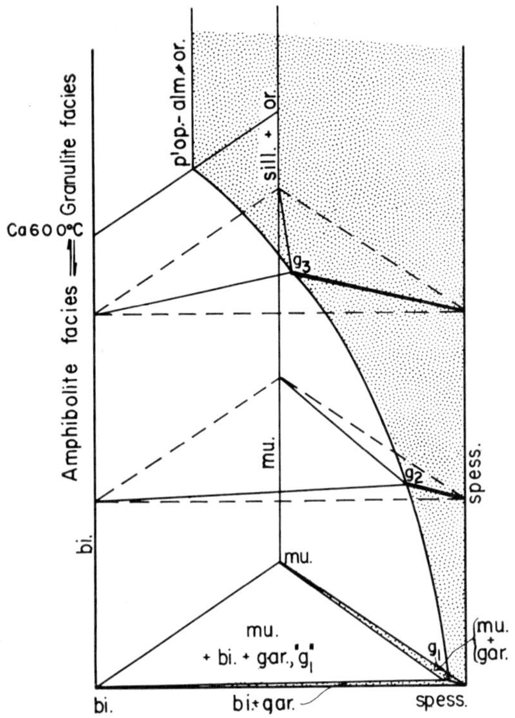

FIG. 29.—Tentative and simplified diagram, showing how the spessartite content in garnet changes with degree of metamorphism. In the space limited by muscovite, spessartite, and the curve through g_1–g_2–g_3, muscovite exists together with garnet, the composition of which can change from spessartite to the curve through g_1–g_2–g_3. In the space limited to biotite, spessartite, and the curve through g_1–g_2–g_3, garnet and biotite exist. In the space limited by biotite, muscovite (or orthoclase and sillimanite), and the garnet curve, the composition of the garnet is given by the curve through g_1–g_2–g_3.

Several other reactions in which biotite is involved and where the manganese content in garnet is pertinent take place in rocks. For example, biotite and kyanite (or sillimanite) give garnet above a certain rock pressure (and temperature):

$$\text{biotite} + \text{kyanite} + \text{silica} \rightarrow \text{potash feldspar} + \text{garnet} + \text{water} .$$

By introducing spessartite, we shall find that the equilibrium curve is parallel to that in Figure 29 and is displaced just a little toward lower temperatures.

Biotite may react with cordierite to form garnet and feldspar, upon increasing pressure. In this case, also, the equilibrium pressure at which this reaction takes place will decrease with increasing concentration of Mn in the garnet, because cordierite is not inclined to take much Mn in its lattice.

CHLORITE, BIOTITE, AND MUSCOVITE

It is well known, from the study of schists, that chlorite often is altered into biotite at rather low degrees of metamorphism. The classic chlorite and biotite zones of regional metamorphism (Barrow, 1893, 1912) are distinguished by this transformation. Since chlorite does not contain the potassium necessary to transform it into biotite, biotitization of chlorite is possible only by reaction with some potashic minerals like muscovite or potash feldspar (we shall here disregard the possibility of the introduction of potassium from external sources). Let us consider a possible reaction between muscovite and chlorite:

$$\underset{\text{muscovite}}{2KAl_3Si_3O_{10}(OH)_2} + \tfrac{5}{3}\underset{\text{chlorite } a}{(Mg,Fe)_6Si_4O_{10}(OH)_8} \rightleftharpoons \underset{\text{biotite}}{2K(Mg,Fe)_3AlSi_3O_{10}(OH)_2} + \underset{\text{chlorite } b}{(Mg,Fe)_4Al_4Si_2O_{10}(OH)_8} + \tfrac{14}{3}SiO_2 + \tfrac{8}{3}H_2O\,, \quad (65)$$

where the right-hand side, with biotite and aluminum-rich chlorite, is the stable assemblage at high temperatures. The important thing in this reaction is that the Al content in chlorite is changeable within wide limits, because Al can enter the $(Mg,Fe)_6$, as well as the Si_4, positions, so that chlorite *a* (antigorite) actually is miscible with chlorite *b* (amesite). In muscovite the Al content is stoichiometrically fixed; and in biotite, only limited variations of the Al content are possible.

Thus equilibrium can be established between muscovite, biotite, and a chlorite of a certain Al content at a given P,T-combination (in excess of H_2O and SiO_2). If P and/or T changes, the chlorite may continue to be in equilibrium with muscovite and biotite by changing gradually its Al content. We have fairly good reasons to assume that the aluminum content in the chlorite implied in this equilibrium system will increase gradually with rising degrees of metamorphism, provided that the Mg/Fe ratio is constant. This statement certainly does not imply that Al-poor chlorite (antigorite) is unstable at somewhat elevated temperatures. It applies only to the chlorites which are in equilibrium with muscovite and biotite. Figure 74 (p. 142) gives the general trend of the system.

GARNET AND CHLORITE

Disregarding the water content of chlorite, the Al-rich members of this mineral series may roughly be considered a low-temperature "modification" of ferromagnesian garnet. It is indicated by numerous petrographic observations from metamorphic complexes that the chlorite side of equation (66) corresponds to a low degree of metamorphism, and the garnet side to a high degree. Our experience also enables us to place the reaction temperature (pressure) of equation (66) somewhere within the epidote-amphibolite facies:

$$\text{Fe,Mg-chlorite} + SiO_2 \rightleftarrows \text{Fe,Mg-garnet} + H_2O\,. \qquad (66)$$

The equilibrium temperature of this reaction is certainly not definitely fixed by the external rock pressure. Not only does the partial vapor pressure of water play a significant role, but also the relative amounts of the mutually exchangeable elements Mg, Fe, and Mn are of the most impor-

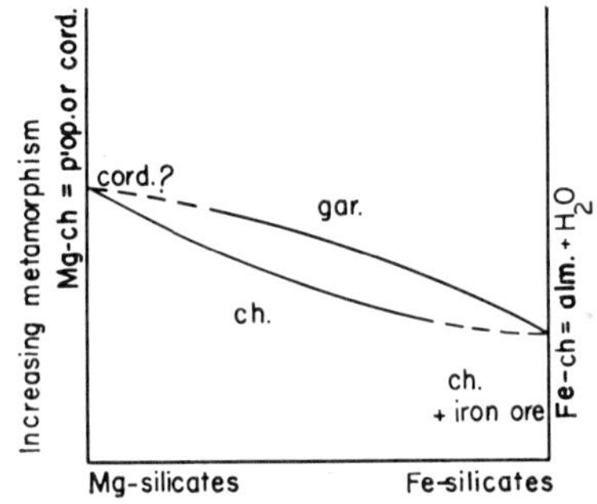

FIG. 30.—Tentative phase diagram, showing the transition: chlorite $\rightleftarrows$ garnet

tant consequence. (We shall disregard the extensive interchange of Al and Si in chlorites.)

Chlorites have a greater Mg/Fe ratio than do coexisting garnets. Such a distribution of Mg and Fe in garnet and chlorite shows, in connection with the fact that chlorite is the low-temperature mineral of the two, that the principal trend of the equilibrium diagram is as given in Figure 30. Unfortunately, the diagram is compiled from only a few observations, and its simplicity is illusive, particularly when the conditions close to the end-members of the system are concerned. Close to the Fe side of the diagram, it seems to be empirically established that chlorite is unstable. The most iron-rich chlorite splits up into iron ore and quartz, together with perhaps some other low-temperature iron silicates (chloritoid, stilpnomelane, minnesotaite, etc.) (see Gruner, 1944*a*, *b*). The almandine garnet, however, being one of the most stable iron silicates known, shows no tendency to split up. Complications also enter the system at its Mg-richer parts. High pressure seems necessary to stabilize more than 80 per cent of pyrope in the garnet structure. Therefore, cordierite and anthophyllite (or enstatite) take the place of garnet under certain metamorphic conditions. In an attempt to clear up the character of the Mg side of the diagram, let us consider equations (67) and (68):

$$\text{chlorite} + SiO_2 \rightleftarrows \text{cordierite} + \text{anthophyllite} + xH_2O\,, \tag{67}$$

$$\text{chlorite} + SiO_2 \rightleftarrows \text{pyrope} + xH_2O\,. \tag{68}$$

Petrographic observations show that both these reactions tend to take place toward the right-hand side with rising degrees of metamorphism. For formation of cordierite and anthophyllite (or enstatite), high tem-

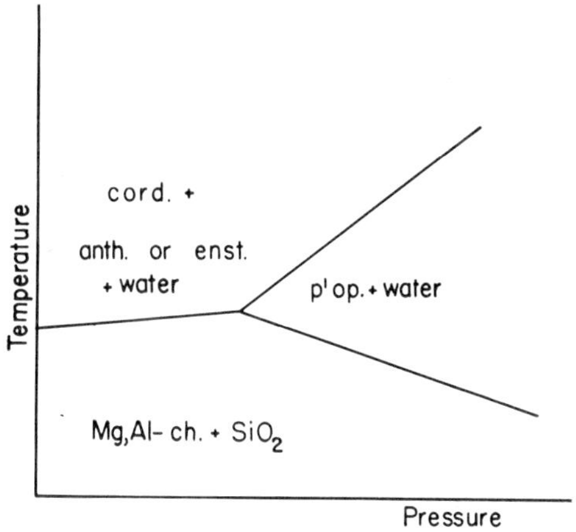

FIG. 31.—The relative stability fields of pyrope, cordierite, and chlorite, based upon field observations.

perature is the most important factor, whereas, in the formation of garnet, high pressure is particularly needed. To visualize the relative effects of pressure and temperature on the pure Mg end-members (eqs. [67] and [68]), Figure 31 has been constructed. We see that it is impossible to visualize the equilibrium conditions of the Mg-richest part of the system under consideration without introducing separate pressure and temperature axes. With regard to the most Fe-rich parts of the system, the relative effect of pressure and temperature is not so important, although it seems that, in order to stabilize almandine, a minimum pressure is needed. In the hornfelses of contact metamorphism, for example, where pressure has been low but temperature very high, cordierite and hypersthene form at the expense of chlorite (Goldschmidt, 1911):

$$\overset{\text{low temp.}}{\text{chlorite} + \text{silica}} \longrightarrow \overset{\text{high temp.}}{\text{cordierite} + \text{hypersthene} + \text{water}}\,. \tag{69}$$

In regionally metamorphosed rocks, however, garnet forms very commonly, as indicated by reaction (66), although a high Mg/Fe ratio in the rocks will favor the formation of cordierite together with enstatite or anthophyllite, even in regional metamorphism (Eskola, 1914, 1915).

By introducing MnO into the system, as is done in the case of most natural rocks, complications arise because manganese will not distribute itself evenly among the several minerals, but, instead, the garnet usually almost completely absorbs the manganese in the form of spessartite. Applying reasoning similar to that in the case of biotite and garnet (p. 57),

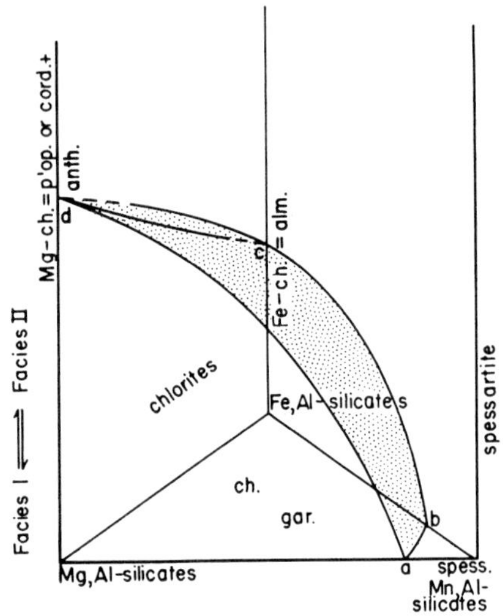

FIG. 32.—Diagram, showing schematically the stability relations of (Fe, Mg, and Mn)-garnets and (Fe,Mg)-chlorites. Below the surfaces *a*, *b*, *c*, and *d*, chlorites are stable together with garnets of compositions given by the *a*, *b*, *c*, and *d* surface. Above this surface only garnet is stable, in part together with anthophyllite and cordierite. Facies *I* is the green schist facies; facies *II* is the epidote amphibolite facies.

we can construct a three-component equilibrium diagram, as shown in Figure 32. According to the previous discussion, the diagram can be valid only as far as intermediate and rather high Fe/Mg ratios are concerned.

TALC, CALCITE, AND ACTINOLITE

In many marbles the following reaction is typical at a somewhat elevated temperature (corresponding to the upper parts of epidote amphibolite facies):

$$4SiO_2 + \underset{\text{talc}}{5Mg_3Si_4O_{10}(OH)_2} + 6CaCO_3 \underset{\text{high temp.}}{\overset{\text{low temp.}}{\rightleftarrows}} \underset{\text{tremolite}}{3Ca_2Mg_5Si_8O_{22}(OH)_2} + 2H_2O + 6CO_2 \,. \qquad (70)$$

At a given rock pressure, the equilibrium temperature is determined if the partial H_2O and CO_2 pressures are known. Calcite and talc are unable to mix with corresponding Fe compounds to any appreciable extent. Tremolite, however, is able to mix extensively with a hypothetical ferro-tremolite, thus forming the natural actinolites. It is evident that such a solution of Fe-silicate in the tremolite must lower the equilibrium temperature of reaction (70) according to the general rule discussed on pages 22–29.

Figure 33 shows the principal equilibrium diagram thus appearing. It may be of interest to compare this diagram with the biotite-garnet diagram (Fig. 29, p. 58), for a moment. In the latter case it does not seem impossible that the pure spessartite is stable and able to form down to the low temperatures of cementation of sediments (Pettijohn, 1949). The pure ferrotremolite, $Ca_2Fe_5Si_8O_{22}(OH)_2$, however, may perhaps be entirely unstable; and, even if this compound is stabilized by some minor content of Mg, it is empirically known that such an Fe-rich actinolite is unstable at the low temperatures of cementation (Pettijohn, 1949). It is not fully known what minerals the Fe-rich actinolite changes into at a low degree of metamorphism. The most probable possibility seems to be that actinolite (or a theoretically pure ferrotremolite) changes into calcite, siderite, or dolomite, plus iron ore and quartz, or some stilpnomelane or minnesotaite and quartz. Because of our ignorance on that point, the Fe side of Figure 33 is uncertain.

An excess of calcite in the system represented by equation (70) will not affect the transition temperature of this reaction; but, upon a certain rise in temperature, diopside will now form by interaction between calcite and tremolite:

$$3CaCO_3 + \underset{\text{actinolite}}{\overset{\text{low temp.}}{Ca_2(Mg,Fe)_5Si_8O_{22}(OH)_2}} + 2SiO_2 \overset{\text{high temp.}}{\rightleftarrows} \underset{\text{diopside}}{5Ca(Mg,Fe)Si_2O_6} + 3CO_2 + H_2O\,. \qquad (71)$$

Hedenbergite ($CaFeSi_2O_6$) will form by a corresponding reaction between ferrotremolite (or actinolite) and calcite. Experience from studies of metamorphic rocks shows that the reaction temperature of equation (71) is only a little higher than that of equation (70). In this connection it should be noted that the splitting-up of tremolite into diopside and pure Mg-silicate (enstatite, anthophyllite, or cummingtonite) according to equation (72) requires a considerably higher temperature than does the formation of diopside according to equation (71):

$$\underset{\text{tremolite}}{\overset{\text{low temp.}}{7Ca_2Mg_5Si_8O_{22}(OH)_2}} \rightleftarrows \underset{\text{diopside}}{14CaMgSi_2O_6} + \underset{\text{anthophyllite}}{\overset{\text{high temp.}}{3Mg_7Si_8O_{22}(OH)_2}} + 4SiO_2 + 4H_2O\,. \qquad (72)$$

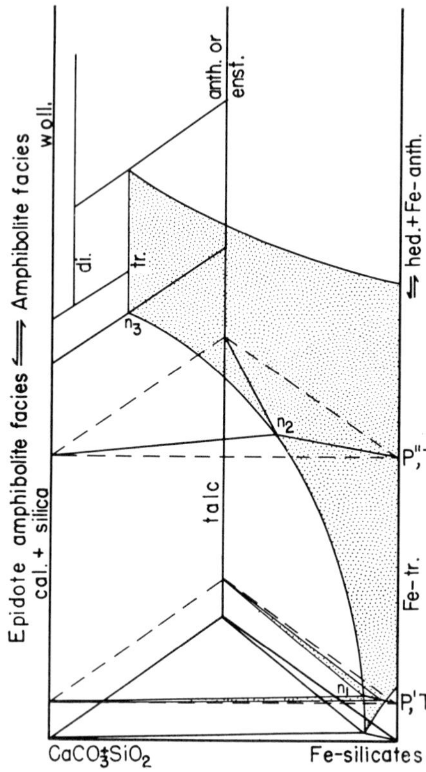

a

FIG. 33*a*.—Tentative diagram, showing some phase relationships between tremolite, ferrotremolite, and several other minerals. Based on field observations. The field limited by Fe-tremolite, tremolite, and the curve n_1–n_2–n_3 covers the mixed crystal tremolite-ferrotremolite. The curve n_1–n_2–n_3 gives the composition of actinolite in equilibrium with talc and calcite and silica. (The minerals tremolite and diopside are placed closer to the Ca corner than would correspond to their composition.)

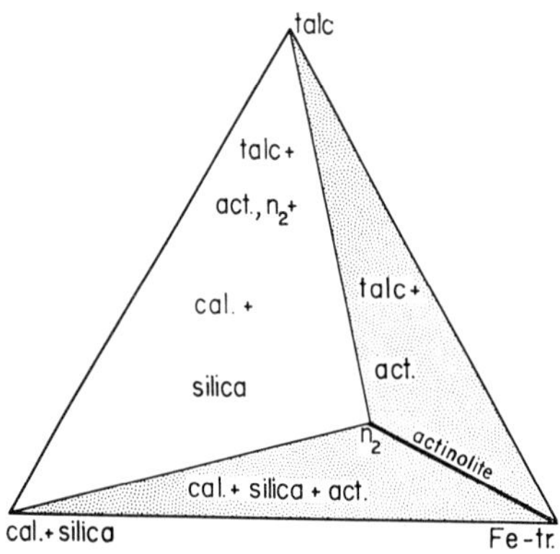

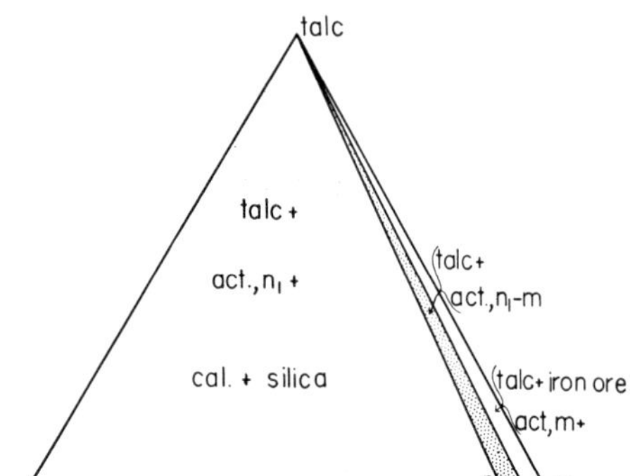

FIG. 33*b* and *c*.—Sections at *P''* and *T''* and *P'*, *T'*, respectively, in Fig. 33*a*

Little is known about the relative temperatures of equations (71) and (72), so that the main trend of the equilibrium curves of the upper parts of Figure 33 is uncertain.

ALUMINOUS HORNBLENDE IN RELATION TO SEVERAL OTHER COMMON MINERALS

Scattered observations from rocks containing hornblende indicate that some regularity exists between the content of Al in the hornblende and the degree of metamorphism (Eskola, 1939; Foslie, 1945; Ramberg, 1944*a*, 1948*c*). We have already discussed several aspects of the stability of hornblende, where the emphasis was put upon the Mg/Fe ratio in hornblende or the Na/Ca ratio in coexisting plagioclase.

According to Warren (1930), the hornblendes are built up of double Si_4O_{11} chains oriented parallel to the *c*-axes. In between the Si_4O_{11} chains are Ca^{2+}, Mg^{2+}, Fe^{2+}, Fe^{3+}, Ti^{4+}, some Na^{+} and K^{+}, etc., placed in certain positions. Per unit of Si_4O_{11}, there are $\frac{3}{2}$ positions, which preferably are occupied by Ca^{2+} and Na^{+} or K^{+}, and $\frac{5}{2}$ positions, which Mg^{2+}, Fe^{2+}, Fe^{3+}, and some other ions find particularly suitable. As $\frac{1}{3}$ of the (Ca^{2+},Na^{+}) positions may remain unoccupied without collapse of the lattice, the empirically established chemical formula for actinolite-tremolite is justified:

$$(Na,Ca)_{2\text{-}3}(Mg,Fe)_5Si_8O_{22}(OH,F)_2.$$

Hornblende belongs to those silicates in which Si can be rather extensively replaced by Al. To compensate for the valency change thereby produced, Al may enter the $(Mg,Fe)_5$ positions, or some of the empty $(Ca,Na)_{2\text{-}3}$ places may be occupied by sodium or potassium, or, eventually, part of the $(Mg,Fe)_5$ positions may be occupied by trivalent iron and/or tetravalent titanium. Let us consider an imaginary and highly unstable aluminum-rich hornblende molecule which we shall call molecule "A" in this discussion: $Ca_2Al_5(Si_3Al_5)O_{22}(OH,F)_2$. It is well known that no natural hornblende is so rich in aluminum, but we may consider some of the natural hornblendes to be mixtures of actinolite-tremolite and "A." So, for example, a mixture of 4 actinolite and 1 "A" will have the following formula, which is stable under certain conditions:

$$Ca_2(Mg,Fe)_4Al(Si_7Al)O_{22}(OH)_2\,.$$

Although mixtures of molecule "A" and actinolite are simplifications of the natural hornblendes, we shall use these mixtures as models in our discussion.

We have seen that pure tremolite can be formed from talc and calcite (or dolomite) by rising temperature:

$$\underset{\text{talc}}{5Mg_3Si_4O_{10}(OH)_2} \overset{\text{low temp.}}{+} 6CaCO_3 + 4SiO_2 \overset{\text{high temp.}}{\rightleftarrows} \underset{\text{tremolite}}{3Ca_2Mg_5Si_8O_{22}(OH)_2} + 2H_2O + 6CO_2\,. \tag{73}$$

It is established that the right-hand side of equation (73) represents the high-temperature assemblage. Suppose that a small amount of the "A" molecule is introduced into the system at the equilibrium temperature of equation (73) at a given pressure. We know from experiments and petrographic studies that neither calcite nor talc is able to dissolve appreciable amounts of Al or any Al compound in their lattices. Therefore, the small amount of the "A" molecule introduced will mix with the tremolite to form a tremolite with a certain small content of Al. By necessity (p. 19), this Al-carrying tremolite must have somewhat less partial molal free energy with reference to the pure tremolite "molecule" than an Al-free tremolite at the same P,T-conditions, and the equilibrium temperature of equation (73) must be depressed somewhat by the introduction of molecule "A" into the hornblende.

This conclusion is, at first glance, in direct opposition to the general experience of a gradually increasing content of Al in hornblendes with increasing degrees of metamorphism. However, this discrepancy is not so serious as it may appear to be. Consider now the A, C, and F diagrams (see definition, p. 138) (Fig. 34). It is then plainly seen that hornblende, with some content of Al, may form by reaction between calcite and chlorite:

$$\text{calcite} + \text{chlorite} \underset{}{\overset{\text{low temp.} \quad \text{high temp.}}{\rightleftarrows}} \text{aluminous actinolite} + \text{zoisite} + CO_2 + H_2O\,. \tag{74}$$

It is also known that the hornblende-zoisite association forms at the expense of chlorite and calcite when the temperature exceeds that of the green schist facies (p. 139). The equilibrium temperature of reaction (74) appears to be somewhat lower than that of reaction (73) in the sequence of regional metamorphism (Ramberg, 1944*a*). Because the aluminum content of both chlorite and hornblende is changeable, the equilibrium temperature of equation (74) will also depend upon the Al content. As chlorite is more able to accept Al in its lattice than is hornblende at these low temperatures, rising Al content in the system should expand the stability field of chlorite and hence raise the equilibrium temperature of equation (74). We cannot be certain about what will happen in detail, but it seems likely that, if calcite is present in excess, the equilibrium temperature of equation (74) would reach a maximum when the aluminum content in the chlorite has reached its limit. At this temperature-pressure

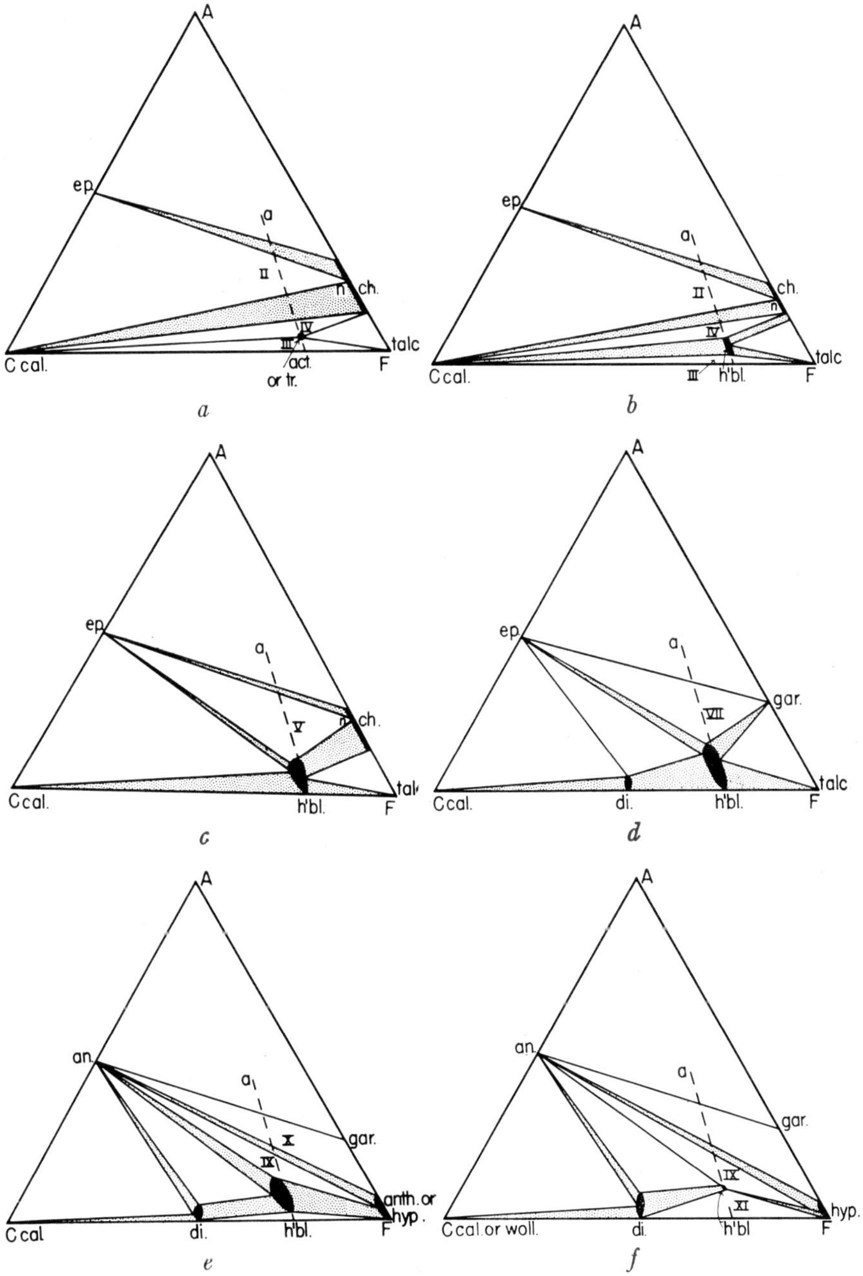

FIG. 34*a–f*.—ACF diagrams, showing the principal changes of Al content in hornblende at changing degrees of metamorphism. *a, b, c, d, e,* and *f* represent increasing grades of regional metamorphism, from the green schist facies up to the granulite facies (see Fig. 35).

condition, it appears likely from field observations that chlorite starts changing into garnet (Fig. 78). These relations are shown roughly in the equilibrium diagrams, Figures 34 and 35.

Above the temperature represented by Figure 34, *c*, chlorite (or garnet) and calcite are unstable together. At these high temperatures, garnet and zoisite interact and form the molecule "A," which, upon increasing metamorphism, will be "adsorbed" in the hornblende present, as shown in Figure 35.

At a sufficiently high degree of metamorphism, hornblende is entirely unstable. With reference to the pure tremolite, the following reaction takes place:

$$\underset{\text{tremolite}}{\overset{\text{low temp.}}{7Ca_2Mg_5Si_8O_{22}(OH)_2}} \rightleftarrows \underset{\text{diopside}}{14CaMgSi_2O_6} + \underset{\text{anthophyllite}}{\overset{\text{high temp.}}{3Mg_7Si_8O_{22}(OH)_2}} + 4H_2O\,. \quad (75)$$

The degree of metamorphism at which reaction (75) takes place is rather high in the higher parts of amphibolite facies (see field *XI*, Fig. 35).

Let us now see what should most reasonably happen to the Al-rich hornblendes under a high degree of metamorphism. One of the most important reactions is the following:

$$SiO_2 + \underset{\text{hornblende}}{Ca_2(Mg,Fe)_3Al_2Si_6Al_2O_{22}(OH)_2} \rightleftarrows \underset{\text{anorthite}}{2CaAl_2Si_2O_8} + \underset{\text{enstatite}}{3(Mg,Fe)SiO_3} + H_2O\,, \quad (76)$$

showing that rhombic pyroxene and anorthite are formed from hornblende at high temperatures. According to the considerations presented hitherto, together with data discussed elsewhere, the principal diagram showing the stability of hornblende of the formula $Ca_2(Mg,Al)_5(Si,Al)_8O_{22}(OH)_2$ should generally be as shown in Figure 35.

This diagram will be greatly modified when applied to natural rocks because rocks are very rarely purely composed of CaO, MgO, Al_2O_3, SiO_2, and H_2O. As hornblende is a "sponge" able to absorb almost any element, it is evident that the presence of Fe^{2+}, Fe^{3+}, Na^+, Ti^{4+}, etc., may change the diagram drastically. We shall now consider the trend of these changes.

1. If Na_2O is introduced into the system as an exchange for CaO, the most important change of the equilibrium curves will be due to the occurrence of plagioclase instead of pure anorthite. The main effect will be a narrowing of the stability field of the hornblende.

2. The introduction of FeO and Fe_2O_3 at the expense of MgO will result in several interesting changes. First, the lowermost equilibrium curves will be depressed considerably as a result of the conditions already considered on page 63. Second, the upper curves must be displaced up-

ward, as shown on page 56. That is: the introduction of FeO makes the stability field of hornblende expand toward higher, as well as lower, temperatures. Increase of the ratio Fe/Mg will stabilize garnet at the expense of cordierite, a fact which most reasonably will affect the Al-richest parts of the hornblende field.

3. As hornblende is able to absorb an appreciable amount of Ti in the lattice (often up to 3–4 per cent TiO_2, more rarely 7–8 per cent), the in-

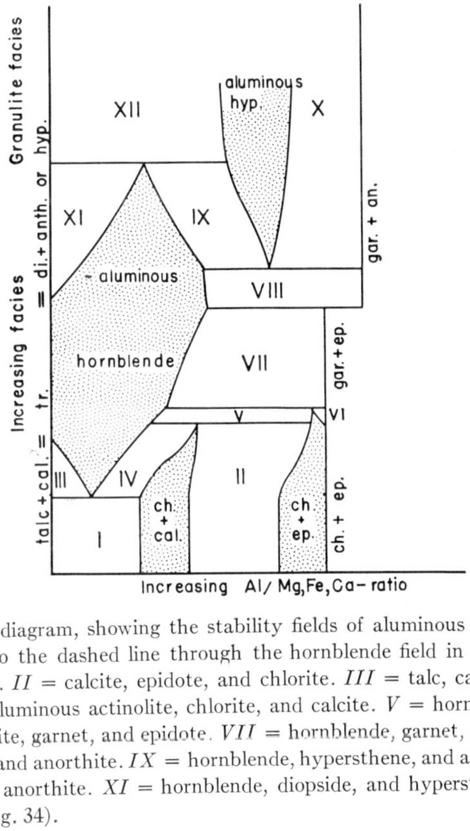

FIG. 35.—Phase diagram, showing the stability fields of aluminous hornblende. The abscissa corresponds to the dashed line through the hornblende field in Fig. 34*a–f*. *I* = talc, calcite, and chlorite. *II* = calcite, epidote, and chlorite. *III* = talc, calcite, and actinolitic hornblende. *IV* = aluminous actinolite, chlorite, and calcite. *V* = hornblende, chlorite, and epidote. *VI* = chlorite, garnet, and epidote. *VII* = hornblende, garnet, and epidote. *VIII* = hornblende, garnet, and anorthite. *IX* = hornblende, hypersthene, and anorthite. *X* = hypersthene, garnet, and anorthite. *XI* = hornblende, diopside, and hypersthene (see the corresponding fields in Fig. 34).

troduction of TiO_2 in the system will expand the stability field of hornblende. That is true only where the uppermost parts of the hornblende field are concerned, because low-temperature (-pressure) hornblendes show only a small tendency to dissolve titanium.

4. The existence of potassium is important because it tends to change chlorite, garnet, and cordierite into biotite. Without further discussion here, it can be shown that this fact indicates strongly that potassium (for example, as potash feldspar) helps to minimize the stability field of hornblende, in particular, to render Al-rich hornblendes unstable.

5. The diagram (Fig. 35) is supposed to represent mineral associations containing SiO_2 in excess as quartz. This is of significance because the hornblendes, which are stable at high temperature and pressure, have two of the eight silicon positions occupied by aluminum (Foslie, 1945). These hornblendes, therefore, behave as if they were unsaturated, with reference to silica in the same way as do most olivines; and a deficit of SiO_2 in the rock will expand the stability field of hornblende toward a higher degree

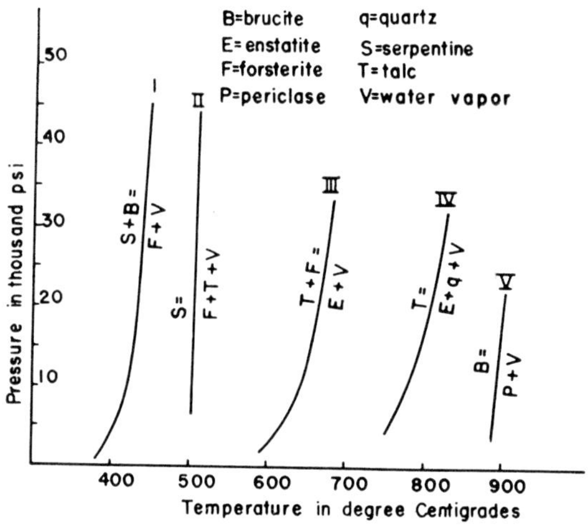

FIG. 36.—The equilibrium temperature as function of pressure for several reactions among magnesium minerals. (After Bowen and Tuttle, 1949.)

of metamorphism. In other words, reaction (77) takes place at a higher temperature than does reaction (76):

$$\underset{\text{hornblende}}{Ca_2Mg_3Al_4Si_6O_{22}(OH)_2} \underset{\text{low temp.}}{\overset{\text{high temp.}}{\rightleftarrows}} \underset{\text{enstatite}}{MgSiO_3} + \underset{\text{forsterite}}{Mg_2SiO_4} + \underset{\text{anorthite}}{2CaAl_2Si_2O_8}\,. \qquad (77)$$

THE SYSTEM $MgO\text{-}SiO_2\text{-}H_2O$

The system $MgO\text{-}SiO_2\text{-}H_2O$ is one of the very few systematically and quantitatively investigated silicate systems which are of significance for metamorphic petrology (Bowen and Tuttle, 1949). Figure 36 summarizes the equilibrium and stability relations established by Bowen and Tuttle. Of particular importance to note are: (1) forsterite and enstatite can form under stable conditions at a surprisingly low temperature (400° C.) under certain circumstances; (2) serpentine is completely unstable above about 500° C.; (3) talc is likewise unstable beyond about 750°–800° C.; and (4) anthophyllite and/or Mg-cummingtonite are not present in the equilibrium diagrams for the system $MgO\text{-}SiO_2\text{-}H_2O$.

We shall discuss point 1 a little further. From the diagrams it may be concluded that, at low temperature, forsterite and enstatite are unstable together with water vapor because talc and/or serpentine is situated between enstatite and/or forsterite and the water-vapor corner. It is evident, however, that this situation only shows that enstatite and forsterite are unstable together with water vapor of the same pressure as the total pressure at which the experiment was performed. The two anhydrous minerals mentioned may still be stable with water vapor of a partial pressure lower than the total pressure. It is easy to show that they really are, because enstatite is stable together with talc, and forsterite together with serpentine, two hydrous minerals which must have finite partial water-vapor pressures. This leads to the conclusion that enstatite and forsterite are stable below curve *I* (Fig. 36), provided that the water-vapor pressure corresponds to—or is lower than—the dissociation water-vapor pressure of talc and serpentine, respectively.

Therefore, if ultrabasic rocks contain water vapor in cracks and microfissures at a partial pressure below the dissociation pressure of talc or serpentine, then enstatite and/or forsterite is still stable at temperatures below curves *I* and *II* in Figure 36.

As a further consequence, the curves representing the following reactions:

$$\left.\begin{array}{c} S + B = F + V\,; \qquad S = F + T + V\,; \qquad T + F = E + V\,; \\ T = E + Q + V\,; \qquad B = P + V\,, \end{array}\right\} \quad (78)$$

are all displaced toward lower temperature if the partial water pressure in the rock is less than the total rock pressure. We do not know the partial water-vapor pressure which has existed in a recrystallizing rock; hence we do not know the equilibrium temperatures of the several equations above, even if the rock pressure can be calculated. The curves given by Bowen and Tuttle represent maximum temperatures at which the hydrous minerals or mineral assemblages are stable.

The metastable character of anthophyllite as indicated by the experiments is rather surprising, in view of the abundance of this mineral in some metamorphic rocks. As suggested by Bowen and Tuttle, however, the obvious stability of anthophyllite in natural rocks may be explained by the presence of Al^{3+}, Fe^{2+}, etc., in the molecule. It is, in this connection, easy to show that the admixing of Al^{3+} in the anthophyllite lattice will make it more stable relative to talc, enstatite, and forsterite than a "pure" Al-free anthophyllite would be. That is so because anthophyllite is more capable of replacing Si and Mg by Al than is talc, enstatite, or forsterite.

It is interesting to notice that olivine, with some content of iron, was not serpentinized by water vapor at the P,T-conditions given by curve I, Figure 36. The temperature had to be dropped to 340° C. at a pressure of 15,000 lb/in^2 in order to serpentinize a natural olivine with 10 mol per cent of Fe_2SiO_4. This is exactly what we should expect on the grounds that serpentine does not show the same affinity for iron in its lattice as does olivine. The situation is shown diagrammatically in Figure 37.

THE SYSTEM MgO-Al_2O_3-SiO_2-H_2O

The system MgO-Al_2O_3-SiO_2-H_2O contains many minerals of great significance for metamorphic rocks. It has been studied by Yoder (1952),

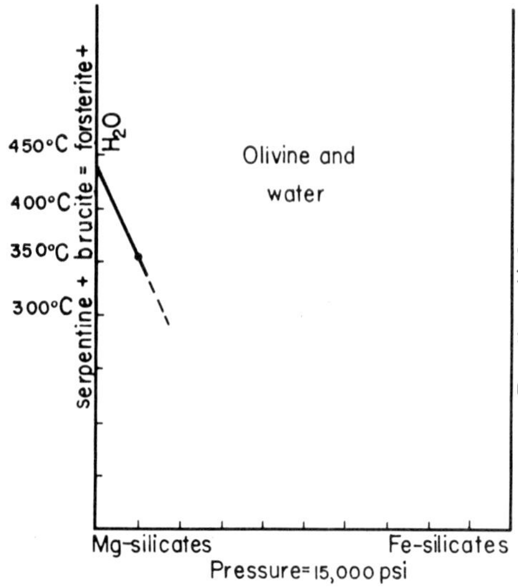

FIG. 37.—The effect of iron in olivine upon the equilibrium temperature for the reaction olivine + $H_2O \rightleftharpoons$ serpentine, brucite, and iron ore. The inclined line to the left gives the composition of olivine, which is in equilibrium with serpentine and brucite. (Based on data from Bowen and Tuttle, 1949.)

but Yoder's detailed discussion was not yet published when this book went to press. However, Dr. Yoder has been kind enough to let the writer publish some of the crucial phase relationships found by him in his experiments (Fig. 38).

SYSTEMS IN WHICH SPHENE IS INVOLVED

Sphene, although a minor constituent of rocks, plays a rather important role in the study of metamorphism. As an isolated mineral, sphene is stable from the lowest temperatures of metamorphism to the high temperature of melting; furthermore, sphene is stable at the highest pressures recorded by rocks—the pressure of granulite facies. However, at a high degree of metamorphism—in the border field between amphibolite facies and granulite facies—sphene undergoes several reversible reactions with common rock-forming minerals under liberation of rutile and/or ilmenite, so that the mineralogical environment under which sphene can exist stably at high pressure and temperature becomes very restricted. The following reactions are of particular importance:

$$\underset{\substack{\text{sphene}\\56.0}}{CaTiSiO_5} + \underset{\substack{\text{kyanite}\\44.9}}{Al_2SiO_5} \rightleftarrows \underset{\substack{\text{anorthite}\\100.8}}{CaAl_2Si_2O_8} + \underset{\substack{\text{rutile}\\19.0}}{TiO_2}\,, \tag{79}$$

Mol volumes: 56.0, 44.9, 100.8, 19.0

$$\underset{\substack{\text{sphene}\\56.0}}{CaTiSiO_5} + \underset{\substack{\text{muscovite}\\138.3}}{KAl_3Si_3O_{10}(OH)_2} + \underset{23.0}{SiO_2} \rightleftarrows \underset{\substack{\text{anorthite}\\100.8}}{CaAl_2Si_2O_8} + \underset{\substack{\text{potash feldspar}\\109.1}}{KAlSi_3O_8} + \underset{\substack{\text{rutile}\\19.0}}{TiO_2} + H_2O\,, \tag{80}$$

Mol volumes: 56.0, 138.3, 23.0

$$\underset{\substack{\text{sphene}\\4\times56.0}}{4CaTiSiO_5} + \underset{\substack{\text{cordierite}\\225.7}}{Mg_2Al_4Si_5O_{18}} \rightleftarrows \underset{\substack{\text{anorthite}\\2\times100.8}}{2CaAl_2Si_2O_8} + \underset{\substack{\text{diopside}\\2\times66.1}}{2CaMgSi_2O_6} + \underset{\substack{\text{rutile}\\4\times19.0}}{4TiO_2} + \underset{23.0}{SiO_2}\,, \tag{81}$$

Mol volumes: 4 × 56.0, 225.7

$$\underset{\substack{\text{sphene}\\4\times56.0}}{4CaTiSiO_5} + \underset{\substack{\text{garnet}\\114.8}}{Mg_3Al_2Si_3O_{12}} + \underset{23.0}{SiO_2} \rightleftarrows \underset{\substack{\text{anorthite}\\100.8}}{CaAl_2Si_2O_8} + \underset{\substack{\text{diopside}\\3\times66.1}}{3CaMgSi_2O_6} + \underset{\substack{\text{rutile}\\4\times19.0}}{4TiO_2}\,, \tag{82}$$

Mol volumes: 4 × 56.0, 114.8, 23.0, 100.8

$$\underset{\substack{\text{sphene}\\56.0}}{CaTiSiO_5} + \underset{\substack{\text{hypersthene}\\31.6}}{MgSiO_3} \rightleftarrows \underset{\substack{\text{diopside}\\66.1}}{CaMgSi_2O_6} + \underset{\substack{\text{rutile}\\19.0}}{TiO_2}\,. \tag{83}$$

Mol volumes: 56.0, 31.6, 66.1, 19.0

The foregoing reactions go to the right at a high degree of regional metamorphism (granulite facies), as indicated by the fact that sphene is not found with any of the minerals on the left-hand side of the several reactions at the conditions of granulite facies. The molal volumes are given in these equations in order to indicate the possible effect of high pressure. We realize that none of these reactions is fixed by a single curve in the P,T-field because of the mixed crystals taking place in the reactions.

Let us first consider reactions (79) and (80). It is plainly seen that these reactions are of the same general type as the "saussurite" reactions (56) and (57) (pp. 50–54). Both reactions go to the right-hand side above certain P,T-combinations, according to experience with metamorphic rocks. If albite is introduced into the systems, we arrive at equilibrium diagrams similar to Figure 39. The equilibrium temperature and/or pressure will in either case be depressed gradually by increasing concentration of albite in the plagioclase. It should be noticed that the equilibrium temperatures of equations (79) and (80) are not known. So far, it may even

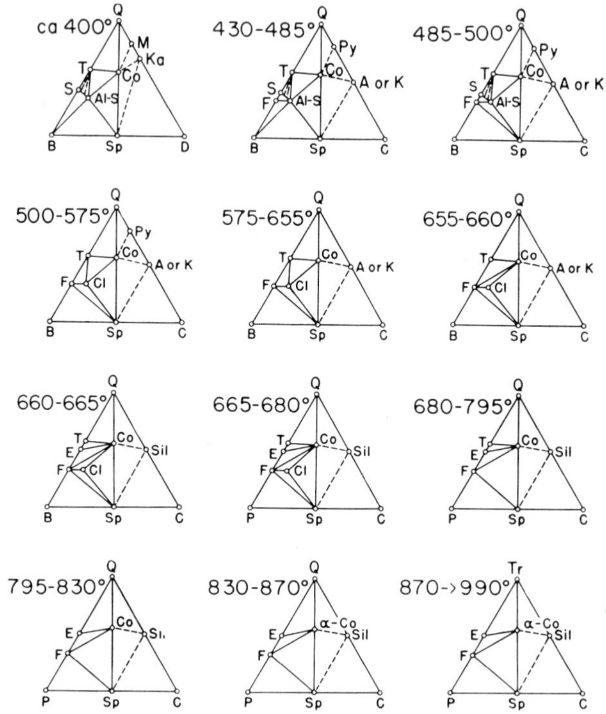

FIG. 38.—Phase relationships of the MgO-Al_2O_3-SiO_2-H_2O system at various temperatures and at 15,000 psi water-vapor pressure (after Yoder, 1952, and personal letter of March 13, 1952). *A* = andalusite; *Al-S* = aluminous serpentine; *B* = brucite; *C* = corundum; *Cl* = clinochlore; *Co* = cordierite; *D* = diaspore; *E* = enstatite; *F* = forsterite; *K* = kyanite; *Ka* = kaolinite; *M* = montmorillonite; *P* = periclase; *Py* = pyrophyllite; *Q* = quartz; *S* = serpentine; *Sil* = sillimanite; *Sp* = spinel; *T* = talc; *Tr* = tridymite.

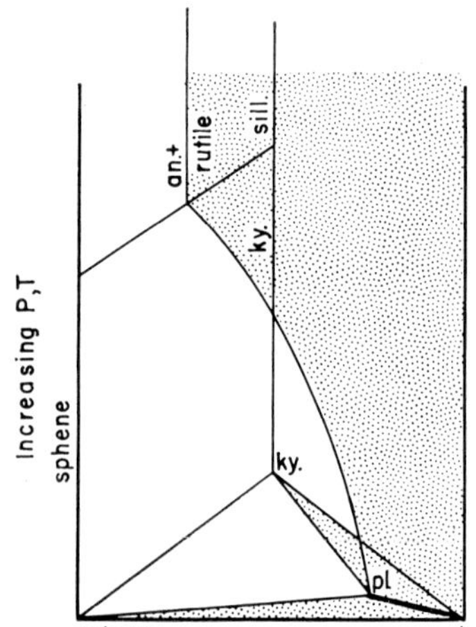

FIG. 39.—Tentative diagram, showing the composition of plagioclase (the curve from point *Pl.*) in equilibrium with sphene and kyanite.

be that the equilibrium temperature is situated beyond the melting temperature, so that the upper part of diagram (39) represents an unstable equilibrium.

Only intermediate and acid plagioclases are recorded from rutile-carrying rocks. These parageneses may correspond to equilibrium in the lower part of Figure 39.

In reactions (81), (82), and (83), not only the composition of the plagioclase but also the Mg/Fe ratios of the ferromagnesian minerals control the temperature-pressure curves of equilibrium.

In natural rocks, the conditions are not so simple as those pictured here. Very often, for example, rutile is not liberated, but the TiO_2 expelled by the reactions may combine with some iron oxide under formation of ilmenite. In other cases the titanium from the reacting sphene enters into biotite or hornblende, both of which are specially enriched in TiO_2 under the conditions of high-grade regional metamorphism.

Many of the stability conditions of minerals as discussed in this chapter are in excellent agreement with the theory of crystal energy and can be predicted to a certain extent by means of theoretical crystal energetics. This science is therefore a valuable tool in the interpretation of petrology. The reader can find some information on crystal energetics in the Appendix.

CHEMICAL KINETICS OF METAMORPHISM

METAMORPHIC rocks usually contain mineral assemblages corresponding approximately to chemical equilibrium at a certain P,T-combination. These states of equilibrium are definitely determined by the bulk chemical composition of the rock and by the P,T-relations. The fact that states approaching equilibrium are generally reached in rocks subjected to change of pressure or temperature shows that rearrangement of matter within rock bodies takes place rather extensively under certain circumstances which are met with in the earth's crust.

To explain the fact that solid rocks undergo recrystallizations (and neomineralization) of different kinds, long ago a "pore fluid" was called upon as the conducting medium, in the same way that the physicists called upon an ether to transfer light waves. It is now known that the artificial ether is nothing but the fields of forces surrounding the most minute constituent particles of matter as well as huge heavenly bodies. In many respects the pore fluid in metamorphism is nothing but an infantile materialization of the reactive ions, atoms, and molecules in rocks.

We are familiar with the mobility and chemical reaction within solid lattices, of which perthitization of feldspar provides an example. If the perthitization by unmixing had had to be explained in the days of orthodox belief in the axiom *corpora non agunt nisi fluida*, it would have been assumed that the fluid pore solution penetrated bodily into the crystalline lattice, there to perform the chemical process of unmixing. Today this kind of process is explained more simply, because it has been revealed that the constituents of crystals are not entirely fixed in their sites inside the lattices.

CRYSTAL KINETICS

Fundamental to a discussion of the kinetics of any system of condensed atomistic particles is Maxwell's law of distribution of kinetic energies or velocities in such systems. In a gaseous, as well as liquid or crystalline, body at any finite temperature, the fraction of molecules (ions, atoms) for which the velocity is between v and $v + dv$ is:

$$4\pi \left(\frac{m}{2\pi kT}\right)^{3/2} v^2 e^{-mv^2/2kT} dv , \tag{84}$$

where m is the mass of the particle and k is Boltzmann's constant, which is equal to the gas constant R divided by Avogadro's number. Plotting the Maxwellian function (Fig. 40) shows that it starts at zero for zero velocity, reaches a maximum at the most probable speed at the given temperature ($v = \sqrt{2kT/m}$), and asymptotically approaches zero again when v approaches infinity (see Slater, 1939, pp. 56–58).

The mean kinetic energy of the particles is $\frac{3}{2}kT$. Now it is of particular interest for our purpose to note that it is possible by some simplifications and integration to determine the fraction of particles having a kinetic energy greater than a certain limit, say ϵ_0. Thereby can be determined the percentage of activated particles in or on a condensed phase which are able to undergo chemical reaction or activated diffusion. The percentage of particles in a liquid, gaseous, or crystalline assembly having kinetic energy greater than ϵ_0 is proportional to $e^{-\epsilon_0/RT}$, where R is the gas constant, so that, if ϵ_0 is the average bonding energy per mol of the phase,

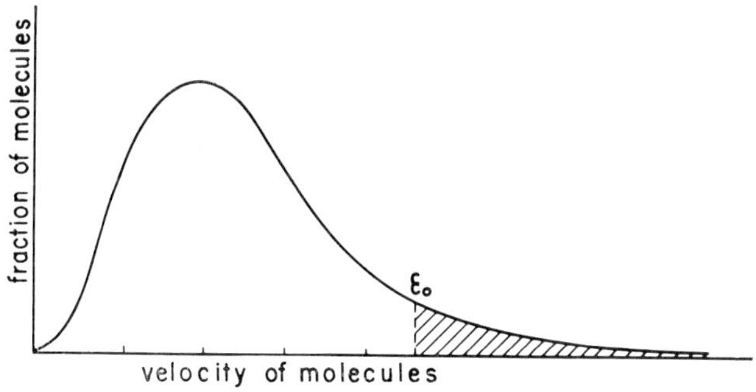

Fig. 40.—The distribution of velocities in an assemblage of molecules

the percentage of activated or reactive particles is given by the $e^{-\epsilon_0/RT}$ function.

It is rather interesting for petrologists to note that, regarding the common physicist's viewpoint of the crystalline state, the well-known physicist, J. Frenkel, writes, as late as 1946: "Crystalline bodies are usually described as possessing a perfect regular structure. The only deviation from perfect regularity which is taken into account consists in the vibration of the atoms about their equilibrium positions, which are assumed to constitute an ideal three-dimensional crystal lattice. . . . From this point of view, however, such processes as the evaporation of a crystal, its dissolution in a liquid medium, or the mutual diffusion of two crystalline substances would be utterly unintelligible." Frenkel had in mind here the fact that the so-called reversible "imperfections" in crystals are neglected and overshadowed by the great amount of knowledge about the static structural building of crystals achieved with the aid of X-ray equipment.

Regardless of the structural architecture of a crystal, it is evident that its constituents are not kept together by infinitely large bonding energies. At a temperature beyond absolute zero, there consequently must be a certain fraction of the building constituents which possess large enough energy to loosen from their regular sites.

It is easy to imagine that energy-rich particles will loosen from the surface layer of a crystal and escape into the surrounding vacuum or gaseous or liquid media in which the particles can dissolve. It is not so obvious, however, that, inside close-packed crystals, atoms or ions can escape from their sites of equilibrium about which they oscillate. There simply seem to be no holes or free spaces to escape into. But even in theoretically close-packed lattices, particles may happen to escape from sites of equilibrium for different reasons. First, a particle escaping from the surface layer leaves an open site which may be filled by a particle from deeper inside the crystal. By this mechanism, the hole itself migrates into the volume of the crystal. At thermodynamic equilibrium at a given temperature, a particular kind of crystal, with an exactly defined composition, possesses a definite number of holes per volume. These holes will, under equilibrium conditions, migrate back and forth in the crystal in such a manner that no net one-way transfer of holes takes place. Evidently, the migration of a hole in one direction is identical to the migration of a particle in the opposite direction.

Suppose that a perfectly ideal close-packed crystal exists at absolute zero temperature. This crystal completely fills a container with movable walls. Let us increase the temperature and keep the pressure on the expanding walls of the container just equal to (or somewhat greater than) the vapor pressure, so that no vapor develops. Under thermodynamic equilibrium at a given temperature, the crystal possesses a certain concentration of holes or vacancies. As no particles have evaporated, the vacant sites have developed simply by expansion of the total volume of the crystal. That is, the formation of vacant sites in a close-packed crystal is not necessarily due to the escape of surface particles into surrounding solvents or vacuum. Once one realizes that holes will form at somewhat elevated temperatures in close-packed lattices, one also realizes the possibility of migration of particles via these holes.

One may imagine that displacement of ions or atoms takes place in a close-packed crystal even if no hole exists, namely, by partial rotation of small clusters of atoms (Zener, 1950). If the three atoms *a*, *b*, and *c* receive energy impulses in the proper direction and of proper magnitude simultaneously, then a partial turning of the cluster will take place (Fig. 41). Atom *a* will come to *b*'s old site, and a later turning of groups *a*, *d*, and *e*, may move *a* still farther away from its original place, and so on.

That this picture of a close-packed crystal is true in principle is evidenced by the phenomenon of self-diffusion in close-packed metals (see Jost, 1937; Barrer, 1941). A certain small fraction of the atoms in a monatomic metal crystal are statistically jumping from place to place inside the crystal.

The mean displacement, X_m, of an atom (ion) in unit time is determined by the self-diffusion coefficient, D, according to the following equation:

$$X_m = \sqrt{2Dt}\,, \qquad (85)$$

where X_m is the displacement in time, t. Table 2 gives the diffusion coeffi-

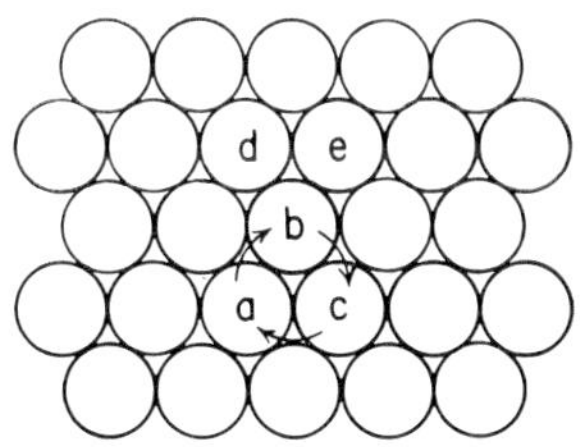

FIG. 41.—Schematic visualization of the "rotation-of-clusters" idea of diffusion in close-packed lattices (see text).

TABLE 2

DIFFUSION COEFFICIENT OF SOME METALLIC AND SALT SYSTEMS (AFTER BARRER, 1941)

SYSTEM	D_0 (CM^2SEC^{-1})	E (CAL/MOL)	D_T (CM^2SEC^{-1})			
			At 400° K.	At 500° K.	At 800° K.	At 1,000° K.
Pb in Pb	5.1	27,900	2.96×10^{-15}	3.32×10^{-12}	1.22×10^{-7}	4.64×10^{-6}
Au in Au	1.26×10^{2}	51,000	1.76×10^{-26}	1.76×10^{-21}	1.38×10^{-12}	9.07×10^{-10}
Cu in Cu	1.1×10	57,200	6.38×10^{-31}	1.1×10^{-24}	2.64×10^{-15}	3.52×10^{-12}
Cd in Cu	3.5×10^{-9}	8,200	1.16×10^{-13}	9.10×10^{-13}	2.07×10^{-11}	5.6×10^{-11}
C in Fe	4.9×10^{-1}	36,600	4.9×10^{-21}	4.9×10^{-17}	4.9×10^{-11}	4.9×10^{-9}
Mo in W	6.3×10^{-4}	80,500	6.93×10^{-48}	4.28×10^{-39}	6.3×10^{-26}	1.64×10^{-21}
U in W	1.0	100,000	2.4×10^{-55}	2.0×10^{-44}	4.9×10^{-28}	1.4×10^{-22}
Ag in Pb	7.5×10^{-2}	15,200	3.75×10^{-10}	1.73×10^{-8}	5.25×10^{-6}	3.60×10^{-5}
Au in Pb	4.9×10^{-1}	13,000	3.92×10^{-8}	1.03×10^{-6}	1.37×10^{-4}	6.86×10^{-4}
Th in W	1.0	120,000	2.80×10^{-66}	3.60×10^{-53}	1.70×10^{-33}	6.20×10^{-27}
Pb in PbCl	7.7	36,800	6.16×10^{-20}	6.39×10^{-16}	6.85×10^{-10}	7.11×10^{-8}
Cu in Ag_2S	46×10^{-5}	3,180	6.44×10^{-6}	1.89×10^{-5}	5.98×10^{-5}	2.9×10^{-5}
Ag in Cu_2S	32.7×10^{-5}	4,570	1.05×10^{-6}	3.27×10^{-6}	1.83×10^{-5}	3.27×10^{-5}
H_2 in Ni	2.04×10^{-3}	8,700	....	....	....	....
H_2 in Fe	1.65×10^{2}	9,200	....	....	....	....
Na on OW-surface	....	4,400–7,400	....	5.9×10^{-4}	3.3×10^{-3}	....
K on OW-surface	....	~15,300	....	1.1×10^{-5}	3.4×10^{-3}	....

cient for a number of metals. Table 3 gives the mean displacements for some diffusion systems.

The self-diffusion coefficient increases with temperature as follows:

$$D_T = D_0 \cdot e^{-E/RT}, \tag{86}$$

where E is the energy of activation per mol of the activated diffusion process. Table 4 gives some energies of activation of self-diffusion com-

TABLE 3

MEAN DISPLACEMENT OF DIFFUSING PARTICLES IN 10^8 YEARS

Substances	At 500° K. (Cm.)	At 800° K. (Cm.)	Substances	At 500° K. (Cm.)	At 800° K. (Cm.)
Pb in Pb	~100	~20,000	Th in W	$\ll 1$	<1
Au in Au	<1	60	Pb in PbCl	~1	2,000
Cu in Cu	<1	2	Cu in Ag_2S	200,000	400,000
Cd in Cu	40	300	Ag in Cu_2S	100,000	200,000
C in Fe	<1	400	Na on OW	1,000,000	5,000,000
Mo in W	$\ll 1$	$\ll 1$	K on OW	400,000	5,000,000
U in W	$\ll 1$	$\ll 1$			
Ag in Pb	10,000	100,000			
Au in Pb	60,000	600,000	H_2O in heulandite at 75° C.	~10,000	....

TABLE 4

HEAT OF VAPORIZATION, E_{vap}, AND HEAT OF ACTIVATED SELF-DIFFUSION, E_{diff}, IN THE EQUATION: $D_T = D_0 e^{-E/RT}$

Substance	D_0 (Cm^2Sec^{-1})	E_{diff} (Cal/Mol)	E_{vap} (Cal/Mol)
Pb	6.4	28,000	47,000
Au	1.85	51,000	92,000
Cu	11.0	57,200	81,000
Bi	10^{-3}	31,000	47,000

pared with the energy of vaporization of some metals (Frenkel, 1946). Note that the heat of diffusion is roughly proportional to the heat of vaporization. It is important to note that D increases with temperature according to an exponential function, showing that the effect of temperature may be very great. Below a certain temperature interval, the self-diffusion may be entirely negligible, whereas, above the often narrow temperature interval, self-diffusion may be of the greatest consequence. Table 2 shows some calculations of change of D with T for a few substances.

A close packing is very common among monatomic metals. As the metallic bonding forces usually can be distributed evenly between an atom and any number of neighbors, the atoms will tend to surround themselves with as many neighbors as possible on geometrical grounds, and the bulk crystal will be a building of closely packed spheres, provided that the temperature is not too high.

The common minerals are not so simply built. A mineral is usually composed of a complex assemblage of different ions having different ionic radii, different number of charges, and also partly possessing directed valency forces of some homopolar nature (for example, the Si-O bonds [Wells, 1945, p. 74]). The response to the requirements of all these different properties of the constituents cannot usually be a close-packed lattice. A close packing of such assemblages of different constituents may result in a great excess of the positive or negative charge of the lattice and/or in severe bending strain of the directed valency forces. The excess of electric charge or the bending of valency forces decreases the lattice energy to such an extent that it cannot be compensated for by the gain in energy due to the smaller mean distance between the constituents in a close-packed configuration. The stablest lattices are therefore those which are almost completely neutral electrically and possess no severe bending or super- or undersaturation of the valency forces. Only in the case of the simplest ionic crystals may such structures also be close-packed.

In a number of silicate minerals, there exist openings and channels of atomic dimensions which are thermodynamically stable even when the reversible imperfections due to thermal agitation are disregarded. On purely geometrical grounds, therefore, it is possible that individual particles can migrate within the volume of many of the mineral lattices. Along the *c*-axis of quartz, for example, there are open channels so large that silver ions as well as Li^+ and Na^+ can pass rather easily (Chentsova and Vedeneeva, 1950). Similar channels are present in beryl and cordierite. Evidently, also, the feldspars are fairly open, and the micas as well as the hornblendes have open channels in their lattices. However, the geometry and the question of openings in the lattice are only one side of the problem; another more important side is the question of the energy required to loosen and move a particle along possible openings in the silicate structures. The available X-ray data of mineral structures can inform us in many cases as to whether holes and channels are available for migration of ions inside the several minerals; but there exist practically no data of the energy required for such migrations in natural minerals.[1] Although

1. That the question of bond energy is highly significant for the mobility of ions inside silicate lattices is clearly shown by the contrast in mobility in alkali feldspars and in plagioclases. The mixing or unmixing of alkali feldspar requires only that the univalent and hence weakly bonded sodium and potassium move in the lattice. Hence this process goes very rapidly and can readily be observed in laboratories under the microscope at fairly moderate temperatures (F. Laves and J. Goldsmith, oral communications). The homogenization of a zoned plagio-

we do not know the self-diffusion coefficient of common minerals, we must accept the fact that such a mobility of individual particles takes place inside mineral lattices, not only because of the phenomena of unmixing or homogenization of mixed crystals found in natural rocks but also because of the experimentally established fact that order-disorder transitions take place in many minerals (Buerger, 1948). The disordering taking place at increasing temperature and the reverse ordering occurring at lowering temperature are verifications of the theoretical conclusion that a certain migration of energy-rich particles must take place in crystals, even at moderately elevated temperatures.

As long as no direct mobility data are available for common minerals, we shall present values from other crystalline substances so that we may get some information on the order of magnitude to be expected and the limits inside which to operate. Tables 2, 3, and 4 give some data.

The kinetics of crystals, here briefly outlined, is conceivable on grounds of the open structures and the reversible and thus thermodynamically stable lattice imperfections (Hedvall, 1942*b*). These imperfections are determined definitely by pressure and temperature for a given crystal of given composition (and size, provided that the size is below a certain limit). By changing pressure and temperature continuously within the stability range of the given crystal, the "concentration" of imperfections and the amount of movable particles also change continuously, generally being a positive function of temperature. These kinds of imperfections may be parallelized with the dissociation of molecules into atoms in a molecular gas mass or the dissociation of molecules into ions in a liquid solution or a silicate melt. The microscopic differences in state (energy, velocity, position, etc.) of the several particles building up the crystal phase, all certainly correspond to the same macrostate, namely, the state of the particular crystal we consider.

It has long been known that crystals also contain different imperfections which have been referred to as irreversible, or thermodynamically unstable, imperfections ("irreversible," *Gitterauflockerung* [von Hevesy, 1928, 1933]). This sort of irregularity in a given lattice depends upon the history of the crystal and not only upon the pressure, temperature, and composition (and size). The irreversible imperfections, therefore, tend to

clase, however, requires not only that calcium and sodium exchange places, which most likely would be a rather easy process, but also that aluminum and silicon exchange places. Because very large amounts of energy are required to move the trivalent Al or the tetravalent Si out of their oxygen tetrahedra, this process will be an exceedingly sluggish one. For that reason it is very natural that zoned plagioclases are common in rocks because they will find it almost impossible to homogenize when once formed unless a thorough recrystallization takes place, as pointed out by Goldsmith (1952).

disappear in the course of time if the crystal is left to itself for a sufficient period of time at a temperature at which the thermal mobility inside the crystal is appreciable. This immediately gives us an important fact: the irreversible imperfections have increased the free energy of the crystal as compared with a crystal with mere reversible imperfections. (Incidentally, we also note that a completely ordered crystal without any imperfections whatsoever has, at elevated temperature, higher free energy than the corresponding crystal with its required amount of reversible imperfections).

Irreversible imperfections may be of several kinds, of which some examples will be given.

1. A crystal which has existed for a long period of time at high temperature and has thereby acquired the proper "concentration" of reversible imperfections at this temperature. If the temperature is now dropped suddenly, the imperfections, being reversible at elevated temperature, may still remain in the crystal but now as irreversible imperfections.

2. A crystal exposed to stress, rapidly changing P or T, or too rapid growth in a supersaturated system develops small fissures and flaws which do not belong to the reversible structure of the crystal. The imperfect crystals consist of a mosaic of blocks or stringers bounded by minute cracks of "molecular" thickness, and not strictly parallel-oriented. This mosaic or lineage structure apparently develops in most crystals not grown under the most carefully controlled conditions (Buerger, 1934; Barrer, 1941, pp. 313–21). Evidences for the mosaic structure are found by X-ray studies (Bunn, 1946, pp. 362–67) and also by study of the electric conductivity on surfaces of crystals, of diffusion through crystals, and of the ability of some crystals to "dissolve" appreciable amounts of molecules too large to enter the true lattice. So, for example, H_2O, alcohol, and ether could be forced into quartz at 15,000 atm. pressure (Poulter and Wilson, 1932). The shattering of the quartz upon release of the heavy pressure showed that the liquids found their way into cracks such that the expansion of the liquids shattered the crystals. It is interesting to note that large molecules like glycerine and oil could not be forced into the mosaic fissures in quartz, thereby indicating an upper limit for the width of the cracks.

Such irreversible imperfections may, as long as they last, provide channelways for the migrating particles inside the crystalline body. If the irreversible imperfections are not continuously created by some external agencies, as, for example, a continuously acting mechanical stress, they will ultimately tend to disappear by the very diffusion and motion of the particles constituting the crystal. It may often happen that the free-energy difference between the crystal with irreversible imperfections and that with reversible imperfections is so low that no annealing will take place (Kochendörfer, 1941). For that reason, mosaic and lineage structures in crystals are the rule and not the exception.

We have seen, so far, that crystalline individuals are not dead bodies in accordance with the phrase *corpora non agunt nisi fluida*. Indeed, chemical reactions may, under proper temperatures, take place extensively by a migration of activated particles inside lattices, along their surfaces and mosaic fissures. A growing body of experiments proving this statement is now developing, and the reader is referred to Barrer, Jost, Hedvall, and others.

On the other hand, rocks are not composed of one huge single crystal (except some pegmatites); rocks are crystalline aggregates with a great area of crystal interfaces and surfaces. Just as some metamorphic reactions must take place by motion of particles through the volume of minerals (exsolution of mixed crystals, etc.), other petrogenetic processes require extensive motion of particles along mineral boundaries and in the interstices between the mineral grains. One knows that CO_2 and H_2O, for instance, do not usually accumulate in a rock after these compounds have been liberated by some mineral reactions which we know have taken place in the rock during its history. In these instances it seems evident that water and carbon dioxide molecules have migrated along the surfaces of the mineral grains, or partly perhaps along some mosaic fissures, because these particles are too large to penetrate most mineral lattices with sufficient ease. (So far, zeolite lattices are sufficiently open to permit H_2O, NH_3, CO_2, etc., to diffuse inside these lattices rather rapidly [Tiselius, 1934; Hey, 1935]). Table 5 gives some data. One may tentatively assume that the transfer of H_2O and CO_2 does not occur as molecules or HO^- and CO_3^{2-} ions, but rather as H^+ and C^{4+} ions, which are so small that the geometry of the lattices should not provide difficulties. Bugge (1945) suggested that H^+ is moving rather than H_2O. But, when one considers the large bonding energies between H and O as well as between C and O, such a transfer of H^+ and C^{4+} through mineral lattices appears highly improbable as a process of petrologic importance.

There are other evidences, also, of intergranular or interface transfer. Most chemical transport of large particles inside solid rocks must, for structural reasons, extend along the mineral surfaces and interstices. We know of numerous cases of metasomatic ore bodies in which sulfur and oxygen, both large atoms or ions, have exchanged place during the metasomatic evolution of a solid rock. In such cases one cannot assume that lattice diffusion played an important role.

In addition, the occurrence of reaction products along mineral borders in some metamorphic rocks indicates that the intergranular space was an important channelway of transport.

Experiments also show that the intergranular is often the easiest way along which minerals can communicate chemically. The catalytic effect of "fluxes"—particularly water and fluorine—on reaction among solid minerals (Bowen and Schairer, 1935*a;* Griggs, 1940; Fairbairn, 1949) strongly indicates the surfaces or interfaces to be the actual avenue of the processes, because the "fluxes" do not dissolve appreciably in the minerals.

The surface of crystalline bodies provides conditions of the greatest consequence for the phenomenon of reaction in the solid rocks. The surface of crystals has a residual field of force, so that there is a tendency for

TABLE 5

DIFFUSION OF WATER IN HEULANDITE ALONG THE DIRECTIONS [201] AND [001] (AFTER TISELIUS, 1934)

Temperature (° C.)	$D_{(201)}$ (Cm^2Sec^{-1})	$D_{(001)}$ (Cm^2Sec^{-1})	Temperature (° C.)	$D_{(201)}$ (Cm^2Sec^{-1})	$D_{(001)}$ (Cm^2Sec^{-1})
20.0	2.7×10^{-7}	2.3×10^{-8}	60.1	7.6×10^{-7}	1.45×10^{-7}
33.8	4.1×10^{-7}	4.5×10^{-8}	75.0	1.11×10^{-6}	2.8×10^{-7}
46.1	4.8×10^{-7}	6.6×10^{-8}			

all crystals to absorb on their surfaces practically any type of atoms or stable molecules (ions) (Weiser, 1939, p. 24; Adam, 1941). In particular, if a crystal is in contact with a gas or vapor, the concentration of the molecules (atoms) of the gas or vapor will be greater on the surface of the crystal than in the gas in bulk. This fact, in connection with the observation that the lateral mobility of adsorbed particles on surfaces is usually (Volmer, 1939) very great, is important for metamorphic petrology. We shall not go into the theory of adsorption in great detail here but merely give a simplified picture of the phenomenon.

There exists a relationship between the pressure of a gas and the concentration of the adsorbed gas on the surface of a given crystal at constant temperature (Freundlich's and Langmuir's "adsorption isotherms"), such that the concentration on the surface increases with increasing gas pressure. On the other hand, the effect of rising temperature at constant pressure is to decrease the concentration of adsorbed particles (the "adsorption isobar"). If the pressure of the adsorbed gas increases at constant temperature, the surface concentration of adsorbed particles also increases, and the surface may be supersaturated in such a way that either (1) there is formed a continuous and stable unimolecular (two-dimensional) layer on the surface (Langmuir) or (2) three-dimensional crystallites or liquid (amorphous) clusters of the adsorbed substance develop on the surface of the adsorbent.

The pressure of the gas in equilibrium with a stable adsorbed layer one molecule (atom) thick—a monolayer—is usually only a small fraction of the vapor pressure of a macrobody of the adsorbed substance in its condensed state. By increasing the pressure of the gas, another layer may form on top of the monolayer, and so on until the multilayer becomes so thick that its thermodynamic properties (e.g., vapor pressure) approach closely those of a condensed macrobody of the adsorbed substance.

The pressure of a gas in equilibrium with three-dimensional crystallites or liquid clusters on an adsorbent is approximately the same as the vapor pressure of the condensed adsorbed substance in bulk.

TABLE 6

STABLE FILMS (AFTER BARRER, 1941)

	Heat of Adsorption of Some Stable Films (Cal/Mol)
Cs on W.	40,800– 65,100
Th on W.	172,000–178,000
Na on OW.	17,000– 32,000

An approximate criterion on whether a monolayer or three-dimensional crystallite (or liquid cluster) of an adsorbed substance is stable is provided by the heat of adsorption as compared with the heat of condensation in bulk. When considering adsorption of metals on metals, glass, quartz, mica, etc., it has been found that, if stable monolayers form, the heat of adsorption is greater than the heat of vaporization of the adsorbed metal; if crystallites (or unstable films) are formed, the heat of adsorption is much less than the heat of vaporization of the adsorbed metal (Barrer, 1941, p. 341). Table 6 gives some thermochemical data from stable films on crystals.

There exist several qualitative observations indicating great lateral mobility of adsorbed metals in cases of unstable films, but, unfortunately, no measured diffusion coefficients are given in the literature. For stable monolayers, on the other hand, quite a number of mobility data are available. Table 7 gives the surface diffusion coefficient of Na and K on WO, of Cs on W, and of Th on W. It is interesting to note that the mobility of Na and K along the WO surface is of the same order of magnitude as in watery solutions. Using the formula $x = \sqrt{2Dt}$ (p. 79), we may calculate that the average Na particle travels along a WO surface some 10^5 cm. (1 km.) in the course of 10^6 years at $t = 600°$ K.

One has every reason to expect the surface mobility in the cases of unstable films to be considerably greater than the values given above for the stable films. It is found that the activation energy of surface diffusion, in the investigated cases, is some one-fourth to one-eighth of the heat of adsorption. The heat of adsorption of the unstable films given in Table 8 is known, and, if we assume that the heat of activation of surface diffusion, in these cases also, is some one-fourth to one-eighth of the heat of adsorption, we may conclude that the lateral mobility of such unstable films is considerably greater than that of stable films (Barrer, 1941).

TABLE 7

SURFACE DIFFUSION ON OW (AFTER BARRER, 1941)

SUBSTANCE	*D* (CM²/SEC)				
	At 500° K.	At 600° K.	At 700° K.	At 800° K.	At 1,535° K.
Na	5.9×10^{-4}	1.77×10^{-3}	2.78×10^{-3}	3.3×10^{-3}	....
K	1.1×10^{-5}	1.9×10^{-4}	1.26×10^{-3}	3.4×10^{-3}	....
Cs	1.5×10^{-7}	1.6×10^{-6}	8.5×10^{-6}	3.0×10^{-5}	....
Th	....	....	....	....	1.8×10^{-9}

TABLE 8

UNSTABLE MONOLAYERS: HEAT OF ADSORPTION OF SOME METALS ON FOREIGN SOLIDS, H_1, COMPARED WITH THE HEAT OF CONDENSATION OF THE METALS ON THEMSELVES, H_2 (AFTER BARRER, 1941)

Substances	H_1 (Cal/Mol)	H_2 (Cal/Mol)	Substances	H_1 (Cal/Mol)	H_2 (Cal/Mol)
Cd on glass	3,500	28,000	Cd on Ag	5,000	28,000
Cd on Cu	3,000	28,000	Hg on Ag	2,500	18,500

The surface mobility, as well as the heat of adsorption and heat of activation of surface diffusion, changes appreciably with concentration of adsorbed particles in the surface layer. The higher the surface concentration, the greater the mobility.

Little is known about the nature and type of particles which exist on mineral surfaces and interfaces in rocks under conditions favorable for recrystallization. There can be no doubt, though, that such adsorbed particles are present wherever space is available. Space is available not only between mineral grains but also in the mosaic fissures, which, indeed, may be wide enough to permit entrance of even large molecules, as we have seen above.

It is not reasonable that continuous films of the adsorbed particles (H_2O, CO_2, etc.) cover the interfaces completely, but it is more likely that along parts of the interfaces minerals are directly contacting and grown together.

Although no experimental information exists as to the character of the adsorbed particles along the intergranular borders and mosaic fissures in rocks, some estimations may be made. It seems evident that the most abundant adsorbed substance is simply H_2O. It is natural to assume that water would constitute a stable monolayer on most minerals even at rather low water-vapor pressure in the rock, because silicates are "wetted" by water, thus showing that the bonding forces between water molecules and mineral surfaces are rather strong. (Glass is always covered by an H_2O layer if not evaporated carefully; see also Frederickson, 1951.) It is not known, though, which water-vapor pressure is required to form such monolayers on minerals at different temperatures, but it must evidently be much less than that required to condense liquid water in bulk. Besides H_2O molecules, all the elements constituting the several minerals must be present in one form or another, adhering to the surfaces. As fluorine and chlorine are almost never completely lacking in rocks, it is reasonable to assume that molecules like metal fluorides and chlorides are present among the adsorbed surface particles. It is also likely that oxides and hydroxides are present among the adsorbed molecules.

Whatever the conditions in the intergranular, they are by no means identical to an ordinary liquid watery solution. In order that such a solution may exist, a minimum free space is needed, so that a truly condensed liquid body, and not only individual molecules, can exist in the space available. Furthermore, the water-vapor pressure in the rock must be so high that ordinary liquid water will condense.

The water film on the mineral surfaces may be as dense as ordinary liquid water, and it would be in equilibrium with a water vapor wherever free space is available for gaseous phases. The concentration of mineral-building dispersed particles in this adsorbed water film cannot be directly predicted from the available data of solubility of minerals in watery solutions. It is to be expected that the concentration of individual mineral-building ions on the mineral surfaces is very much greater than the concentration in the bulk of a saturated watery solution (Weiser, 1939, pp. 89 ff).

The previous discussion may be summarized in the following picture of a metamorphic rock. The main constituents are the crystalline minerals, which are grown together tightly, but with some "free" space at places along the grain boundaries and in the mosaic fissures. Adhered to the surfaces along the "free" spaces are water, carbon dioxide, fluorine, and other "fugitive" substances, together with the ions and molecules constituting the minerals in bulk, the relative concentration in the "film" being greatly different from the relative concentrations in the minerals as well as in the true pore fluid, if such is present. When the "free" space between the mineral lattices is too great to be entirely filled with the adsorbed film, the excess space must be occupied either by a true liquid solution or by a vaporous solution. Assuming mutual chemical equilibrium throughout such a rock, the ionic constituents will be distributed among the several phases in a certain manner, every element constituting a certain fraction of any phase, depending upon the partition function. The partition function would often be such, however, that some elements, which constitute the major part of one phase, enter other phases to only a negligible extent. For example, while water constitutes the major part of a liquid pore solution, some of the minerals may contain only traces of water (or the elements constituting water). In a gaseous pore solution, CO_2 may be the major molecule; it may constitute a lesser fraction of the adsorbed films; and eventually some minerals contain only weak traces of the constituents of carbon dioxide. Evidently also, the compositions of the adsorbed films on the several minerals are different. These rather complicated circumstances are controlled by the surprisingly simple requirement that the corresponding partial molal free energies are the same for all phases, as well as at a minimum when equilibrium is reached. It is therefore seen that the direction of the chemical reaction, as long as no great amount of fluids occurs, can be determined by considering only the relative free energies of the several solid minerals.

We are now in a position to consider some aspects of the mechanism of recrystallization of a rock. As an example, we shall assume that tremolite forms in a rock containing talc and calcite. Prior to the start of this reaction, the rock has been exposed to low temperature, so that the combination of calcite and talc is stable. Upon rising temperature, tremolite is stabilized. The rock is supposedly a quartzite, with scattered grains of calcite and talc.

What is needed to make tremolite start and continue to grow? First, all the elements constituting tremolite, $Ca_2Mg_5Si_8O_{22}(OH)_2$, must be available at the very point at which the nucleus of tremolite is formed. Second, the elements must be continuously supplied to the place of growth as long as the growth takes place.

We have seen that, at equilibrium, the partial molal free energy of a component is the same throughout the rock at any point inside or outside the minerals, showing that all the constituent elements exist anywhere in the rock. That is, some Ca, Mg, Si, and O exist everywhere in the calcite- and talc-carrying quartzite, but it is certainly not true that the concentration of these elements is so great at any one point in the rock that a nucleus of tremolite could form there. The probability of forming a nucleus would depend upon the energy of nucleus formation—which is different in different places in the mineral aggregate—the concentration of the relevant elements, and the mobility of the elements forming the nucleus.

Controlled by these circumstances, a nucleus of tremolite is formed, for example, close to a talc crystal (Fig. 42*a*). This nucleus absorbs the elements Ca, Mg, Si, O, and H from the surroundings. Any of these elements but Ca occurs in great concentration in the very place at which tremolite is formed, so the minimum factor is Ca, which occurs in low concentration

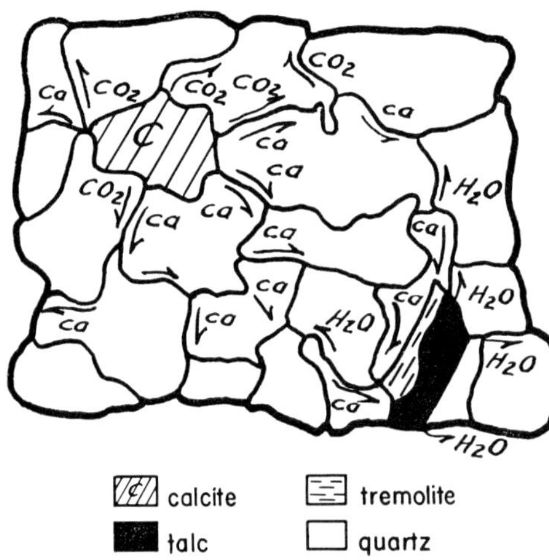

FIG. 42*a*.—The possible directions and paths of migration of Ca, CO_2, and H_2O when talc is changed into tremolite by "interaction" with a distant calcite grain in a quartzite.

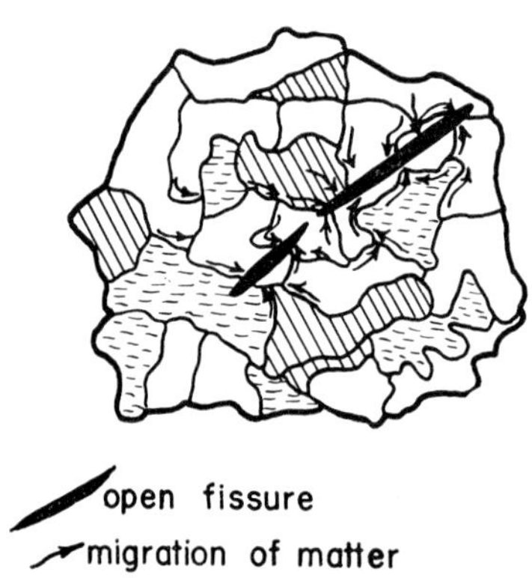

FIG. 42*b*.—The direction of migration of disperse constituent in a rock with newly opened tension cracks.

scattered in the surroundings. The formation of the first small nucleus will, therefore, give rise to a severe relative drop in the calcium concentration in the immediate surroundings. A further growth of the tremolite is now largely determined by the ease with which Ca is transferred to the nucleus from the surrounding sources for Ca, namely, calcite. The very formation of the tremolite nucleus made the partial free energy of Ca drop a certain amount just in the proximity of the nucleus, such that a free-energy gradient of Ca is created pointing out from the growing nucleus. The free-energy gradient will flatten out slowly, so that the free energy of Ca must drop throughout a gradually larger area. Stationary free-energy gradients will be maintained between the tremolite and the closest calcite grains, the calcite grains gradually dissociating and supplying Ca to tremolite. From now on, the velocity of supply of Ca to the tremolite also depends upon the velocity of dissociation of calcite and the velocity of escape of CO_2. Likewise, H_2O is liberated from talc during the alteration to tremolite, so that the growth of the tremolite also is determined by the rate of migration of H_2O from the dissociating talc.

Indeed, the recrystallization of rocks is not a simple chemical process but, instead, a complex interplay of dissociating, moving, and consolidating ions and molecules. It is rare that the process is so simple that a stabilized mineral forms along the common border between two minerals constituting all the elements necessary to build up the newly stabilized mineral. Some constituents are usually lacking in the closest surroundings and must be supplied from some distance, whereas other elements or molecular compounds are superfluous and likely to disappear from the site of mineral growth.

It is rather amazing, under such circumstances, that mutual equilibrium is commonly approached in recrystallized rocks. This fact can be explained only by assuming that the mobility in rocks is large as compared with the time of metamorphic alteration.

HOW GROWING METAMORPHIC MINERALS PROVIDE SPACE FOR THEMSELVES IN ROCKS

So far, we have considered the mechanism of recrystallization of rocks in some detail. We shall now consider the "space problem," i.e., how do growing minerals provide space for themselves when imbedded in a crystalline matrix? Three possibilities seem to exist (Wegmann, 1935): (1) the mineral may grow in open fissures or minute cavities in the rock—*secretionary growth;* (2) the mineral may grow by replacing, so to speak, ion by ion, an old mineral or a cluster of minerals occupying the same volume as the new mineral—*chemical replacement sensu stricto;* (3) the mineral may provide space for its own body by pushing the surrounding grains aside: *concretionary growth sensu stricto.*

SECRETIONARY GROWTH

A mineral existing in a small cavity in a rock has lower molal free energy than the same mineral existing in the bulk of the rock, provided that the rock is exposed to some pressure. The pressure in the cavity is usually less than the pressure acting on the mineral in the bulk of the rock. Hence, according to the law $dF/dP = V$ (p. 13), a mineral in the cavity has somewhat less free energy than the same mineral in the rock. Not only does a small cavity provide a favorable place for formation and growth of newly stabilized minerals, but it also represents a place to which the surrounding minerals would be transferred by chemical transport (activation, migration, and consolidation) (Fig. 42*b*). *The mere existence of a cavity or minute crack, in many cases, is sufficient to explain the free-energy gradient driving the secretionary growth of minerals filling cavities or small cracks in rocks.* Concerning this question see also page 216.

REPLACEMENT *sensu stricto*

In order that replacement *sensu stricto* may take place, chemical instability among the minerals must be created by a P,T-change. (We disregard for a moment the introduction of elements from external sources.) By this type of replacement, an old unstable mineral alters into a new mineral by the gain of some elements and the loss of others. A very simple example of such a process is the interdiffusion of two metals able to form mixed crystals, e.g., Ag and Au. Silver diffuses into Au at the same time as Au diffuses into Ag. If these two processes proceed with exactly the same velocity, we have an ideal example of chemical replacement.

In the example above, there exists originally a free-energy gradient of Ag tending to transfer the Ag atoms into the gold body, and a free-energy gradient of Au tending to move the Au atoms from the pure gold metal into the silver body. This is actually the true criterion on chemical replacement: there must exist partial free-energy gradients so directed that some elements tend to enter the site of an unstable host mineral at the same time that other elements tend to migrate out of the unstable structure. In the case of ideal volume-by-volume replacement, these free-energy gradients, velocities of migration, etc., must be so delicately intercontrolled that no change in volume takes place. It is certainly not easy to see how this complex of processes can result in constant volume in the sense of Lindgren (1933), even if, for example, enstatite is formed by simple interaction between forsterite and silica: $Mg_2SiO_4 + SiO_2 \rightarrow 2MgSiO_3$. We shall assume that the two minerals are in direct contact and that a thin layer of enstatite is formed along the interface (Fig. 43). If the newly forming enstatite replaces forsterite volume by volume, it means that the space originally occupied by forsterite will eventually be occupied by enstatite. The molal volume of forsterite is $V_f = 44.1$ cc., and of enstatite, $V_e = 31.6$ cc. If 1 mol of silica is introduced into the forsterite, 2 mol of enstatite are formed, and a volume expansion of $\Delta V = 19.1$ cc. has accompanied the process. To keep the volume constant, therefore, silicon (or SiO_2) is introduced at the same time that magnesium (or MgO) is removed, and it can be calculated that, for each mol of silicon introduced, about 0.9 mol of magnesium must be removed, provided that the volume is to remain unchanged.

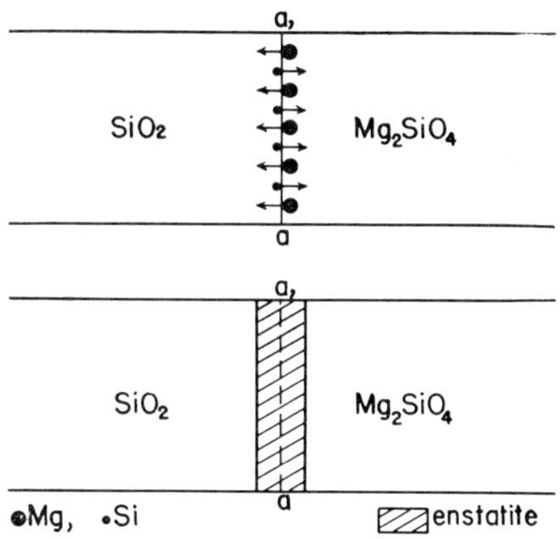

FIG. 43.—The interaction between contacting olivine and quartz. Arrows show the directions of diffusion.

In this particular case of interaction between silica and forsterite there actually exists a tendency to remove some Mg^{2+} (or MgO) from the forsterite, because the free energy of Mg^{2+} (or MgO) is higher in the forsterite grain than in the grain of quartz. Therefore, Mg^{2+} (or MgO) will tend to migrate away from the forsterite through the growing enstatite layer, at the same time that Si (or SiO_2) is migrating in the opposite sense (Fig. 43). But, even so, it seems unintelligible that the removal of magnesium from forsterite should necessarily be of such magnitude that the newly formed enstatite will occupy the same volume as the original olivine. Under other circumstances, however, the newly formed metasomes may occupy exactly the same volume as the unstable host mineral. Let us, in a further study, suppose that the minerals yielding the constituents which are building up a newly forming mineral are some distance apart in the rock. We may, for example, consider the formation of muscovite in the following reaction:

$$\underset{\substack{\text{potash feldspar}\\ 109.1}}{KAlSi_3O_8} + \underset{\substack{\text{sillimanite}\\ 50.0}}{Al_2SiO_5} + H_2O \rightarrow \underset{\substack{\text{muscovite}\\ 138.6}}{KAl_3Si_3O_{10}(OH)_2} + \underset{23}{SiO_2}\,. \tag{87}$$

Mol volume: 109.1, 50.0, 138.6, 23

Suppose that the reaction takes place directly in or on the sillimanite cluster, potassium, water, silicon, etc., being transported to the place of reaction through intergranular and lattice diffusion from surrounding parts of the rock. When 1 mol of sillimanite is combined with the required amounts of potash, water, and silica to form muscovite, an expansion of some cubic centimeters follows. Since the minerals yielding K, Si, and H_2O are some distance away, this increment in volume will not take place easily by a compensative volume shrinkage in the surroundings due to removal of K, H_2O, etc. The formation of a small muscovite nucleus consequently causes a rise of pressure in and around itself. This local pressure increment causes the free energy of muscovite as well as of sillimanite to rise, and three things may happen: (1) the potassium, silica, and water particles in the surroundings may now be unable to produce more muscovite by interaction with sillimanite; (2) the free energy of sillimanite at the high-pressure place may be so large that Al and Si start to migrate away, thereby permitting more muscovite to be formed by replacement of the remaining sillimanite under constant volume; and (3) the surrounding crystalline matrix may yield plastically as time goes on, thereby providing space for the growing muscovite grain under local swelling. It thus depends upon several circumstances whether replacement at constant volume will take place or not. It is important to note that the reason for constant volume in case 2 above is the fact that the free energy of the replaced mineral—sillimanite—increases with rising pressure, thereby initiating an escape of Al from the scene, so that the lost volume of disintegrated sillimanite may just be compensated for by the volume of the growing muscovite.

Under different circumstances, there are other possible reasons for more or less constant volume under replacement. We may, as an example, consider replacement of hornblende by biotite after the following equation:

$$\begin{aligned} &\underset{\text{hornblende}}{3Ca_2(Mg,Fe)_4Al_{x_1}Si_{y_1}O_{22}(OH)_2} + 4K^+ + H_2O \\ &\qquad \rightarrow \underset{\text{biotite}}{4K(Mg,Fe)_3Al_{x_2}Si_{y_2}O_{10}(OH)_2} + 6Ca^{2+}\,. \end{aligned} \tag{88}$$

The potassium may be transferred to the hornblende from potash feldspar grains situated some distance away. The volume lost by the escape of Ca^{2+} from hornblende may be approximately compensated for by the gain of K^+ in biotite. One may well ask what rules insure that Ca^{2+} continues to leave the unstable hornblende lattice while K^+ enters the growing biotite lattice. The answer is found in the fact that the incoming K, under its reaction with the hornblende, "extracts" Mg, Fe, Al, etc., out of the hornblende lattice to incorporate these elements in the mica lattice. Thereby, the relative concentration of Ca is increased somewhat in the hornblende. That means (p. 18) also that the partial molal free energy of Ca in hornblende is raised so that there is created a potential gradient tending to make Ca migrate away from the place of reaction.

As a matter of fact, the thermodynamics of subsolidus reactions in rocks may, under certain circumstances, be such that replacement proceeds in an approximately volume-by-volume manner. In numerous instances, however, Lindgren's law of constant volume is not valid.

CONCRETIONARY GROWTH

Strict concretionary growth would correspond to the following case: A grain of garnet, for example, starts to grow in intergranular space and continues to grow slowly by supply of material via the intergranular paths, utilizing no elements of the minerals in the immediate surroundings. To provide for the necessary space, the surrounding crystalline matrix has to be pushed away in front of the borders of the expanding garnet. It is more likely, however, that the growing garnet will utilize some of the constituents of surrounding minerals, thus growing partly by replacing these minerals, partly by pushing them aside.

The ability of a mineral to push aside surrounding minerals depends to some extent upon the force of crystallization and upon the interfacial energy between the growing mineral and the minerals of the matrix. The greater the force of crystallization, the more likely is the concretionary growth, because the less will the growth be hindered by the increase of free energy with rising confining pressure. Still considering just recrystallization within a confined part of a rock, it is evident that a mineral growing after the principle here discussed must receive its constituents from minerals lying within rather moderate distances from the growing mineral. The whole rock is "supersaturated" with the newly formed mineral—for example, garnet. A few nuclei are formed at some favorable sites. Supposing that the activation energy of nuclear formation is rather high, the degree of supersaturation may be great without rapid formation of new nuclei. In this case the increased pressure in and closely around the few growing garnets can be rather appreciable before the growth stops.

It seems obvious that the expanding volume of the growing garnet must be approximately compensated for by contraction at other places in the rock. If, for example, the garnet forms at the expense of hypersthene and cordierite, the sites of these minerals are the places where the volume contracts. The rest of the rock aggregate lying between the dwindling hypersthene and cordierite grains and the grain of the growing garnet must then yield by means of plastic flow, as shown in Figure 44.

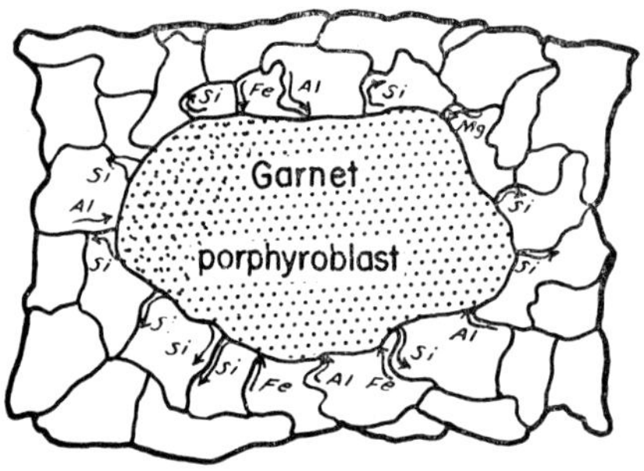

FIG. 44.—The migration of elements under growth of garnet porphyroblast in a quartz-rich rock.

Although we have considered the phenomena of secretionary growth, replacement *sensu stricto*, and concretionary growth separately, this does not indicate that these processes are separate in rocks. To be sure, it is most likely that a single mineral grows by means of all three principles combined.

THE KINETICS OF DEFORMATION OF ROCKS AND MINERALS

OBSERVATIONS in the field and through the microscope have established that the apparently very rigid and brittle crystalline rocks must be able to yield in a plastic manner under certain circumstances (Fig. 45). Several important questions arise in connection with such plastic (and cataclastic) yielding of solid rocks: (1) What is the origin, magnitude, and direction of the stresses needed to create the plastic deformation? (2) How can the plastic yielding of such a brittle substance be understood physically? (3) What is the relationship between the flow structure (preferred orientation of mineral grains, etc.) and the direction of flow of the rock body? Etc.

Not many of these questions can be answered satisfactorily and rigorously as yet, in spite of the numerous field and microscopic observations recorded from tectonites (= mechanically deformed rocks). The vast number of results from the petrofabric analysis introduced, some twenty years ago, by W. Schmidt and B. Sander are interpreted ambiguously in several respects, and the several authors hold opposite, or at least highly diverging, viewpoints regarding the significance of their observations. A typical example may be provided by the interpretations of the subparallel arrangement of prismatic minerals (lineation), which is encountered in numerous hornblende schists, sillimanite schists, etc. Some authors are certain that the lineation is perpendicular to the direction of plastic flow or laminar gliding in the rock body, others find that lineation of mineral grains doubtlessly coincides with the direction of flow (Cloos, 1944, 1946; Anderson, 1948; Fairbairn, 1949). The several explanations of the cause of the preferred lattice orientation of, for example, quartz in tectonites are not less contradictory (see Fairbairn, 1949; Sander, 1948). The reasons for this lack of satisfactory and unambiguous theoretical explanation of the flow structure observed in rocks must again be blamed on the difficulties in performing the proper experiments, which repeat closely enough the conditions under which natural metamorphites are formed. The plastic flow of a solid rock, as it takes place in the earth's crust, is undoubtedly a very slow process, just as slow as the recrystallization of a rock. We can hardly expect, therefore, to repeat experimentally in any detail the plastic deformations of most rocks and minerals, any more than we can repeat the formation of a coarse-grained gneiss by recrystallization of a fine-grained slate. As yet, even the seemingly simple process of formation of a coarse marble from a dense limestone has not been reproduced in laboratories. With reference to the plastic yield of rocks, the experiments by Griggs (1940) and by Fairbairn (1949) are highly promising.

Under these circumstances, it appears that the most valuable aid to our understanding of the mechanical behavior of rocks comes from the theoretical and experimental knowledge achieved during the intense studies of metals and other easily yieldable crystalline substances. Most

FIG. 45.—Plastically co-folded basic and acidic gneiss layers. The basic dark layers in this case have behaved in an unusually ductile manner. Generally the basic components (amphibolites) are rather rigid or competent when deformed together with acidic gneisses, which often are plastic or incompetent (see Figs. 63 and 69). Southeast Bay, West Greenland, 1948. Photograph by M. L. Ramberg.

of this chapter will, as a consequence, be devoted to our general information about the mechanical properties of crystalline substances.

As is well known, crystalline matter may—from a mechanical point of view—be classified into ductile, or plastic, and brittle crystals. Most metals belong to the first group, as do many organic crystals and some ionic crystals under the proper physical conditions. Silicate minerals, as a common rule, are brittle under everyday laboratory tests (low confining pressure and not too long duration of experiments). The behavior of neither ductile nor brittle crystals under application of mechanical stress is simple, and the conclusions to be drawn from the numerous experiments are not yet compiled in general laws with extensive validity. However, some correlations of general importance are evident and will be considered.

If exposed to a mechanical stress with a shearing component acting along translation lines in the lattice, a ductile or plastic single crystal will respond to the stress principally as follows (Fig. 46):

1. A small shearing stress, τ, will strain the crystal elastically: that is,

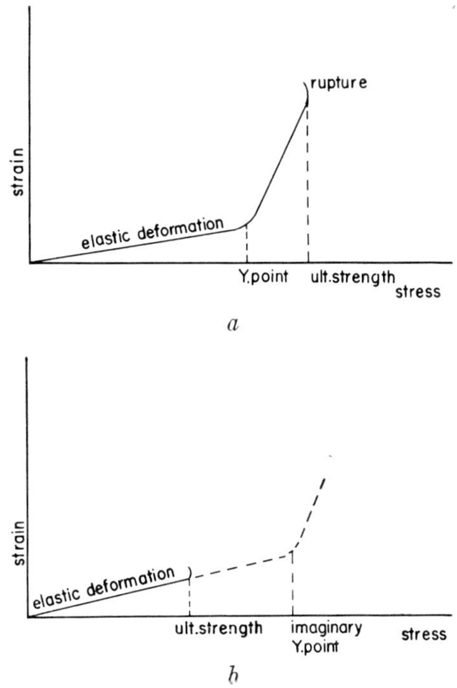

FIG. 46.—General diagram, showing the mechanical properties of crystals under stress: strain, yield point, and ultimate strength. *a* holds for ductile crystals; *b* is valid for brittle crystals.

on complete withdrawal of the stress, the original shape of the crystal is immediately recovered.

2. When the stress surpasses a certain value—the so-called *yield point*, τ_y, characteristic of the particular direction in the given crystal and also determined by temperature and pressure—permanent deformation of the crystalline body in part takes place. On release of the stress, the crystal returns only partly to its original shape, even if left unstressed for an infinite length of time. The difference between the original shape and the new shape of the crystal is due to plastic flow. The elastic part of the deformation which is recovered on withdrawal of the stress may not

always disappear suddenly, as in the case of elastic deformation below the yield value. Instead, a certain part of the recovery of shape may go on rather slowly and perhaps need some elevated temperature to be completed. This slow elastic recovery, which is most prominent in polycrystalline test pieces, is called "elastic aftereffect" (Houwink, 1940, pp. 5–7) and, if elevated temperature is needed, "thermo-recovery" (Fig. 47). Another important phenomenon found by plastic deformation tests of crystals is called "hardening": ductile crystal starts to yield plastically at a certain stress value, say τ_y, but, instead of yielding with a constant rate under the constant stress, as a liquid body would do, increasingly larger stresses are required to maintain the constant rate of deformation. Thus,

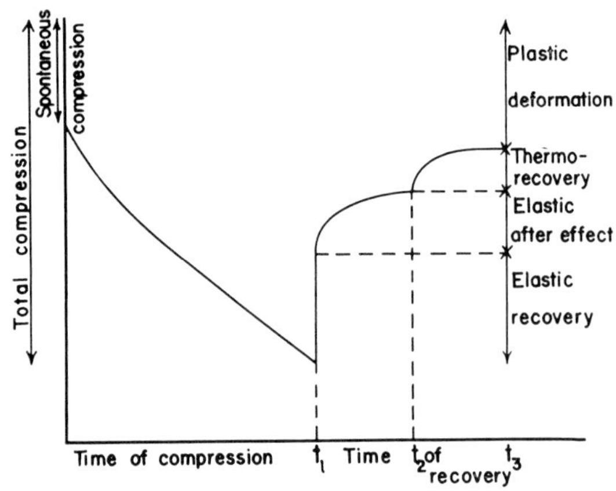

FIG. 47.—Diagrammatic representation of the interrelationships of compression, elastic recovery, and thermo-recovery, plastic deformation, etc., as functions of time. (After R. Houwink, 1940.)

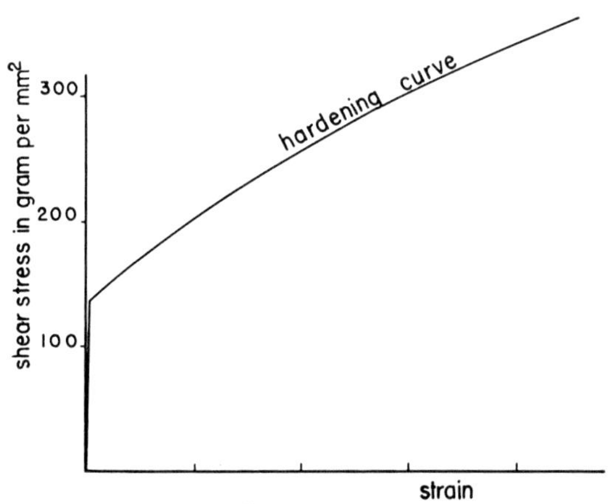

FIG. 48.—General shape of a hardening curve for crystalline substances

provided that the applied stress is not gradually raised, the yielding comes to a stop when the deformation has reached a certain percentage. That is, the yield value of the strained crystal has been increased by the very deformation itself (Figs. 48–50). A gradual increase of the applied stress increases the amount of strain until rupture occurs at a stress value called the *ultimate strength*.

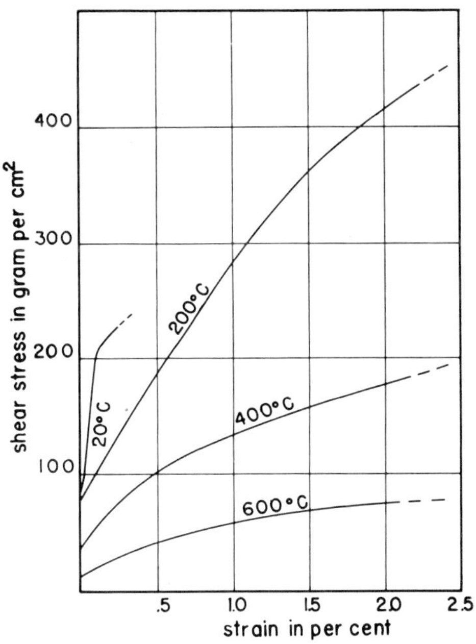

FIG. 49.—Hardening curves at different temperatures for rock salt. (After Theile, 1932)

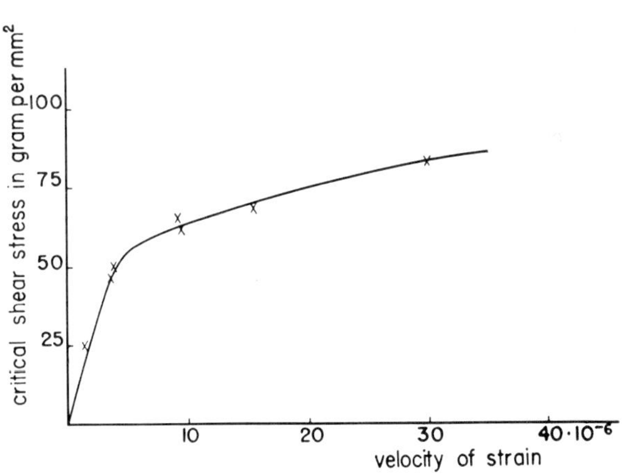

FIG. 50.—Critical shear stress (yield point) of pure Sn crystals as a function of velocity of strain. (After Chalmers, 1936.)

It is emphasized that the behavior here outlined is valid only if the duration of the experiment is not extraordinarily long.

Brittle crystals, on the other hand, exhibit no interval of plastic flow but are subjected to rupture after a certain amount of elastic deformation and corresponding magnitude of stress. One may say that the ultimate strength of such crystals is less than their yield value.

Quartz, feldspars, and most of the other common rock-making minerals (except micas, chlorite, carbonates, and a few more) are to be classified as rigid and brittle under everyday conditions. "In normal laboratory tests and building application most rocks are brittle rather than plastic and possess a well defined strength . . ." (Griggs in G.S.A., Special Paper No. 36, p. 108). However, as it is known that pressure and temperature both have some effect upon the property of plasticity, it may be tentatively assumed that the high pressure and temperature prevailing in the depths of the crust could introduce some plasticity into the brittle mineral aggregates. Experiments and theory, however, agree on the point that temperature and pressure of values reasonably to be expected within the earth's crust do not, by themselves, change the brittle rocks into plastic bodies. The most common rock-forming silicate mineral—quartz, for example—is brittle even at a confining pressure of 25,000 atm. (Bridgman, 1937). (It can be argued that a brittle crystal has an ultimate strength lower than its yield value which, therefore, is not substantiated. Rising pressure raises the ultimate strength more rapidly than it raises the yield value, so that at higher pressures a yield value may occur below the ultimate strength. It is seen, though, that the magnitude of the yield value thus materialized must be greater than the ultimate strength at low pressure [Fig. 51]. Even if some minerals may be plastic at high pressure, they must have a correspondingly high yield value, as seen from the ultimate strength of some minerals [Table 9].) It is commonly agreed that rocks have yielded in their solid state under much less stress than the ultimate strength, as seen in Table 9. In this connection, it may be emphasized that regionally metamorphosed rocks do not commonly contain open cracks or ruptures, although one must necessarily expect that such ruptures have been developed in the course of the history of schists and gneisses. After these rocks have been made accessible by erosion, one finds that the ruptures which once must have been open are closed either by plastic yield in the surroundings or by some mineral fillings. It seems reasonable to conclude, then, that rocks have rather low yield values; otherwise, one would expect to find open cracks and fissures as common features of metamorphic rocks, because only those cracks which were surrounded by stresses greater than the yield values would collapse. Under metamorphic recrystallization, rocks are evidently able to yield under very small stresses, provided that time is sufficiently long.

Most crystals show lower yield points, the higher the temperature is. Accordingly, it appears offhand that natural minerals may become sufficiently plastic at some elevated temperature. This possibility, however, is by no means supported by the experiments by Griggs and Bridgman (see G.S.A., Special Paper No. 36). From metal investigations, it is known that the effect of temperature upon the yield point is rather small. "The critical shear stress (yield point) even near absolute zero is still of the same order of magnitude as at room temperature" (Barrett, 1943, p. 298).

There are certain aspects of the mechanical properties of tested crystals

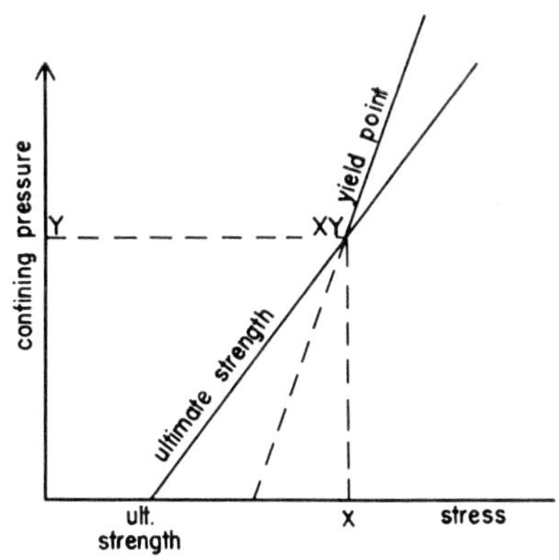

FIG. 51.—The influence of confining pressure upon the yield point and ultimate strength of crystals. Below pressure *y* the crystal is brittle; beyond this pressure the crystal can behave plastically because the yield point is below the ultimate strength.

TABLE 9

CRUSHING STRENGTH OF ROCKS (DATA AFTER *Handbook of Physical Constants*)

Rock Type	Average Strength (Kg/Cm²)	Range of Strength (Kg/Cm²)
Granite	1,480	370–3,790
Syenite	1,960	1,000–3,440
Diorite	1,960	960–2,600
Gabbro	1,800	460–4,700
Gneiss	1,560	810–3,270
Quartzite	2,020	260–3,200
Marble	1,020	310–2,620
Sandstone	740	110–2,520
Limestone	960	60–3,600
Slate	1,480	600–3,130
Serpentine	1,230	630–1,230
Tuff	310	100– 520

which are of considerable geological importance. The yield point of an undeformed (unhardened) crystal in a certain direction is not constant at given P and T but depends importantly upon the duration of the test. The longer a stress is applied, the lower the actual yield point of the crystal. A question naturally arises: Does the yield value approach zero when the duration of the test approaches infinity? The geological importance of this question is evident when the length of geological time is considered. Before we attempt to answer this question, we shall study other aspects of the phenomena of plasticity and elasticity.

The fact that a slow plastic deformation actually takes place below the practical yield value is explained by the so-called "creep," which apparently is attributed to quite different processes in the crystal lattices than the plastic flow occurring at stresses beyond the yield value.

In the same way as a slow creep or relaxation takes place in elastically strained crystals below the yield value, the so-called "recovery" or "annealing" takes place in a hardened crystal stressed beyond the yield point.[1] When a hardened crystal (or crystalline aggregate) is left to itself for some length of time (depending upon temperature, type of crystal, and percentage of hardening), some atomistic rearrangement takes place inside the lattice, so that the crystal reattains the low yield value it possessed in the unhardened, thermodynamically stable, state, although the shape and macrostructure of the crystal (or crystalline aggregate) do not necessarily change. This process is called "recovery" or "annealing" and may be distinguished from recrystallization by the fact that the latter is always characterized by formation of new crystals and disappearance of old ones. It is now evident that the existence of the phenomena of recovery or annealing results in a continuous creep in hardened crystals when subjected to a constant stress of a magnitude equal to, or less than, the "hardened" yield value, but greater than the yield value of the unhardened state.

To explain these main aspects of the mechanical properties of crystalline matter, we shall briefly consider the atomistic character of crystal lattices. In so doing, we shall mainly follow the considerations of Dehlinger (1939).

Figure 52 shows a simple atomistic arrangement which may serve our purpose. The shearing stress acting along the surface is $\tau \text{g/cm}^2$, and the number of atoms (ions) per square centimeter is n. So that, provided that the external stress applied is evenly distributed on the several atoms, each is affected by a force $1/n$ times τ in the direction of the arrows. It is often possible to calculate the energy of bonds which keep an atom in its regu-

1. The "yield point" refers here to the unhardened state of the crystal.

lar site; it is therefore also possible to calculate theoretically the shearing stress capable of sliding one layer of atoms (ions) over another by simple slip (Dehlinger, 1939). (Incidentally, it is interesting to note that such calculations result in yield values and ultimate strength values 500 to 10,000 times greater than the actually observed values [see, e.g., Kochendörfer, 1941]. Furthermore, the simple picture above would require that the shear angle due to elastic strain be about 50 per cent when the yield takes place, because the atoms in a layer must be displaced to the top of the energy barriers before they will jump freely to their new stable positions [Barrett, 1943, p. 332]. Actually, the shear angle is much less. It is

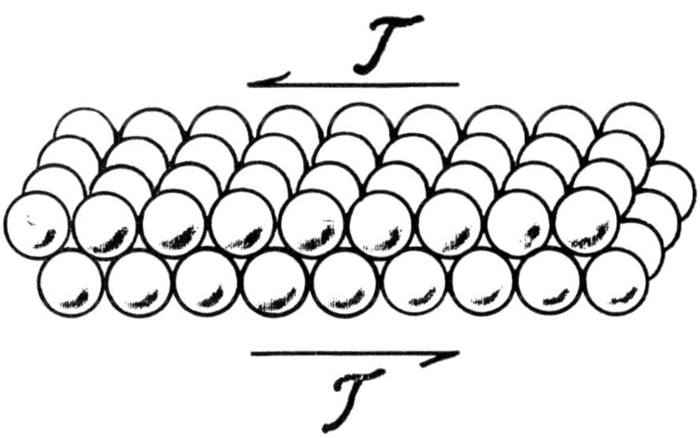

FIG. 52.—A double layer of atoms in a crystal acted upon by a shear of magnitude τ

of particular significance for structural petrology that but minor admixing of "impurity" elements in a crystal may, in cases, change the yield point rather drastically [Barrett, 1943, p. 298].)

It is known that the plastic yield at stresses greater than the yield value is due to mutual slip of layers or blocks of atoms along certain planes and directions in the lattices (so-called "slip planes" and "slip directions"). Because the atoms or ions are bound together with forces of a certain magnitude, a characteristic minimum stress is also required to create slip and corresponding plastic deformation. The Maxwellian distribution law requires, as we have seen, that a certain small fraction of the constituent particles in a crystal be able to exchange places in the lattice by individual jumps. By applying a stress on a crystal, thus putting the lattice in an elastic strain, the energy barriers in the lattice will be unsymmetric, as shown in Figure 53. Hence the energy required to move an atom (ion) in the direction of mechanical stress will be somewhat less than that required to move it in the opposite direction. As a consequence, the number of particles jumping in the direction of stress will be somewhat greater than the number jumping in the opposite direction. The result will be a net transportation of matter along the mechanical stress component and, therefore, a slow creep of the crystal in this direction. This corresponds rather closely to a directed self-diffusion in the crystal.

We now have a very simplified and qualitative picture of the difference between the plastic flow by translations or slip in the lattice and the creep by atomistic jumps. Theoretically, these minute jumps would take place in any stressed crystal unless the temperature is not too close to 0° K. As a consequence, one may conclude that an absolute yield point does not exist except at absolute zero. The yield point determined by most experiments is actually the critical stress at which the crystal starts to yield by slip in the lattice. However, the yield by microcreep below the experi-

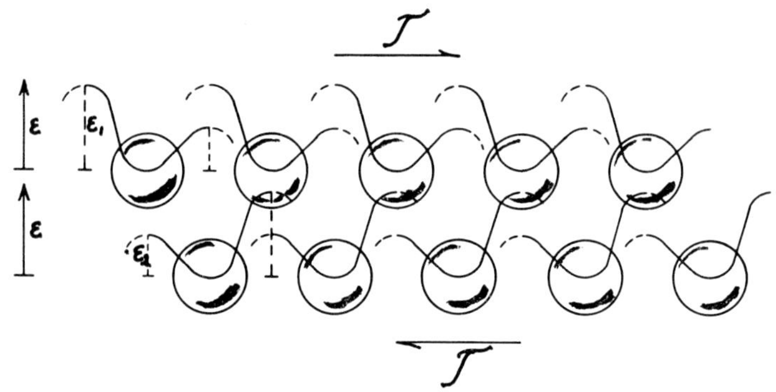

FIG. 53.—The distortion of the energy "wave" in a sheared lattice. For motion from left to right in the upper layer, a small energy, ϵ_2, has to be overcome; for motion in the opposite direction, a large energy, ϵ_1, has to be overcome. In the lower layer the situation is the opposite.

mentally determined yield point may be so infinitely slow that, even during geological periods, no appreciable flow will result because the viscosity coefficient of a crystal at stresses below the yield point is much greater than at stresses beyond the yield point.

According to the theory of solids, there is a rather fundamental difference between the mechanism of slip or translational gliding in unhardened crystals beyond the elastic limit and the mechanism of creep at smaller stresses. Moreover, there are also important differences between the mechanisms of creep below the yield point and the creep caused by recovery or annealing of hardened crystals beyond the yield limit. Below the yield value, the creep is caused, as it is believed, by migrating individual atoms (ions), that is, by a kind of self-diffusion. The "creep" of hardened crystals stressed beyond the elastic limit is more likely due to en bloc motion of layers or blocks of atoms (ions) much similar to the translational gliding in unhardened crystals. As distinguished from the yield in unhardened crystals, however, a translation in a hardened crystal cannot appear before the thermal agitation of the atoms has recovered the lower strength characteristic of a less hardened crystal. Hence the creep by recovery, annealing, or recrystallization of crystals stressed beyond the yield value is not so rapid as the simple translational gliding in an unhardened crystal, but it is much more rapid than the creep below the elastic limit, where block motion by gliding is theoretically impossible or at least exceedingly unlikely.

With reference to flow of brittle rocks, it is of some consequence to remember that the slow microcreep taking place below the yield value will occur irrespective of whether the crystal is ductile or brittle. This kind of creep must occur in brittle silicate rocks rather commonly, although the quantitative significance of it in the process of flow of rocks is not yet known.

The resistance against plastic flow by translation gliding above the yield value is different in different directions in the lattices. Indeed, along some directions a crystal may be brittle, whereas in other directions—i.e., along translation lines in translation planes—the same crystal may be highly plastic. Hence the viscosity of a crystal depends greatly upon the direction in the space lattice along which the translation extends. This matter is of importance for the lattice orientation in strained rocks (pp. 121–31). It is also of consequence for the understanding of flow structure in rocks to know that the slip lines in some crystals are different at different temperatures (Barrett, 1943, p. 298). The slip lines in a lattice may also be different at different pressures, although the latter has not yet been investigated. These circumstances may, in part, explain the great variety of, for example, lattice orientation of quartz found in different metamorphosed rocks.

In a similar manner, one must also assume that the microcreep taking place below the yield value extends differently in different directions in a lattice, simply because the self-diffusion coefficient often depends greatly upon the direction along which it is measured in crystals of lower symmetries. In metallic Bi, for example, it was found (Seith, 1933) that the diffusion coefficient along the *c*-axis was 10^6 times that perpendicular to this direction in the lattice (Fig. 54). This fact will certainly also cooperate in controlling the orientation of mineral lattices in deformed rocks or tectonites. A similar anisotropic effect is found in the elasticity of crystals: the elasticity modulus is different in different directions in the lattice (Sonder, 1948).

We have, so far, considered the behavior of single crystals under mechanical stress. The behavior of rocks as polycrystalline aggregates can be constructed only within certain limitations, by integration of the properties of the individual grains. One principal new effect is introduced when the test piece changes from a single crystal to a polycrystalline aggregate, namely, the possibility of plastic flow by recrystallization and differential transportation of matter along the grain boundaries and through pore fluids—indirect componential movements, as it has been termed by Sander (1930).

First, it is plain that the pure translational part of the plastic flow in a crystalline aggregate with randomly oriented crystal units would meet

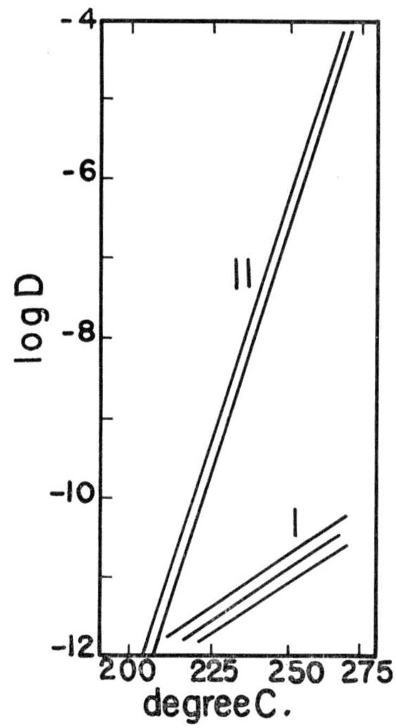

FIG. 54.—Anistropic self-diffusion in Bi; D is the diffusion coefficient. I = parallel to c-axis; II perpendicular to c-axis. (After Seith, 1933.)

greater resistance than in a single crystal favorably oriented relative to the mechanical stress. In the polyaggregate, the random orientation of the individual grains provides that only a small fraction of the grains are oriented with their glide lines (and glide planes) parallel to the direction of shearing stress. Only these few grains offer a minimum resistance against the plastic flow. Most grains have their translation lines inclined to the direction of shearing stress, thus yielding correspondingly greater resistance to the flow. Those grains with their glide lines oriented perpendicular to the shearing stress fail to yield by translation at all.

Supposing that there exists a certain shearing stress which is homogeneous within a certain volume of the rock and exceeding the yield value of the favorably oriented mineral grains, then one would expect that a translation gliding took place first in the more favorably oriented grains, thereby tending to turn the less favorably oriented grains into the proper position, with their glide lines approximately parallel to the flow lines. However, one would also expect rupture to take place in grains with an unfavorable orientation. Suppose that the shearing stress is by no means sufficient to start rupture if all (or the majority of) crystals are oriented with their glide lines perpendicular to the direction of stress. If only a small percentage of the grains are so oriented, however, the rest of the crystalline aggregate will yield plastically, thereby more or less releasing itself from the stress, which thus will ultimately concentrate in the rather small cross-section of the transversely oriented grains. Therefore, these unfavorably oriented crystals may easily break by rupture, and the flow can continue.

We should consider the effect of recrystallization and intergranular transportation of material as the agency of plastic yielding of a polycrystalline test piece. To start and maintain these processes, which are definitely chemical by nature (activation, diffusion, consolidation), some chemically acting forces are necessary. As a matter of fact, a mechanical stress acting on a crystal induces chemical potential differences in such a way that the chemical processes (diffusion) thereby initiated tend to release mechanical stress. We have already considered the phenomena of microcreep and annealing, which also are of a certain "chemical" nature, in contradiction to the purely mechanical translational gliding, although the processes of microcreep and the true annealing (recovery of a hardened crystal) are not chemical in the respect that new phases always are formed and old phases disappear. The chemical process takes place inside the crystal, so that atomistic transportation releases the stress without necessarily forming new nuclei of crystals.

By recrystallization due to stress, the chemical nature of the flow is more striking as far as the macroscopic result is concerned, because new crystals grow at the expense of unstable and intensively strained crystals. Under these circumstances, chemical transportation of matter extends along one path or another from the stressed area to the unstressed or less stressed areas.

As these processes, when discussed in geological literature, have caused quite common confusion, it appears to be worth while to consider the matter in some detail. Let us take the simple system of a single crystal and its liquid solution. We shall assume at the beginning that the entire transfer of substances takes place via the solution. An experiment, as pictured in Figure 55, is performed. Some pressure is applied on the plate *g*, and it is assumed that the solution, as a minutely thin layer, can exist between the crystal surface and plate *g*. If we suppose that the saturated concentration of the solution increases with increasing pressure, then, plainly, the concentration of the dissolved crystal is greater underneath the plate *g* than it is close to the unpressed surfaces. It seems straightforward to conclude that the dissolved particles will diffuse away from underneath the plate *g*, where the concentration is great, to the bulk part of the solution, where the concentration is lower. The stressed crystal face will therefore dissolve gradually, whereas the free crystal face will grow. This is nothing but the well-known *Riecke's principle*.

If, on the contrary, the saturation concentration decreases with rising

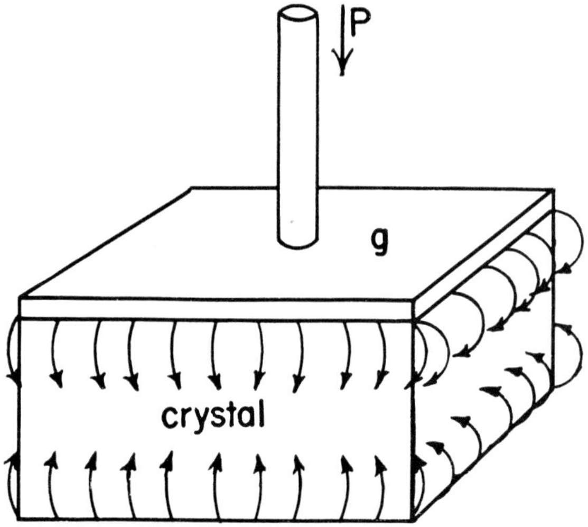

FIG. 55.—Illustration of Riecke's principle. The crystal is exposed to directed pressure, the arrows indicate the migration of dissolved material.

pressure, then the concentration underneath the plate *g* must be somewhat less than close to the free surfaces of the crystal.

At a first casual glance one may consequently conclude that the free face will dissolve and the dissolved particles will diffuse toward the stressed face and precipitate upon this face, so that the growth will extend against the greatest pressure, as has been assumed by Rosenquist (1942). Being rather embarrassed by the fact that this conclusion evidently contradicts LeChattelier's principle, as well as the second law of thermodynamics, we hasten to investigate the situation a little further. Then it appears that the hasty conclusion drawn from case 2 is not valid because, entirely independent of the relationship between pressure and saturation concentration, the molal free energy of the crystal and the partial molal free energy of the solute in the solution at high pressure underneath the plate *g* is greater than at the free surfaces, where pressure is low. Therefore, the chemical transfer by diffusion in the solution in both cases takes place from the stressed surface toward the free surface. In the second case, however, it happens that the concentration gradient is directed against the free-energy gradient. But the latter, which always is the driving gradient of any isothermal chemical process, is in either case directed in such a way that the transfer extends from the stressed to the free surfaces. Considering, now, the material transfer within the mineral grain (self-diffusion), it is evident that this will also extend from the high-pressure parts of the crystal to the low-pressure parts.

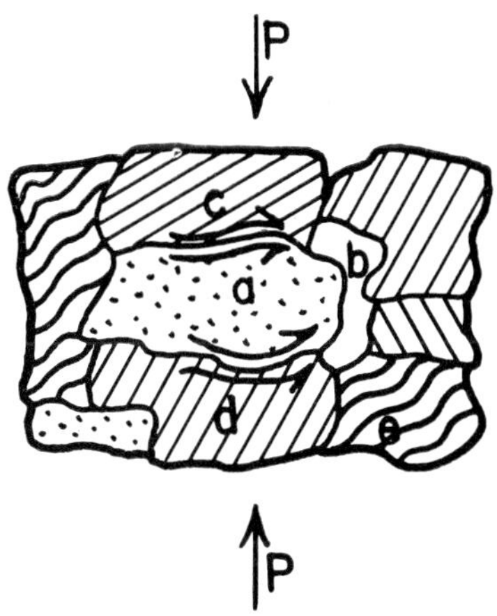

FIG. 56.—Riecke's principle in rocks. The one-sided pressure makes matter "dissolve" along the contacts between *a* and *c* and *a* and *d*. Matter diffuses along the arrows and consolidates again in the open space, *b*.

We can, with the greatest confidence, conclude that the deformation of the crystal due to the chemical transportation of matter within the lattice itself, along the mineral surfaces, or within a surrounding solution will extend in such a direction that the crystal yields under the mechanical stress. Whether the solubility in the particular solvent used increases or decreases with rising pressure is entirely irrelevant to the direction of the process. It is here to be understood that we are considering situations in which the unstability is due solely to mechanical stress. It is a priori not impossible not only that precipitation on the free surfaces of the stressed crystal takes place but also that nuclei of new crystals are created in the somewhat supersaturated solution which is formed outside the liquid layer underneath plate *g* (Fig. 55). This is probable, in particular, when we consider a polycrystalline aggregate such as that in Figure 56. The solution in the small space *b* will be supersaturated with respect to the stressed crystal *a*. As the solution body *b* is surrounded by several minerals, it may happen that the nuclei of *a* start to grow on any of the surfaces of the surrounding minerals. This small modification of the idealized case just discussed would undoubtedly be characterized as recrystallization, and the continuation of this process would be classified as recrystallization flow.

Like any other chemical process, recrystallization flow will, with respect to rate, be greatly influenced by catalysts. The character of the solution present would significantly change the rate of recrystallization flow and hence the corresponding viscosity. As the power of recrystallization is greatly influenced by temperature, recrystallization flow would also

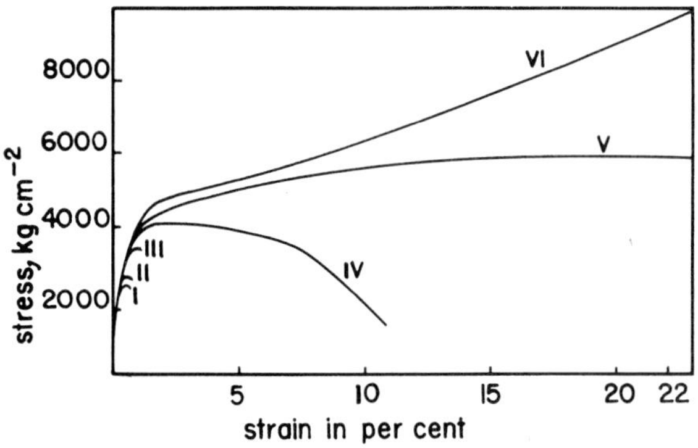

FIG. 57.—Stress-strain relationships for Solenhofen limestone. Curves *I*, *II*, *III*, *IV*, *V*, and *VI* correspond, respectively, to 1,000, 2,000, 4,000, 6,000, 8,000, and 10,000 atm. (After Griggs, 1936.)

increase much more steeply with temperature than would flow by pure translational gliding. There will, thus, be a number of factors influencing the viscosity of natural rocks, factors the effect of which is hardly known, even in principle.

Some of the controlling factors found by experiments by Griggs and others are listed in Tables 10 and 11 and demonstrated in Figures 57–59. In addition to these factors, the rate of recrystallization flow will be influenced by the same factors as those which control the general chemical mobility in rocks (see pp. 76–91).

Before we can synthesize the happenings in a rock exposed to some kind of mechanical stress, we shall need some further information about details. One important question is concerned with the orientation effect of recrystallization flow and microcreep. Would we expect to find synkinematic preferred orientation of minerals (p. 124) if the flow was solely a recrystallization flow or a microcreep? Certainly, on rotation, prismatic or flaky minerals would be oriented by shape during recrystallization

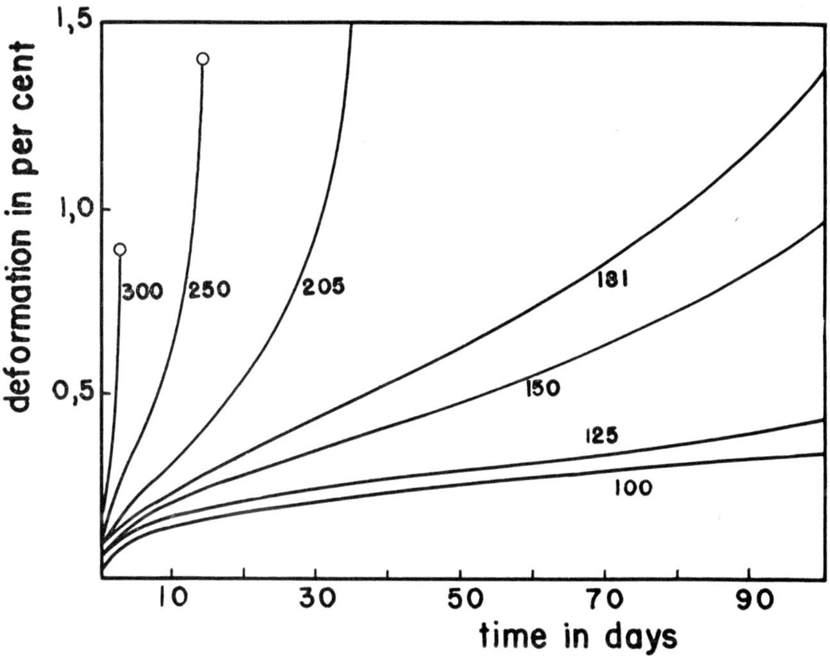

FIG. 58.—Creep of alabaster in H_2O at 24° C. The stress values in kg/cm^2 are indicated on each curve. (After Griggs, 1940.)

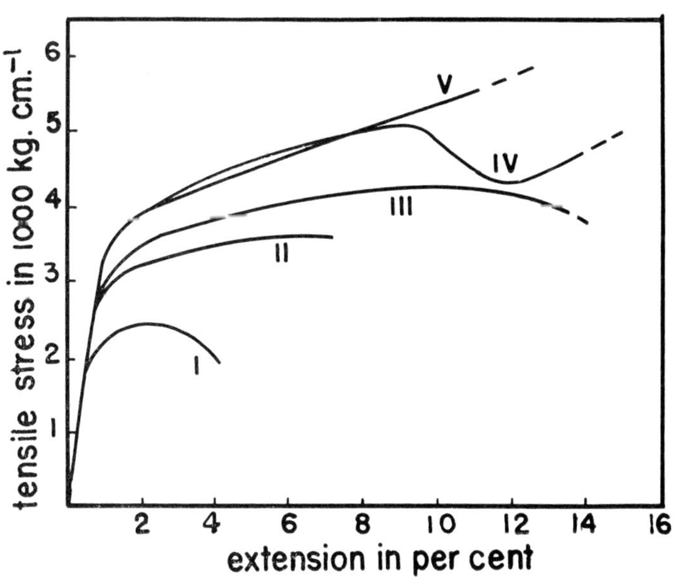

FIG. 59.—Deformations of limestone under tensile stress at different confining pressures. Curves *I*, *II*, *III*, *IV*, and *V* correspond to 2,830, 4,500, 6,000, 8,000, and 10,000 atm., respectively. (After Balsley, 1941.)

TABLE 10*

RESISTANCE TO SHEARING UNDER HIGH CONFINING PRESSURE
(AFTER *Handbook of Physical Constants*)

Mineral	Confining Pressure (Kg/Cm^2)					Remarks
	10,000	20,000	30,000	40,000	50,000	
Andalusite	2.4	6.7	10.9	13.3	15.4	Radial fibers; snaps beyond 15,000
Anorthite	2.2	7.7	11.0	14.0	14.0	
Augite	1.1	3.0	6.3	10.0	13.0	Radial fibers; snaps
Bronzite	1.8	3.4	9.0	13.0	16.0	
Diopside	2.5	5.9	9.0	12.0	14.0	
Garnet	......	......	......	......	15–19	Radial fibers; snaps
Hornblende	1.1	3.1	6.4	10	13.0	
Pyroxene	1.4	4.5	8.0	11	15.0	
Sericite	......	......	......	......	13.0	Snaps; orientated
Sillimanite	2.1	6.4	10.1	11.2	11 8	Snaps beyond 12,000
Graphite	2.9	4.7	6.9	9.0	9.9	Violent snapping beyond 10,000
Calcite	2.8	5.1	6.4	7.5	8.6	Smooth
Magnesite	2.0	4.6	6.9	8.2	9.0	Quiet motion
Quartz	......	......	......	......	14.5	Snaps
Rock salt	0.9	1.8	2.2	2.4	3.0	Quiet motion
Galena	0.5	0.97	1.29	1.44	1.61	Smooth motion
Hematite	2.3	6.9	10.9	14.0	16.7	Snaps

* Figures in columns give average shearing stress in 1,000 kg/cm^2.

TABLE 11

EQUIVALENT VISCOSITY OF SOME ROCKS (AFTER GRIGGS, 1936, 1939, AND 1940)

Material	Confining Pressure (Kg/Cm^2)	Differential Pressure (Kg/Cm^2)	Remarks	Viscosity Poises
Solenhofen limestone	1	1,400		$>2 \times 10^{22}$
	10,000	4,400		3.1×10^{15}
	10,000	5,500	Dry	9.3×10^{14}
	10,000	6,760		2.2×10^{14}
Halite	1	61		3.0×10^{17}
Alabaster	1	420		$>2.42 \times 10^{19}$
	1	103		*ca.* 38×10^{16}
	1	125		14.4×10^{16}
	1	165		6.04×10^{16}
	1	205	Wet	2.64×10^{16}
	1	300		0.42×10^{16}
	1,000	250		1.47×10^{16}
	1,000	250		0.27×10^{16}
	1,000	300		0.064×10^{16}
Average earth's crust	....	....		*ca.* 10^{21}–10^{22}

flow or microcreep in an easily reacting matrix, just as well as by any other type of plastic flow in that matrix. However, should one expect preferred lattice orientation (p. 124), which is independent of the shape of the crystals, to develop during recrystallization flow or creep? As an introduction to the discussion of this matter, we may start with the following picture.

Suppose we consider two rock bodies containing only hornblende. In the one rock body, all the hornblende needles are oriented parallel to the flattening stress, P. In the other rock, the needles are perpendicular to the flattening stress, P. We shall assume that the stress is well below the yield value, so that the plastic flow is taking place partly by diffusion inside the hornblende lattices and partly along the interfaces. We shall also assume that the diffusion coefficient of the lattice, as well as the surface diffusion, is much greater parallel to the c-axis than transverse to this axis (see the concept of growth anisotropy, below). Under these circumstances, it is evident that the velocity of creep will be greatest for the body with hornblende needles oriented transverse to the flattening pressure. This fact, in combination with the principle of the path of least resistance, should indicate that there would be a certain tendency for hornblende or other minerals with anisotropic diffusion coefficients (or growth anisotropy) to orient themselves parallel to the direction of recrystallization flow or microcreep in a yielding rock body.

To pursue this question further, let us consider the situation pictured in Figure 60. A and B are two single crystals of the same mineral exhibiting a typical prismatic habitus (hornblende or sillimanite, for example). A is oriented with the axis of elongation in a 90° angle to the flattening pressure P, whereas B's axis of elongation is parallel to P. The crucial question is now whether there is any difference in thermodynamic stability of grains A and B, so that the one grain would tend to grow at the expense of the other by means of recrystallization via some matrix. (It is evident that none of the grains are really thermodynamically stable because one-sided pressure or stress is in itself an unstable situation. Nevertheless, it is adequate to consider the relative stability between the two grains.) To discuss this question, we need to introduce the concepts of growth and dissolution anisotropy. Most anisotropic crystals growing in a supersaturated and homogeneous milieu will grow more rapidly in some directions than in others, thus providing flaky or prismatic crystals. The basic explanation of this phenomenon is evidently to be sought in the low symmetry of the lattices of such crystals. Hence the bonding energies are different in different directions, and the densities of the atomic planes change greatly with orientation. The number of solute particles which precipitate per unit of surface area on a crystal in a supersaturated solution depends upon which crystal face we consider. Let the concentration of solute in the supersaturated solution be homogeneous; then the number of solute particles which collide with the crystal surface per unit of time and unit of area must be the same for all surfaces. The fraction of the striking particles which really enter the surface layer, thus building up the crystal, however, are different at the different crystal faces, so that the growing crystal will expand with different rates in different lattice directions. The phenomenon of surface migration (Volmer, 1939) usually acts in the same direction because particles condensing on surfaces which

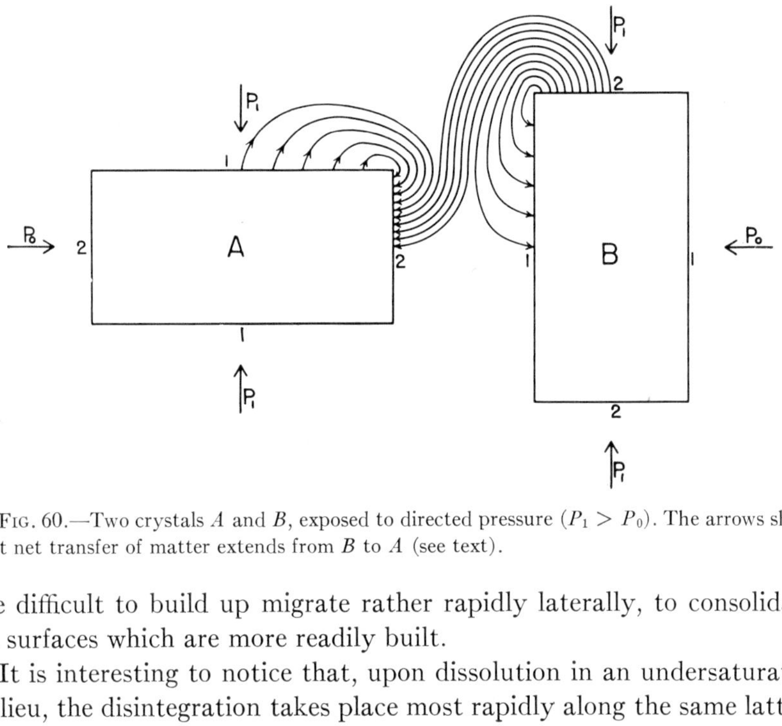

FIG. 60.—Two crystals A and B, exposed to directed pressure ($P_1 > P_0$). The arrows show that net transfer of matter extends from B to A (see text).

are difficult to build up migrate rather rapidly laterally, to consolidate on surfaces which are more readily built.

It is interesting to notice that, upon dissolution in an undersaturated milieu, the disintegration takes place most rapidly along the same lattice directions as the most rapid growth takes place in a supersaturated milieu (the law of reciprocal growth and dissolution [Aminoff, 1927*a*, *b*]).

We shall now assume that the solution is just saturated with reference to crystals A and B under a confining pressure amounting to P_0. The excess flattening pressure, P_1, is applied by means of a permeable plate through which the individual solute particles can migrate freely. Because of the pressure P, the molal free energy of the crystal along faces A_1 and B_2 has risen by a value $\Delta F = \mathrm{V} \cdot (P_1 - P_0)$ relative to the free energy at faces A_2 and B_1. The crystals will, therefore, dissolve along faces A_1 and B_2, but with different velocity: B_2 will dissolve more rapidly than A_1.

The concentration of solvent will increase above the permeable plate on A_1 and B_2, so the solution actually is supersaturated here. Assuming that the supersaturation is not of a value capable of creating nuclei within a reasonable lapse of time, we shall find that the concentration will be slowly homogenized, so that the whole solution ultimately is somewhat supersaturated. Hence precipitation takes place on faces A_2 and B_1. In harmony with the law of growth anisotropy previously mentioned, A_2 must grow faster than B_1. Thus we have found that the faces which dissolve on grain A (A_1) dissolve more slowly than the dissolving faces on grain B, whereas the faces of A which grow (A_2) must expand more rapidly than the growing faces of grain B. If we follow this process somewhat further, we shall find that the chances for a complete dissolution of grain B to the benefit of growth of grain A are very great. Hence we have proved that, during recrystallization under one-sided pressure of a rock with flaky or prismatic minerals (minerals with growth anisotropy), the grains which are oriented with the direction of most rapid growth parallel to the one-sided pressure are unstable relative to the grains oriented with the direction of most rapid growth parallel to the direction of lowest pressure in the rock body.

The preferred lattice orientation thus developed by pure recrystallization under stress may be easily mistaken for lattice orientation caused by translational gliding. It is plain that the preferred lattice orientation caused by recrystallization of anisotropic minerals under stress will also coincide with a preferred shape orientation, because minerals with prominent growth anisotropy tend to develop a tabular or prismatic shape.

It thus seems impossible to create preferred lattice orientation of equidimensional minerals by recrystallization flow, as pictured here. Before concluding that such orientation is due solely to translation gliding in the lattice (p. 104), we should recall the work and ideas of Sonder (1948). He demonstrated that the elastic compressibilities of rocks and minerals are different in different directions. In particular, he found that in many cases the compressibility was greatest along the mineral axis (quartz, calcite), which often is oriented perpendicular to the schistosity in rocks. Evidently, there will be a gain in free energy if a mineral grows with its axis of greatest compressibility along the direction of largest one-sided pressure in a schist. Such an effect would certainly also create preferred lattice orientation by recrystallization flow without the help of translation or twin gliding.

STRESS-ANTISTRESS MINERALS

CLOSELY connected with the question of relative thermodynamic stability of crystals with different orientation relative to mechanical stress is the question of stress-antistress minerals. A stress mineral in the sense of A. Harker and others is a mineral which can be formed only under mechanical stress. An antistress mineral, on the other hand, becomes unstable as soon as stress is put upon it. There is no common acceptance as to whether stress and antistress minerals really exist or not. P. Eskola, for instance, considers the concepts of stress and antistress minerals to be invalid. The matter has drawn much attention from the physicochemical-minded petrologists, but neither experiments (Larsen and Bridgman, 1938) nor theory has succeeded so far in proving or disproving the general validity of the concepts, to say nothing of the question of which particular minerals should belong to the groups of stress or antistress minerals, respectively.

Quite recently, however, a series of unpublished experiments performed at the University of Chicago indicated that, generally, there may well exist crystalline substances with structures which are created under only one-sided pressure. It was shown, in an experiment performed for quite a different exploration, that when stress acted upon some iron-carbon mixed crystals, then the carbon atoms moved into particular sites, thus creating a new crystal structure. Upon release of the stress, the carbon atoms returned to their original sites, hence recovering the original crystal structure—the "antistress" structure. If this process proceeded very slowly, one could easily imagine that the stressed structure could persist in a "quenched" form for a long period of time under homogeneous confining pressure.

A closer consideration of the question suggests that stress minerals may probably exist. Evidently, no minerals are thermodynamically stable under stress because stress in itself is an unstable condition. As was done under the discussion of the stability of differently oriented grains (p. 115), we may also ask whether under stress some minerals are more likely to form than others. We concluded that grain *A* was, because of its orientation, more stable than grain *B* (Fig. 60, p. 116). Here *A* and *B* were two grains of the same mineral. Our question now becomes whether there exists a new modification of the mineral which is still more likely to develop than grain *A*.

We shall, to this end, consider two modifications, *A* and *B*, of a certain substance. Supposing that modification B is isotropic, whereas A is strongly anisotropic regarding velocity of growth and dissolution, B being the stable modification under homogeneous confining pressure.

A and B exist in a solution under a confining pressure P_0. When no stress is applied, the stable B will grow at the expense of the unstable A. Let a one-sided pressure P_1 be applied to both modifications. Clearly, in order that B may start to dissolve along the stressed face (B_2), a minimum pressure P_1 is needed, because the solution was originally somewhat supersaturated with respect to B. When P_1 exceeds the minimum value, B will start to dissolve along the stressed face B_2. Now we shall assume that the rate of growth of any face of B at a given supersaturation is intermediate in value between the rate of growth of faces 1 and 2 on A at the same degree of supersaturation. At some high P_1 therefore, it is not impossible that the rate of dissolution of face B_2 becomes greater than the rate of dissolution of face A_1, in spite of the fact that B would grow along face B_2 at $P_1 = P_0$. The solution thus becomes supersaturated with respect to face A_2. However, the degree of supersaturation is greater about face B_1 than about face A_2. As the velocity of growth on B_1 is less than on A_2 (at the same degree of supersaturation), it might well occur that face A_2 grows more rapidly than face B_1. Hence we may conclude: If a substance can build two crystalline modifications with closely similar lattice energy but with great difference in growth anisotropy, that with the smallest growth anisotropy being slightly more stable than the other under nonstress conditions, the modification with the greatest growth anisotropy is likely to develop under stress conditions.

A similar conclusion should hold for more complex systems: Suppose that X is a crystal with prominent growth anisotropy and is related to crystals Z and Y as follows: $X = Z + Y$, so that, under any confining pressure, the combination $Z + Y$ is stable. Then it does not seem unlikely that X will develop at the expense of Z and Y under stress conditions.

These conclusions may, in particular, be of importance when the extreme and rare compositions of some mixed crystals are concerned. If the mixed crystal has a pronounced anisotropic structure, then it is likely that intense stress enables it to absorb chemical components from minerals with less pronounced anisotropic structure. It is not unlikely, for example, that the sodium component of glaucophane hornblende is absorbed under stress at the expense of the surrounding albite grains.

In conclusion, we may say that the existence of stress minerals does not contradict the laws of thermodynamics. It has yet to be proved, however, that any of the suggested stress minerals really are such. Antistress minerals are in a similar position. It cannot be disproved theoretically that some minerals with less striking growth anisotropy become completely unstable at intense stresses, and thus change into stress minerals.

Petrographic experience indicates strongly that neither stress nor anti-stress minerals are, if they really exist, very numerous. Most micas, horn-

TABLE 12

MOLAR VOLUMES OF SOME MINERALS COMPARED WITH THE VOLUMES OF THE CONSTITUTING OXIDES (AFTER ESKOLA, 1939)

MINERAL NAMES	DENSITIES (GM/CC)	MOLAR VOLUMES OF CONSTITUTING OXIDES		MOLAR VOLUMES OF MINERALS
		Max.	Min.	
Nepheline	2.62	53.3	46.4	54.4
Leucite	2.5	85.9	76.3	87.5
Cordierite	2.6	209 6	186.2	235.6
Andalucite	3.1	53 3	48.1	59.4
Sillimanite	3.24	53.3	48.1	50.2
Potash feldspar	2.56	111.9	98.9	109.0
Albite	2.60	105.3	91.6	101.0
Forsterite	3.216	51.0	44.8	43.8
Diopside	3.275	87.9	72.7	66.3
Enstatite	3.175	38.5	33.7	31.7
Kyanite	3.6	53.3	48.1	45.1
Grossularite	3.53	175.5	142.5	127.9
Almandite	4.25	148.8	136.8	117.3
Pyrope	3.51	142.8	126.6	114.8

blendes, chlorites, kyanite, sillimanite, etc., with obvious growth anisotropy are commonly found in rocks where stress was unimportant or lacking (see Eskola's discussion, 1939, pp. 331–34). Most so-called "stress" minerals have a relatively small molal volume; they are minus minerals[1] in the sense of Loewinson-Lessing. So-called "antistress" minerals are usually plus minerals,[1] as shown in Table 12, compiled by Eskola. This indicates that most stress minerals simply are relatively high-pressure minerals, antistress minerals are stable only at lower pressure.

1. A minus mineral has a mol volume less than the sum of the mol volumes of the constituent oxides; a plus mineral has a larger mol volume than the oxides.

BRIEF SURVEY ON STRUCTURAL PROPERTIES OF METAMORPHIC ROCKS

THE RELICT STRUCTURES

PARTS of the structure of metamorphosed rocks are relicts from the premetamorphic history of the rocks, while other structural properties have been gained during the metamorphism and metasomatism. The relict structures of metamorphites or metasomatites reveal for us the premetamorphic history of the rocks. We may read the evolution through the newly impressed structure because the synmetamorphic alterations have not managed to destroy completely the original character of the rock (palimpsest structure, Sederholm, 1899).

Rhythmic layering in sediments like sandstone-limestone complexes, conglomerate-sandstone, "varvig" slate, and the like often preserve the layering also in the metamorphosed state, or pillow lavas may preserve their typical structure in recrystallized amphibolites. Calcite concretions in shale may be recognized in the form of diopside-scapolite-skarn inclusions in mica schist or sillimanite schist (Figs. 61 and 62). Doleritic (diabasic) sills or dikes which once injected schists may be found as deformed amphibolites in crystalline schists or gneisses. In such meta-igneous rocks the original prophyritic structure or also often the typical pyrogenetic ophitic diabase structure may be present as relicts in the metadiabase. The original pyroxene phenocrysts are uralitized (=altered to secondary hornblende), the plagioclase crystals may be saussuritized (=altered to zoisite, epidote, muscovite, etc.), etc.

Some "augen" gneisses may possibly represent relict structures after igneous porphyritic granites, but most commonly the feldspar "augen" have grown during the metamorphism, i.e., they are porphyroblastic. Tectonic deformation during the growth or after the growth may give the false impression that the porphyroblasts are relicts from phenocrysts.

THE CRYSTALLOBLASTIC FABRIC

We shall consider briefly some of the essential concepts of structural petrology grown from the study of deformed rocks. The science of structural petrology has, in recent years, grown rapidly to the stature of an independent branch of petrology, much too large to cover in a book on metamorphism. The reader is referred to the works of Sander (1930–48), Schmidt (1932), Turner (1948), and Fairbairn (1949), the last, in particular, being a clear and comprehensive text. It will serve our purpose in this section to give some definitions.

Consider a rock recrystallized or metasomatosed thoroughly under the static conditions of confining pressure. The mutual lattice and shape orientation of all the mineral constituents in such a body may be completely randomly oriented in space. The fabric is said to be "isotropic" or "undirected." Although some metamorphic-metasomatic rocks may

FIG. 61.—Skarn inclusions in sillimanite-garnet gneiss. The rocks belong to the granulite facies. The skarn inclusions are probably remnants from limestone layers in shale which has been changed to sillimanite-garnet gneiss by recrystallization under granulite facies conditions. Nagssugtoq, West Greenland. Photograph by H. Ramberg, 1948.

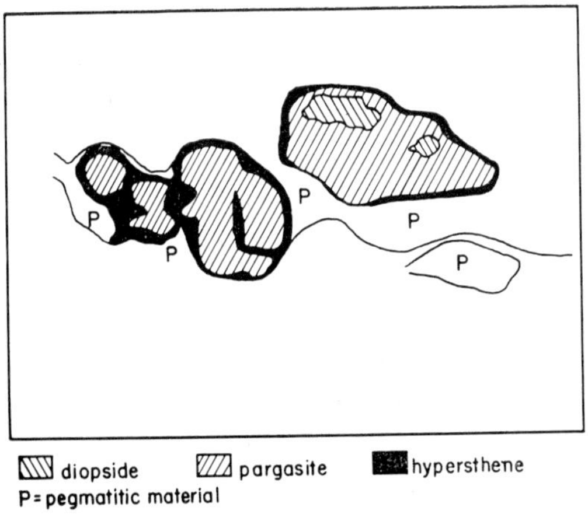

FIG. 62.—Key sketch to photograph, Fig. 61

exhibit such undirected fabric, most metamorphosed rocks develop an anisotropic, or directed, fabric. The anisotropic, or directed, fabric of metamorphic rocks is most commonly caused by deformation and by crystal growth.

A directed growth fabric originates when newly formed minerals grow in some kind of subparallel orientation for reasons different than mechanical strain of the rock body. This kind of directed structure is not uncommon in metamorphic rocks in the broadest sense of the word used in this book. In four cases, such structure is common.

1. In recrystallized or metasomatically replaced sediments with an original anisotropic deposition fabric, the neominerals may orient themselves parallel to the old clastic grains which are being replaced or may function as nuclei of crystal growth.

2. In polymetamorphic gneisses or crystalline schists, the minerals grown during the last recrystallization or metasomatism would often tend to orient themselves parallel to the old unstable minerals, thus giving the new rock a directed fabric, although the rock perhaps formed during static confining pressure.

3. During growth of replacement veins—dilation veins, as well as concretionary veins (pp. 248–56)—the newly formed minerals like biotite, hornblende, muscovite, feldspars, quartz, etc., often orient themselves subperpendicularly to the border surface of the vein (the "comb" structure of many veins).

4. Minerals formed during metamorphism by reaction between chemically incompatible rock units—for example, calcite concretions in shale, or dunite inclusions in anorthosites (Eskola, 1921*b*)—often also tend to orient themselves with their major axes perpendicular to the interface between the two rocks. That is, minerals with growth anisotropy tend to orient their axis of most rapid growth parallel to the direction in the rock along which the growth meets least resistance. A surface perpendicular to the direction of preferred growth is called "blastetrix" by Sander.

The majority of cases of anisotropic structure in metamorphic rocks are caused by deformation under mechanical stress of the rock bodies, the mechanism of which is discussed in a previous chapter (pp. 97–117). During this process some of the structure elements or the fabric elements achieve more or less parallel orientation of either a planar or a linear character, or both.

A fabric element is a single crystal or a group of crystals (superindividual) (Fairbairn, 1949, p. 3) which behaves as a unit with respect to the mechanical forces applied to it. The shape of such units may be planar, linear, or equant (*ibid.*). There are no limitations in size of these fabric elements, which, however, should more properly be called "structure elements" when the size exceeds 1 foot or so in diameter. The common boudins of amphibolite in quartzo-feldspathic gneisses (Fig. 63) or the boudins of pegmatite in micaceous quartzite in Figure 64, as well as the quartz-feldspar veinlets in Figure 65, are all structure elements helping to define the deformation and crystalloblastic structure of the rocks in which they are found.

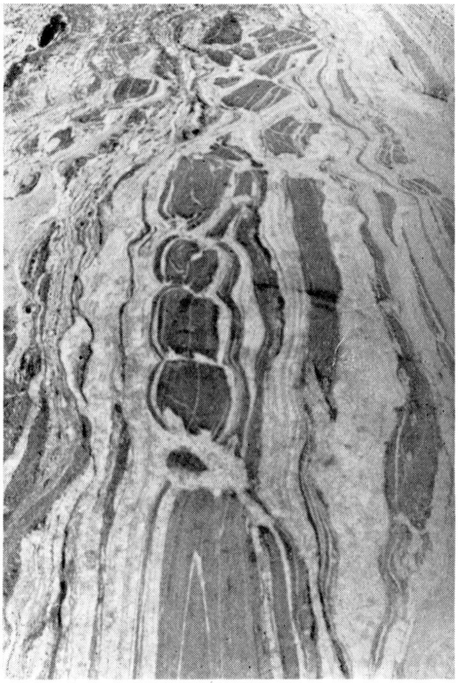

FIG. 63.—Boudins of amphibolite in granodioritic gneiss. Godthaab Fiord, West Greenland. The boudins are about 1 foot thick. Photograph by M. L. Ramberg, 1951.

The preferred orientation of mineral units in deformed rocks may be dimensional, or crystallographic = lattice orientation. The dimensional orientation refers to the external shape of the mineral grain only, whereas the lattice orientation refers to the space lattice of the mineral grain. As we have discussed in the preceding chapter, rocks yield under stress partly by different kinds of plastic flow and creep, partly by rupture or cataclasis. Usually some cataclasis is found in any tectonite (undulatory quartz, fractured crystals, minor cracks and joints, etc.), and some plastic flow has accompanied any mylonitization. Figure 66 shows typical plastic deformation in a gneissic rock; Figure 67 shows a typical mylonite zone; and Figure 68 shows an intensely sheared amphibolite.

FIG. 64.—Photograph of boudins of pegmatite in micaceous quartzite. Near Atlantic City, Windriver Mountains, Wyoming, U.S.A. Photography by H. Ramberg, 1949.

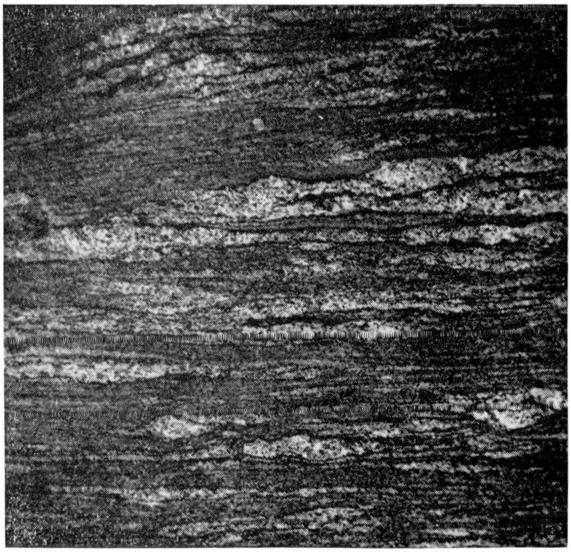

FIG. 65.—Typical veined gneiss. Note concentrations of mica along the borders of some of the pegmatitic veinlets. Caledonian gneiss, West of Trondhiem, Norway. Photograph by H. Ramberg, 1943.

Fig. 66.—Ptygmatic folds in gneiss. Typical example on plastic flow in rocks. West Greenland, 1949. Photograph by H. Ramberg, 1949.

Fig. 67.—Zone of intense shear movement, partly cataclastic. Godthaab Fiord, West Greenland, 1949. Photograph by M. L. Ramberg.

The preferred orientation of fabric elements (mineral grains) gives rise to foliation in which planar structure elements are subparallel, or lineation in which linear structure elements are subparallel.

According to Fairbairn (1949), one may distinguish between the following kinds of lineations: (1) parallel dimensional orientation of elongate single crystals, like hornblende, mica, quartz, etc., or of composite larger fabric units, such as elongate conglomerate pebbles, fossils, etc.; (2)

FIG. 68.—Intensively sheared amphibolite. The unusually well-developed elongation is vertical. West Greenland, 1949. Photograph by M. L. Ramberg.

parallelism of lines of intersection of surfaces of varying orientation, e.g., bedding planes and secondary schistosity; and (3) parallelism of fold axes on a small or large scale.

It has been found convenient to introduce a three-dimensional axial system to describe the orientation of fabric elements in rocks (Sander). These fabric axes are denoted by *a*, *b*, and *c*, *a* being supposed to be parallel to the "direction of tectonic transport,"[1] *b* is parallel to the fold

1. This very point is a matter of intensive discussion among structural petrologists (see p. 97). It appears that, very commonly, tectonic transport or shearing is parallel to that type of lineation which is substantiated by parallelism of elongate mineral grains. In any case, if the lineation is due to elongate pebbles in deformed conglomerates or to elongate mineral grains, it means that the rock itself has been elongated in the direction of the lineation and hence that plastic flow has taken place in this direction, which usually is parallel to the fold axis (see Balk's discussion of salt domes, 1949).

axis and hence usually also parallel to the lineation, and *c* is just normal to the *a–b* plane.

An *s*-surface is any structurally significant plane or surface in the deformed rocks, such as bedding planes, planes of layering in banded gneisses, foliation planes, surfaces of rupture, etc.

The most common fracture surfaces in crystalline schists and gneisses are the tension joints, which are more or less perpendicular to the lineation and elongation, and the shear-rupture surfaces, which are usually parallel

FIG. 69.—Competent amphibolitic layer with tension cracks. This is a very typical feature from gneissic areas and shows the embryonic stage of the common amphibolitic boudins in quartzo-feldspathic gneisses (see Fig. 63). West Greenland, 1948. Photograph by H. Ramberg.

to the elongation. Actually, rupture surfaces, often accompanied by mylonitization, may have any inclination to the tectonic axes system. But "textbook" diagonal joints developed in well-defined pairs are rather rare.

It is interesting to note that very commonly in regionally metamorphosed complexes the brittleness of the rocks is a function of the mineralogical composition in a general way. The more basic the rock, the more subjected to rupture it is. The quartzo-feldspathic gneisses, as well as the quartzites and the mica schists, are very mobile, tectonically speaking (incompetent), whereas the diopside skarn rocks, the amphibolites, and the ultrabasites are usually rigid and immobile (competent), showing little ability to yield in a plastic manner under stress (Fig. 69). This is valid only at a medium and high degree of metamorphism. Although this statement holds true in a general way, exceptions may certainly occur because the mechanical properties of rocks are not well defined but are likely to change with time and chemical milieu, as well as with pressure and temperature. Rocks which are mechanically competent at low-temperature deformation (quartzites, sandstones, limestones) are very incompetent indeed at intermediate and high degrees of metamorphism.

The established fact that, under medium and high regional metamorphism, quartzo-feldspathic minerals and rocks are more easily deformable than the basic and ultrabasic rocks is rather interesting from a crystal-chemical point of view. In dry melts the state of affairs is exactly the opposite: the more acidic the silicate melt, the more viscous it is (Table 13). This is explained chiefly by the higher concentration of

TABLE 13

VISCOSITY OF SOME ROCK MELTS AT 1,400° C.
(AFTER *Handbook of Physical Constants*)

Melt Type	Viscosity in Poises	Melt Type	Viscosity in Poises
Obsidian	1.7×10^6–4.4×10^6	Andesites	150–1,500
Hornblende granite	2×10^6	Diabases	15–400
Hornblende–mica andesite	1.6×10^4		

–O–Si–O– bonds of great energy in acidic silicate melts. One would, offhand, expect a similar situation to exist in the crystalline state. The three-dimensional network of rigid –O–Si–O– bonds in silica and feldspars should offer greater resistance to plastic slip than should the silicates consisting of sheets, bands, smaller groups, or individual islands of silicon-oxygen tetrahedra. In conformity with such structural considerations, it is well known that the chlorites, talc, serpentine, and most micas are rather ductile, owing mainly to slip along the (001) plane. Although the pyroxenes and hornblendes exhibit good cleavage parallel to (110) and partly to (010) and (100)—that is, parallel to the –O–Si–O– chains—it is less certain whether slip takes place along these planes (see Fairbairn, 1949, p. 101). Olivine may, according to Turner (1942) and Chudoba and Frechen (1950), have a glide plane (010). Some data are compiled in Table 14.

Bowen and Tuttle (1949) have, in harmony with similar lattice structural considerations, suggested that dunites may easily be injected as a solid aggregate (in order to account for their experiments, which show that many dunites cannot crystallize out of an injected melt, see p. 70).

As we have seen, however, the field observations clearly demonstrate that basic rocks are usually rigid and immobile as compared with acidic rocks. The present writer thinks that the explanation for these apparent discrepancies lies in the fact that the relative plasticity (crystalline viscosity) of rocks and minerals under medium- to high-grade metamorphism is not governed principally by the ease at which slip takes place in their lattices. The mobility of the rocks is mainly determined by their power to undergo recrystallization or by the ease with which chemical rearrangements can proceed in the rocks and their minerals. The crystalloblastic structure in most metamorphic rocks shows that recrystallization flow is the most important type of yield in silicate rocks, provided that they do not contain mainly ductile minerals like chlorites, talc, and some micas. It appears, then, that the ionic, atomic, or molecular mobility in quartzo-

TABLE 14

EXPERIMENTAL DEFORMATION OF MINERALS*

Mineral	Translation Plane	Translation Direction (Glide Line)
Halite	(110)	[1$\bar{1}$0]
Fluorite	(001)	[110]
Calcite	(01$\bar{1}$2)	[*r*:*r*]
Dolomite	(0001)	[*r*:*c*]
Bronzite	(010)?	[001]?
Anhydrite	(001)	[010]
	(012)	[100]?
Gypsum	(010)	[001]
	(010)	[301]?
Mica	(001)	[100]?
	(001)	[110]?
	(135)	...
	(205)	...
	(104)	...
Augite	(100)?	[001]?
Kyanite	(100)	[001]

* Reproduced in part from Fairbairn (1949).

feldspathic rocks during regional metamorphism is, as a common rule, considerably greater than this type of mobility in basic rocks. In this connection, it is interesting to recall the experiments by Bridgman and Larsen (1938), which showed that quartz failed to yield plastically by translation in the lattice, even at very high pressure and reasonable temperature. Only rupture and cataclasis were produced. In opposition to this result, Fairbairn reports (1949) that quartz sand would yield rather easily and form a compact and deformed true quartzite when a weak Na_2CO_3 solution was present in yield experiments performed at confining pressures of 5,000–50,000 psi and temperatures of 245°–450° C. The effect of solutions as catalysts in the experiments by Griggs also shows that the chemical part of the tectonic deformation of rocks is the most prominent during ordinary thermodynamic metamorphism.

In this chapter and the preceding ones we have discussed the fundamental aspects of the kinetics of deformation and the deformation structure of rocks. We have, perhaps, paid too little attention to the description of structures and deformation fabrics of tectonites. As this branch of metamorphic petrology is still mainly in the empiric-descriptional stage, we consider it outside the scope of this book to go deeper into the subject. For better illumination, the reader is referred, in particular, to the well-balanced book of Fairbairn so frequently cited through this chapter.

THE IDIOBLASTIC SERIES OF MINERALS

The crystalloblastic fabric of rocks is partly controlled by the fact that the several minerals exhibit different tendencies to develop euhedral shape. On empirical grounds, Becke established the following mineral series, showing decreasing tendency to grow with their own crystal shape in metamorphosed or metasomatosed rocks: sphene, rutile, magnetite, hematite, ilmenite, garnet, tourmaline, staurolite, kyanite; epidote, zoisite; pyroxenes, hornblende; brennerite, dolomite, albite; micas, chlorites; calcite; quartz, plagioclase; orthoclase, microcline.

This sequence is valid only statistically; minerals early in the sequence may not invariably develop as euhedral metacrysts, neither may minerals situated late in the series always fail to exhibit euhedral crystal shape. Garnet, for example, certainly is likely to form well-shaped dodecahedrons in schists, but not infrequently one finds anhedral garnet porphyroblasts in metamorphites. On the other hand, microcline usually shows a very weak tendency to develop a euhedral crystal shape in most recrystallized rocks, but occasionally this mineral exhibits exceedingly well-shaped euhedral crystals when growing in gneisses.

The ability of a mineral to form metablasts with well-shaped crystal faces has been attributed to the so-called "form energy" ("shape energy") of the mineral (Eskola, 1939). That is, there is a free-energy difference between a crystal with well-developed crystal faces and the same crystal of the same size with an irregular, faceless shape, the latter being less stable than the former. This energy difference should correspond to the form energy of the mineral. The magnitude of this energy difference between a well-developed crystal face and an irregular surface is different not only for the several minerals but also for the different lattice planes of one particular mineral. The faces (001) in micas and (110) in hornblendes, for example, show a much stronger tendency to develop than any other face of these minerals. Because of interfacial energy between contacting minerals, it is evident that the tendency of a mineral to develop crystal faces depends significantly upon the mineralogical milieu in which it grows.

The ability of metacrysts to form crystal faces may be explained by the mechanism of growth of minerals in recrystallizing rocks. To form an embryonal nucleus of a mineral under given conditions, a certain degree of supersaturation is always required. The growth of the nucleus in a homogeneous milieu usually goes on by completing one surface layer after another. It is considerably easier to continue the growth of an uncompleted surface layer than to start a new layer. Actually, the start of a new embryonal-layer nucleus on a surface may be to some degree compared with the formation of a new crystal nucleus—a certain activation energy is required. The crystal planes on which embryonal-layer nuclei form with the most difficulty but expand laterally most easily, therefore, tend to develop as crystal faces. On the other hand, crystal planes on which embryonal-layer nuclei form rather easily but show less tendency to expand laterally along the plane will fail to develop as crystal faces on the metacryst. It is natural that these conditions should depend upon the anisotropic nature of the interionic bonding energies in the crystalline lattices.

In the case of mica, for example, it is known that the interionic forces across the (001) plane (which is parallel to the Si_4O_{10} sheets) are considerably weaker than the forces acting along this plane. For that reason, it is clear that there would be less tendency to form a small nuclear layer on top of the (001) surface than to form a small nucleus on any surface in the prism zone. However, once the small nuclear layer is formed on top of a (001) surface, new constituents will rapidly be attached to the strong unsaturated valency forces around the edge of the nuclear layer, and this will easily be completed to a new (001) surface. A lateral growth along the surfaces of the prism zone will not take place so easily. Figure 70 shows the conditions schematically.

Now conditions will not be so simple in rocks, because a growing mineral grain is surrounded by several different minerals and hence the interfacial tension varies along the entire surface of the growing grain. This situation controls to a great extent the shape which the growing grain ultimately achieves. Consider again a mica grain which grows in a milieu of, for example, quartz and some ore minerals, etc. (Fig. 71*a*). The interfacial tensions between the mica and the several adjacent mineral grains may vary considerably, because the bonding forces across the border against the mica may be different for the several other minerals. Let us, for example, assume that the interfacial tension between the mica and some oxide ore minerals is low, whereas the interfacial tension between mica and quartz is higher. Then the joint between the ore grain and the (001) face of mica will act as a probable starting place for a new surface layer on the (001) face, whereas the quartz-mica joints do not behave thus. The quartz grains may therefore be pushed away or dissolved (replaced) during the growth of the mica flake, whereas the ore grain will be incorporated as an inclusion (Fig. 71*b*).

Garnet metacrysts may often contain numerous inclusions of quartz grains at the same time as the fabric indicates that biotite flakes have

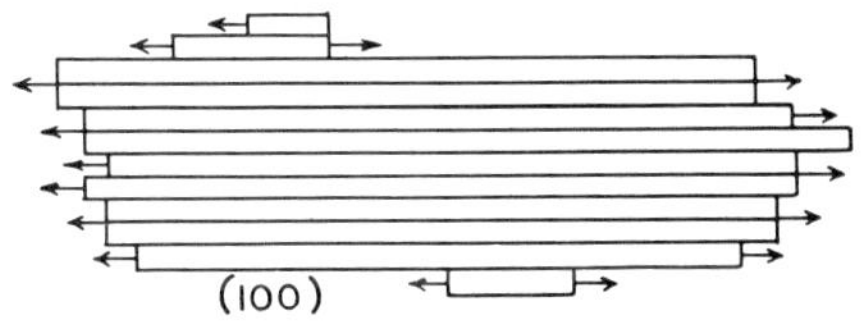

FIG. 70.—Schematic picture of growth of mica. The arrows indicate the most rapid displacements of the lattice during growth from a supersaturated milieu. Most likely, small nuclear layers of a certain minimum size form on the (001) surfaces, and these nuclear layers grow by lateral expansion (see text).

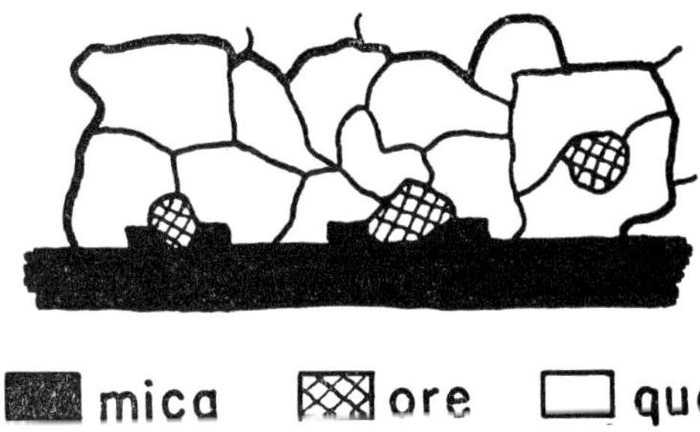

FIG. 71*a*.—The growth of mica in a quartzite with scattered grains of ore. The surface nuclei of mica form most readily in the contact between mica and ore; therefore, the growing mica will incorporate the ore as inclusions but will replace quartz, as shown in Fig. 71*b*.

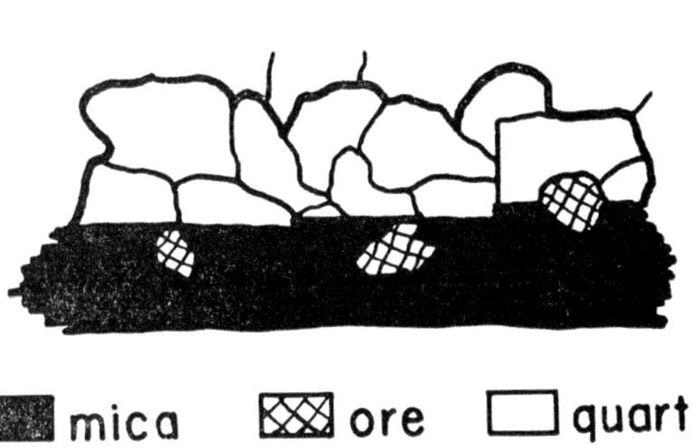

FIG. 71*b*.—Mica with incorporated ore grains (see Fig. 71*a*)

been pushed away (Fig. 72). The reason may logically be that the interfacial energy between garnet and (001) biotite is great, as compared with the garnet-quartz interfacial energy. Here the joints of garnet-quartz may frequently function as birthplaces for the new surface layers to be built on the growing garnet, such that gradually garnet will completely inclose the quartz grain.

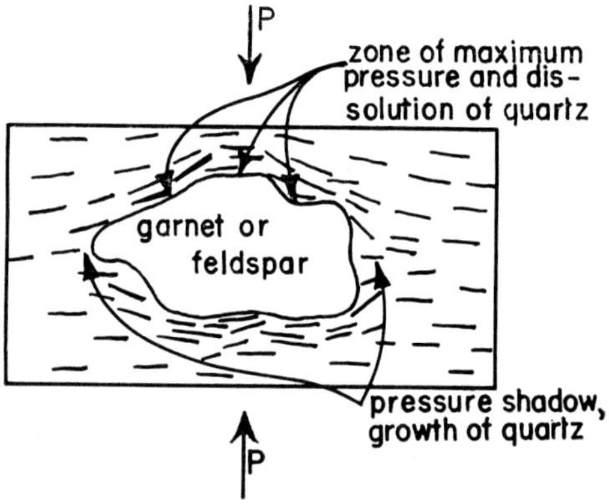

FIG. 72.—Growth of garnet porphyroblast in a micaceous quartzite after the principle of concretionary growth (p. 95). Mica is relatively concentrated in the areas of maximum pressure because the easily mobilized quartz is removed from this place to the pressure shadows.

There is a wide-open field for studies of these conditions in metamorphic petrology today. Very little has been done as yet in an attempt to study the relative interfacial tensions among minerals. These brief remarks may, however, suffice to show the general idea of form energy of minerals and likewise to show that this concept is a relative one—one mineral having varying form energy, depending upon the P,T-conditions as well as the types of minerals among which it grows. The concept of force of crystallization or pressure of growth has been touched upon elsewhere (p. 14).

THE MINERAL FACIES OF METAMORPHIC ROCKS

PETROGRAPHIC experience has established that a close approach to chemical equilibrium among the mineral species is rather common in metamorphic rocks. That is, the mineral assemblages correspond to approximate equilibirum at certain P,T-combinations which may vary from place to place and from rock to rock, and it is believed that the mineralogical evolution of the rocks was "quenched" just at those P,T-conditions.

This fact, perhaps expressed clearly by Goldschmidt (1911) for the first time, enabled Eskola (1921*a*) to introduce his important concept of metamorphic and mineral facies. Eskola defines a certain mineral facies as comprising rocks exhibiting identical minerals if the rocks are of identical bulk chemical composition, but exhibiting assemblages which vary after certain rules if the rocks have different bulk chemical compositions (1939, p. 339).

The importance of this principle is based on the experienced fact that equilibrium commonly is developed in metamorphic rocks so that rocks belonging to the same facies have formed or recrystallized at similar P,T-conditions. It is seen from the definition of facies by Eskola, however, that no requirements as to chemical equilibrium or P,T-conditions are given. Eskola states (1939, p. 339): "Massgebend für die Zugehörigkeit zu einer Mineralfacies ist nur, dass die Mineral Zusammensetzung bei demselben Chemismus dieselbe ist." According to the definition of the concept "mineral facies," then, rocks formed or recrystallized at very different P,T-conditions may be classed in the same facies. That is particularly true in the case of monomineralic rocks which are composed of minerals with a large stability field (quartz, calcite, dolomite, potash feldspar, acid plagioclase, biotite, etc.). A pure quartzite formed at room P,T-conditions should exhibit the same mineral "assemblage" as a quartzite formed just below the transition $\alpha \rightleftarrows \beta$-quartz. A monomineralic potash feldspar rock may also exhibit identical mineralogical composition if formed close to room temperature as well as at some 1,000° C. Also a rock consisting of potash feldspar, quartz, and muscovite (a granitic gneiss) would maintain the same mineral assemblage if recrystallized at P,T-conditions varying over a very wide interval. According to the original definitions of mineral facies, these "identical" rocks would be put in the same facies, although possibly formed at any temperature at which metamorphic rocks can form at all. This particular facies would embrace a P,T-field comprising the whole field of metamorphism.

Let us, on the other hand, consider some more ordinarily composed rocks containing biotite, hornblende, garnet, plagioclase, and other more complex minerals. As seen by the equilibrium diagrams in previous chapters, *identical* composition of the complexly composed minerals at identical bulk chemical composition of the rocks can be achieved only at one particular P,T-combination. Thus, if Eskola's definition of facies (see above) is taken too literally, each point in the P,T-diagram must correspond to one particular facies, and there must be an infinite number of facies.

For the reasons discussed above, we shall here use the following definition of mineral facies: *Rocks formed or recrystallized within a certain P,T-field, limited by the stability of certain critical minerals of defined composition, belong to the same mineral facies.* It is not necessary that a rock contain the critical minerals, in order to be ascribed to a certain facies, for the physical conditions of formation may be determined by different indicators.

By this definition, quartzites of identical mineralogical composition may belong to different facies, depending upon the P,T-conditions of recrystallization. Likewise, two mineralogically "identical" granites containing quartz, potash feldspar, and muscovite may actually belong to very different facies; one may be high amphibolite facies, the other low green schist facies. Furthermore, to belong to the same facies, it is not necessary that rocks of identical chemical composition exhibit identical mineralogical assemblages. That is so because each facies must correspond to a *P,T-field*, not a *point*. Within each such facies field, several mineral reactions may take place without thereby requiring a more detailed facies classification. If the mineral composition is defined to be constant at the same bulk chemistry in a given facies, then there will be an infinite number of facies! To define a particular mineral facies, some critical minerals or mineral assemblages are used such that the facies in question just embraces the P,T-area within which the critical minerals or mineral assemblages are physically stable. Such guide mineral associations are chosen arbitrarily, determined by the requirement that they develop at a bulk chemistry commonly encountered in rocks and that their field of stability be neither too narrow nor too wide.

One of the greatest difficulties connected with the facies principle is to determine whether rocks of varying bulk chemical composition belong to the same facies or not. Clearly, the critical mineral associations chosen in one facies will develop only in rocks with a restricted chemical composition; rocks of different composition will not exhibit the guide mineral assemblage, although they may have formed in the same facies or at the same P,T-conditions as the "typical" rocks. Ultimately to be able to place the several metamorphic rocks in their proper facies (or sequence of facies, if the rocks are polymetamorphic), a knowledge of the relative stability conditions of all rock-forming minerals is necessary. At the present time, such information does not exist, although explorations of the mineral associations encountered in rocks and synthesized in laboratories gradually expand our knowledge. Unfortunately, little is known about the abso-

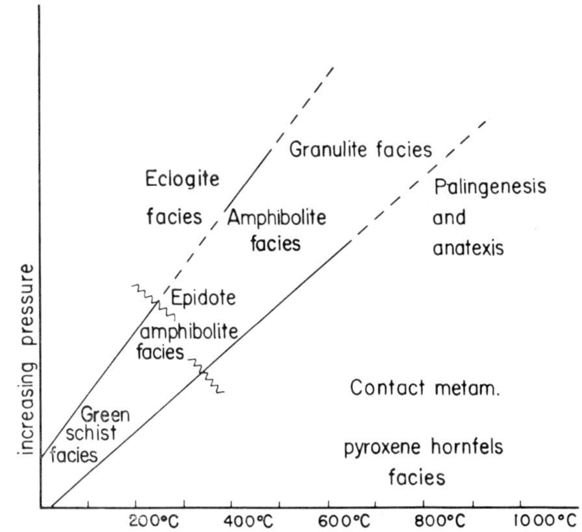

FIG. 73.—Schematic diagram, showing the approximate position of the several facies in a P,T-scheme.

lute temperature and pressure represented by the several facies. We have some information, however, about the relative places of the several facies in a P,T-diagram, as indicated by Figure 73.

The fact that the exact temperature and pressure ranges indicated by a facies are still unknown by no means diminishes the significance of the study of rocks with the important facies principle. The field petrologists can, by mapping and laboratory investigations of the samples, explore finer and finer details as to the relative stability conditions of the mineral assemblages and may gradually succeed in placing all kinds of metamorphosed rocks in their proper relative facies. In the meantime, the study of absolute equilibrium conditions of minerals performed in thermodynamic laboratories may give us enough points in the P,T-diagram to place the several facies absolutely and properly. The present vague estimates of T and P represented for some of the facies will be discussed later.

As to the names of the several facies, the name of the most typical rock formed in the certain facies is used (Eskola, 1921*a;* 1939, pp. 334–45). Usually, also, the chosen rock exhibits mineral assemblages which are sensitive to P,T-changes, as most basic rocks are (for example, green schist facies, amphibolite facies, pyroxene hornfels facies, etc.). That is, the calco-ferromagnesian minerals have much more narrow fields of stability than have quartz, potash feldspar, and acidic plagioclase.

The facies concept can be properly presented only in a polydimensional space, where the several chemical elements (oxides), as well as P and T, are axes. Then it can be shown how the mineral assemblage of a rock would change if the bulk chemistry, as well as pressure and/or temperature, changed. More conveniently, such a polyspace diagram can be divided so that each facies is studied separately, keeping P and T constant but allowing the bulk chemical composition to change. In so doing and in recognizing that the most crucial components in silicate rocks are CaO, Al_2O_3, and (Mg,Fe)O as well as SiO_2, Eskola proposed the so-called "ACF diagrams" (see below), each of which should characterize one single facies. Although the several ACF and ACK (K stands for K_2O) diagrams give a good general view of the several facies, these diagrams fail to give anything but crude information. The reason for that is, in particular, that some of the minerals found in the simplified diagrams are mixed crystals containing elements not shown in the diagrams. Therefore, the pattern of the diagrams will change gradually with changing P and T, and only in rare cases can one definite ACF or ACK diagram be assigned to one facies. Nevertheless, the several facies have been expressed, respectively, by means of one single ACF (ACK) diagram (each) in the previous modern treatises on petrology by Eskola and Turner.

The construction of an ACF diagram of an analyzed rock is simple: First, we note that such diagrams comprise only rocks with free quartz (tridymite), so that all the minerals are saturated with respect to SiO_2. Second, we note that potash feldspar and albite are not recorded in the diagram. These three common rock-forming minerals are physically stable in all facies—they are "through-runners"—so the recording of them is not too crucial for the characterizing of facies. Hence quartz, potash feldspar, and albite may occur together with any one of the stable associations recorded on the diagram. Given the chemical analysis of a rock, the ACF values are calculated as follows: First, the amounts of CaO, FeO, and Fe_2O_3 which go into the accessory minerals—sphene, ilmenite, magnetite, hematite, pyrite, and pyrrhotite—are subtracted from the rock analysis and from the following calculation. To do so, a geometrical determination of the percentage of accessory minerals is needed. From tables or by dividing the weight percentages by the molecular weights, the molecular quotients of the several oxides are found. By definition A′ is then equal to: $(Al_2O_3) + (Fe_2O_3) - (K_2O) + (Na_2O)$; C′ is equal to $(CaO) - (CO_3) - 3(P_2O_5)$; and F′ is $(FeO) + (MgO) + (MnO)$, A′, C′, and F′ being considered simply as three components. The true ACF of the rock appears when A′ + C′ + F′ are recalculated, on the basis of their sum being 100. (Al_2O_3), (FeO), etc., symbolize the molecular quotients of the several oxides.

Some criticism of this procedure may be raised: (1) (Fe_2O_3) is added to (Al_2O_3) because it was assumed that Fe^{3+} replaces Al^{3+} in the silicate lattices, so that the two elements may be considered "identical" when the number of components is concerned. As is well known now, however, Fe^{3+} usually replaces Fe^{2+} and Mg^{2+} in the silicates, so that Fe^{3+} should more properly enter the F value $(FeO + MgO + MnO)$. (2) It is seen that (CO_2) is subtracted from the CaO to give the C value, which means that calcite should not appear in the ACF diagrams. Nevertheless, some very critical mineral associations visualized in ACF diagrams are based upon the presence of calcite in the C corner of such diagrams (Eskola, 1939). There is, therefore, often an inconsistency between the way C is computed from chemical rock analysis and the drawing of the corresponding ACF diagrams, in which calcite as well as dolomite may pose. (In this connection, the reader is referred to the numerous facies diagrams found in the much cited textbooks by Eskola [1939] and Turner [1948].) (3) The F value comprises, by definition, MgO, FeO, and MnO, which then are supposed to be so similar in silicate chemical respects that they can be regarded as one chemical component. This is true only in restricted cases. Commonly, MgO, FeO, and MnO behave as three or at least two components.

For the reasons considered above, and others as well, the ACF diagrams cannot give the detailed story of the mineral assemblages and their changes throughout the sequence of metamorphic facies.

GREEN SCHIST FACIES

Rocks recrystallized at the lowest temperatures of regional metamorphism belong to the green schist facies introduced by Eskola in 1921 and named after the chlorite-rich green stones or green schists so characteristic of the lowest grade of metamorphism. A number of rock-making minerals are stable and able to grow at even the low temperatures which must have prevailed commonly during diagenesis of sediments. The temperature existing under diagenesis of sediments may be in the neighborhood of 100° C. or less.

Yet the following common minerals are known to grow under such circumstances (Pettijohn, 1949): quartz, albite, potash feldspar, muscovite, chlorite, kaolinite, calcite, dolomite, magnesite, tourmaline, rutile, and possibly spessartine-rich garnet. On the other hand, epidote (zoisite), actinolitic hornblende, and serpentine, which are so typical of several of the low-grade green schists, do not belong to the stable minerals of diagenesis (Pettijohn, 1949, pp. 476–507). There seem, therefore, to be some important stability differences among the low-grade minerals upon which may be based a convenient splitting-up of the facies of low-grade rocks. Vogt (1927) has thus distinguished among two (or perhaps three) subfacies within the major green schist facies.

Let us, first, try to straighten out the question of the upper limit of the major green schist facies; i.e., the border line in the P,T-field between green schist and epidote amphibolite facies (see p. 145). Vogt and Eskola consider the lowest P,T-curve at which actinolitic hornblende (or any hornblende) can exist stably to be the highest border line of the green schist facies. Turner (1948, pp. 93–98), however, allows actinolitic hornblende to enter the highest part of his green schist facies. This is just a question of where to place an arbitrary classification border line; more important is the question of the relative stability fields of the minerals characteristic of the green schist facies and its closest neighbors (epidote amphibolite facies, glaucophane schist facies). The following reactions are crucial for the stability fields of several of the most important minerals from the green schist facies:

$$\text{calcite} + \text{talc} + \text{silica} \rightleftarrows \text{tremolite} + \text{water} + \text{carbon dioxide}, \tag{89}$$

$$\text{calcite} + \text{chlorite} + \text{silica} \rightleftarrows \text{actinolite} + \text{epidote} + \text{water} + \text{carbon dioxide}, \tag{90}$$

$$\text{muscovite} + \text{ferromagnesian chlorite} + \text{silica} \rightleftarrows \text{biotite} + \text{aluminous chlorite}, \tag{91}$$

$$\text{chlorite} + \text{silica} \rightleftarrows \text{garnet} + \text{water}, \tag{92}$$

$$\text{potash feldspar} + \text{chlorite} \rightleftarrows \text{biotite} + \text{water} + \text{carbon dioxide}, \tag{93}$$

$$\text{potash feldspar} + \text{kaolinite} \rightleftarrows \text{muscovite} + \text{water} + \text{silica}, \tag{94}$$

$$\text{calcite} + \text{kaolinite} \rightleftarrows \text{epidote} + \text{water} + \text{carbon dioxide}, \tag{95}$$

$$\text{muscovite} + \text{calcite} + \text{silica} \rightleftarrows \text{epidote} + \text{potash feldspar} + \text{water} + \text{carbon dioxide}. \tag{96}$$

It is not convenient to use the chemical formulae in these equations because of the complexity of several of the minerals involved, as, for example, the chlorite and biotite. The reader should study the positions of the minerals involved in the ACF diagrams (pp. 143–49) in connection with these equations, which will give a better understanding of the reactions.

The right-hand sides of these reactions are, as is evidenced by petrographic investigations, stable at higher temperature than the left-hand sides (at constant pressure). Reaction (90) is one of the critical ones as far as the establishment of the uppermost border of the green schist facies goes. As is discussed elsewhere (pp. 62–70), the Mg/Fe ratio, as well as the Al content, affects the equilibrium temperature importantly at given pressure. The reaction is rather complicated: it is, for example, by no means certain that an Al-free actinolite is the type of hornblende which is stable at the lowermost temperature; it seems more logical to presume that some concentration of Al in the actinolite lattice will stabilize hornblende at its lowest possible temperature.

Reaction (91), as suggested by Vogt (1927) and Turner (1948), is particularly interesting because of the fact that members of the same mixed crystal, ferromagnesian chlorite and aluminous chlorite, respectively, occur on either side of the equality sign. This is a system analogous to the system *olivine*, *geikielite*, and *ilmenite*, investigated by Sahama and Torgeson (1949); and hence the two systems should exhibit equilibrium diagrams with similar main patterns. The equation of Sahama and Torgeson is as follows: Mg-rich olivine + ilmenite $\rightleftarrows$ Fe-rich olivine + geikielite (see p. 35).

In equation (91) (Mg,Fe)-rich chlorite corresponds to Mg-rich olivine, Al-rich chlorite corresponds to Fe-rich olivine; muscovite corresponds to ilmenite; and biotite to geikielite (muscovite and biotite do not constitute a mixed crystal; neither do geikielite and ilmenite). In a general way, the principal structure of the chlorite, biotite, and muscovite equilibrium should be as shown in Figures 74, 75, and 76. These diagrams show that the Al/(Mg,Fe) ratio in chlorite in equilibrium with biotite and muscovite should increase with increasing temperature. Furthermore, the formation of biotite after equation (91) seems to require higher and higher temperatures, the more Al there is in the chlorite.

In addition, the ratio Mg/Fe will most likely be of considerable importance, as the difference in the Mg/Fe ratio in coexisting biotite and chlorite is generally appreciable. As far as the writer knows, biotites usually are richer in iron than are coexisting chlorites. That is, a rising Fe/Mg ratio should lower the equilibrium temperature somewhat.

It is obvious that, even at constant rock pressure, the equilibrium temperatures of the mineral equations above are most likely different. Furthermore, because most of assemblages (89–96) are mixed crystals or hydrous and CO_2-carrying minerals, the equilibrium temperature, even for one certain reaction, is not constant at constant rock pressure. How, then, can we define the upper border of the green schist facies? We cannot avoid the fact that we have, at the present time, no way of determining the magnitude of the partial water- and carbon dioxide–vapor pressures which existed during the recrystallization of a given rock.[1] Hence we can never be certain in a specific case whether a rock having the left-hand assemblage of one of the equations above has recrystallized (or formed) at lower or higher temperature than a rock possessing the right-hand association of the same equation. If the water-vapor pressure is low in a rock, an apparently high-temperature anhydrous association may be stabilized at exceptionally low temperature (Yoder, 1952). So, for example, olivine

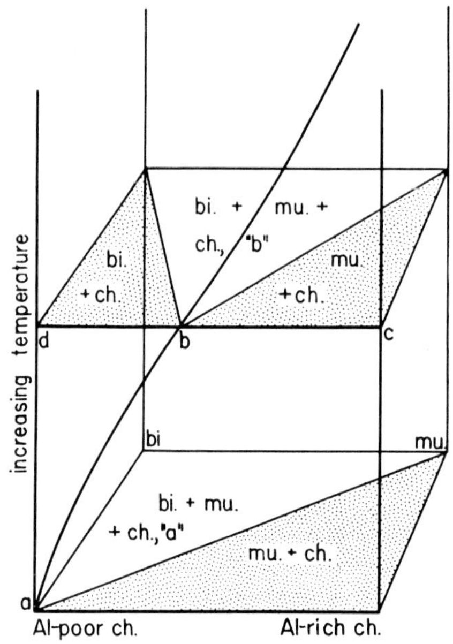

FIG. 74.—Tentative diagram, showing how the Al content of chlorite in equilibrium with muscovite and biotite changes with temperature. The curve *a–b* gives the composition of chlorite when both biotite and muscovite are stably present. The horizontal plane through *d–b–c* is an arbitrary isothermal section.

1. As the magnitude of the partial water-vapor pressure of hydrous minerals is a function of the concentration of HO in the lattice at given T and P, it may be possible in the future to reconstruct the H_2O pressure which has prevailed during crystallization of rocks carrying, say, hornblende or mica.

may be stable at 400° C. at low H_2O pressure (Bowen and Tuttle, 1949). Regarding the effect of the composition of a mixed crystal upon the equilibrium temperature, this is within our ability to account for, inasmuch as the composition can be determined by optical and chemical analysis.

Let us then discuss the several equations somewhat further.

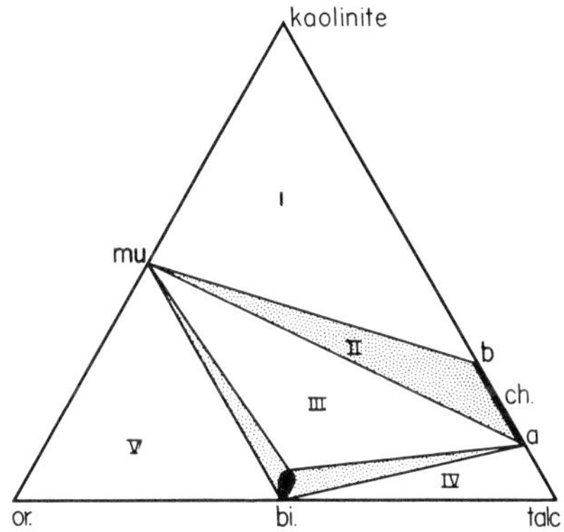

FIG. 75.—AKF diagram for the low green schist facies. The halftone areas are two-mineral fields; the rest are three-mineral fields.

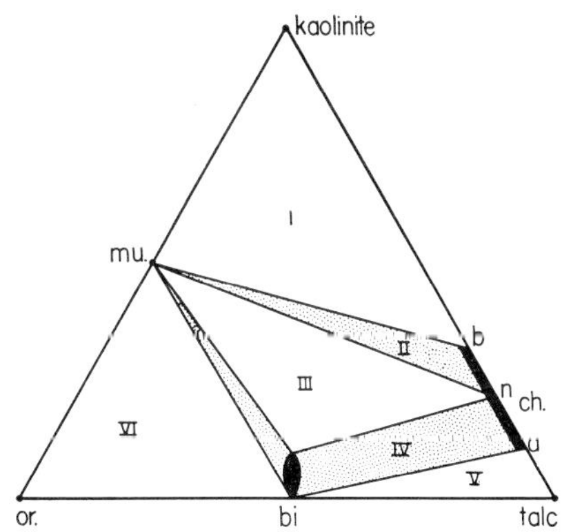

FIG. 76.—AKF diagram for the high green schist facies. The halftone areas are two-mineral fields; the rest are three-mineral fields.

As to equation (89), the equilibrium temperature is determined if rock pressure and the partial water- and CO_2-vapor pressures are given. That is so because talc is, practically speaking, a pure Mg silicate unable to mix with a corresponding iron silicate. Therefore, if no iron-carrying minerals exist together with talc and calcite in a siliceous marble, the hornblende which appears at rising temperature must be an iron-free tremolite. According to petrographic experience, this reaction takes place at considerably higher temperatures than do the other reactions characterizing the green schist facies (Ramberg, 1944*d*). Indeed, talc, calcite, and quartz are stable together, until halfway up in the epidote amphibolite facies. Reaction (89), as it stands, is, therefore, not suitable to define the uppermost border of the green schist facies. Supposing, however, that we consider an actinolite with appreciable amounts of Fe^{2+} (but no Al^{3+}) in the lattice and that this actinolite exists together with calcite and talc to insure a partial water and carbon dioxide pressure of the same magnitude as previously. In this case the temperature will be lower than the equilibrium temperature of equation (89) before the lowest stability temperature of the actinolite is reached. It is not certain into which minerals the iron member of the actinolite series will split up. It is most likely that siderite or some member of the mixed iron-magnesium carbonates is formed:

$$7CO_2 + \underset{\text{actinolite}}{Ca_2Fe_5Si_8O_{22}(OH)_2} \rightleftarrows 2CaCO_3 + 5FeCO_3 + H_2O + 8SiO_2\,. \quad (97)$$

If, for example, hematite is present, the following reaction is likely:

$$\underset{\text{hematite}}{5Fe_2O_3} + 2CO_2 + \underset{\text{actinolite}}{Ca_2Fe_5Si_8O_{22}(OH)_2} \rightleftarrows 2CaCO_3 + \underset{\text{magnetite}}{5Fe_3O_4} + 8SiO_2 + H_2O\,. \quad (98)$$

Reactions (97) and (98) take place at considerably lower temperatures than does reaction (89), thus justifying an equilibrium phase diagram much like Figure 33 (p. 64).

Equation (92) has been discussed thoroughly elsewhere (p. 60). The equilibrium temperature is not fixed unless the Mn/Mg/Fe ratio is fixed. In particular, changing concentration of Mn^{2+} is of great influence as far as the lowermost temperature of formation of garnet goes.

Equations (93)–(96) are rather straightforward and will not be commented upon further.

As a conclusion, we may state that, in a general way, equations (90) and (92) may serve as the uppermost limit of the green schist facies, provided that the compositions of the chlorite and actinolite participating in reaction (90) and of the garnet taking part in reaction (92) are agreed upon. These reactions take place well below the temperature at which kyanite is stabilized. Therefore, neither of the three "dry" aluminum silicates—kyanite, andalusite, or sillimanite—should be present in a stable form in the green schist facies.

Reactions (91), (93), and (96), which generally (provided that the compositions of the mixed crystals are not too extreme) take place below reactions (90) and (92), may serve as finer gradations within the green schist facies. Reactions (94) and (95) most likely have already taken place at very low temperature—(94) during diagenesis and (95) not far above.

Our information about the green schist facies is summarized in Figures 74, 75, and 76 above, and in Figures 77 and 78.

As the green schist facies represents very low temperature and, furthermore, is characterized by an abundance of hydrous minerals, the formation of which is importantly governed by the partial water-vapor pressure, equilibrium conditions are not commonly developed, and, if they are, their interpretation is exceptionally difficult.

In particular, care should be taken in concluding that the typical green schist facies minerals—talc and serpentine—always indicate low temperature when found in rocks. According to Noll and Bowen and Tuttle and others, talc can form stably at temperatures up to 700° C. and serpentine up to 500° C.

The green schist facies is represented by regional metamorphic rocks ranging from muscovite schists over green stones, spilites, soapstones, and talc rocks to pure calcite and dolomite marbles. Some quartzites may also be detected as belonging to the green schist facies because of the presence of minor amounts of green schist minerals (chlorite, epidote, albite, etc.) as "impurities." Such rocks are most commonly encountered in the upper levels of folded mountains and along their marginal zones. The facies is, hence, particularly well developed in the young mountain ranges. Also within the eroded Archean folded mountains (Finland), however, the green schist facies is represented, although more locally.

Many talc rocks and soapstones, as well as even green stones or "green schists" themselves, do not belong to the green schist facies proper, in which they are too often placed; but such rocks may develop an apparent green schist mineral association because of their particular bulk chemical composition. The content of tremolite, anthophyllite (perhaps even pyroxene), and/or garnet with but small amounts of MnO may show the proper place in the facies scheme of such "camouflaged" rocks. Furthermore, numerous "hydrothermal" vein rocks, like alpine veins, ore veins, etc., belong to this facies and are often formed simultaneously with the recrystallization of the host rock, often, as it were, as segregations due to metamorphic differentiation (Königsberger, 1901; Königsberger, Niggli, and Parker, 1940).

EPIDOTE AMPHIBOLITE FACIES

Rocks recrystallized at a somewhat higher grade of metamorphism than the green schists and their facies equivalents often contain common green hornblendes, spessartite-free garnet, and oligoclase—minerals not found in typical low-grade rocks except as unstable relicts. Obviously, the

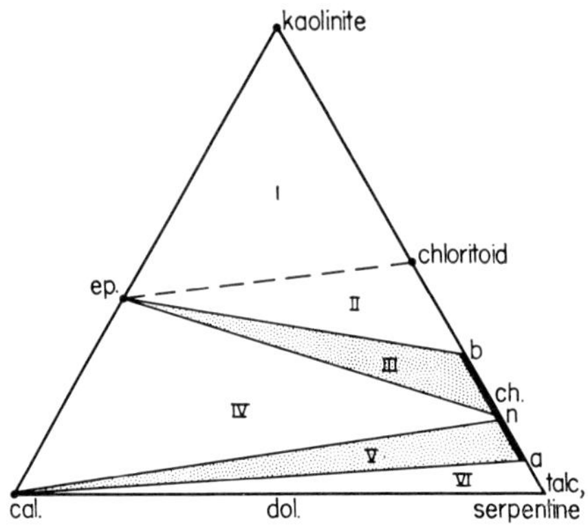

FIG. 77.—ACF diagram of the higher parts of the green schist facies. The halftone areas are two-mineral fields; the rest three-mineral fields.

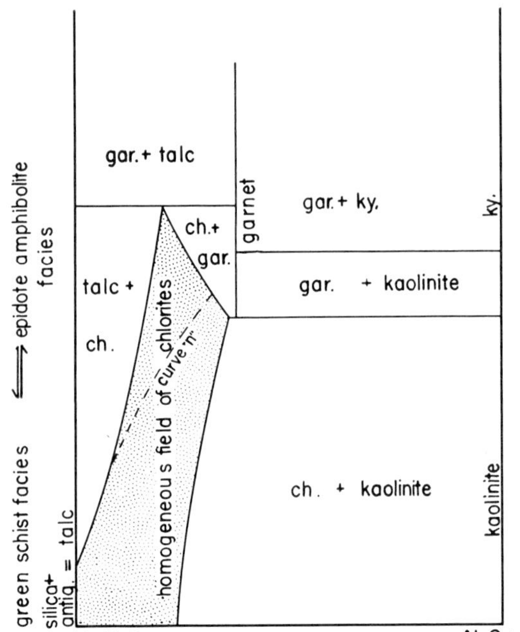

FIG. 78.—Approximate equilibrium situation of the A–F side of ACF diagrams. Curve *n* in the chlorite field shows the composition of chlorite in equilibrium with calcite and epidote; see point *n* in Fig. 77.

proper mineral reactions determining the lowermost boundary of the epidote amphibolite facies should be the same as those fixing the uppermost limit of the green schist facies. The right-hand side of equations (90) and (92) therefore represent the epidote amphibolite facies, provided that some limitation as to the Mg/Fe ratio in the hornblende and to the Mn concentration in the garnet is insured. In addition to these criteria, staurolite and kyanite are usually considered to be typical guide minerals in the epidote amphibolite facies (Vogt, 1927; Eskola, 1939; Ramberg, 1944*d;* Turner, 1948). Now kyanite may be considered a "high"-temperature modification of kaolinite,

$$Al_2Si_2O_5(OH)_4 \rightleftarrows Al_2SiO_5 + SiO_2 + 2H_2O\ , \qquad (99)$$

and staurolite may originate according to one of the following reactions:

$$\text{chlorite-}a + \text{kaolinite} \rightleftarrows \text{staurolite} + \text{chlorite-}b + H_2O\ , \qquad (100)$$

$$\text{garnet-}a + \text{kyanite} \rightleftarrows \text{staurolite} + \text{garnet-}b\ , \qquad (101)$$

where chlorite-*a* and garnet-*a* are richer in Fe than are chlorite-*b* and garnet-*b*.

There is no reason to assume that the three last reactions occur at the same temperature, or at the same temperatures as those of reactions (90) and (92). It seems evidenced by petrographic experience that kyanite needs somewhat higher temperatures to be stabilized than the equilibrium temperature of equations (90) and (92), provided that the Mg/Fe ratio considered in the last two reactions is not too high.

Some very important reactions take place within the epidote amphibolite facies, namely, the reactions in which plagioclase plays a major role:

$$\text{muscovite} + \overset{\text{low temp.}}{\text{epidote}} \rightleftarrows \text{anorthite} + \overset{\text{high temp.}}{\text{potash feldspar}} + H_2O\ , \qquad (102)$$

$$\text{kyanite} + \overset{\text{low temp.}}{\text{epidote}} \rightleftarrows \text{anorthite} + \overset{\text{high temp.}}{H_2O}\ . \qquad (103)$$

The consequences of these reactions have been considered previously (p. 50). The important thing is that plagioclase, because of reactions (102) and (103), becomes more basic, the higher the temperature.

As to the uppermost boundary of the epidote amphibolite facies, different assemblage transitions may be chosen. The writer (1944*a*) has proposed that the lowermost stability border of diopsidic pyroxene should limit the epidote amphibolite facies upward. That is, the following reaction is crucial:

$$2SiO_2 + \underset{\text{tremolite}}{Ca_2Mg_5Si_8O_{22}(OH)_2} + 3CaCO_3 \rightleftarrows \underset{\text{diopside}}{5CaMgSi_2O_6} + 3CO_2 + H_2O\ . \qquad (104)$$

At the temperature corresponding to equilibrium in equation (104) under ordinary regional metamorphic conditions, reaction (102) or reaction (103) would stabilize a plagioclase of composition about An_{30}. The different assemblages characterizing epidote amphibolite facies may be visualized roughly by Figures 79, 80, and 81.

The epidote amphibolite facies limited in the P,T-field, as is done in this book (by reactions [90], [92], and [104]), comprises rather great areas

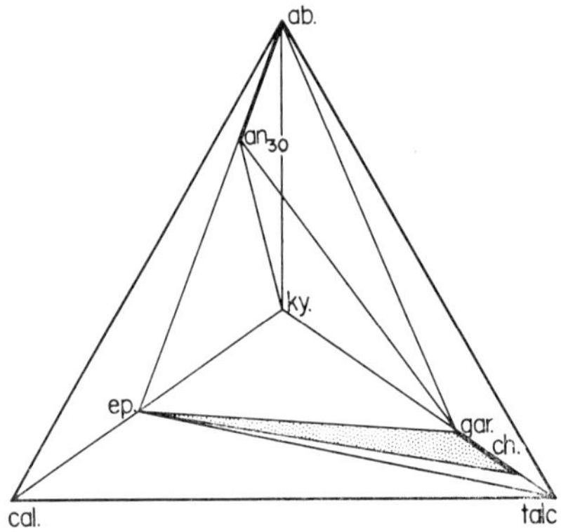

FIG. 79.—Simplified calcite-talc-kyanite-albite tetrahedron diagram of the high epidote amphibolite facies. The base of the diagram is identical to an ACF diagram. No hornblende is present here, because the F component is supposed to contain only MgO (see text and compare with Fig. 80). The important point of the diagram is that, inside the tetrahedron of epidote, garnet, kyanite, and An_{30}, the composition of the plagioclase is fixed (at constant P,T) and independent of the relative amounts of the several minerals mentioned. Inside the tetrahedron of An_{30} albite, garnet, and kyanite, however, the composition of plagioclase is changeable at given P,T-conditions and depends upon the bulk composition of the assemblage. The diagram is considerably simplified for clarity, and several lines are omitted. See, for example, Fig. 27, which corresponds to the calcite-kyanite-albite side of Fig. 79.

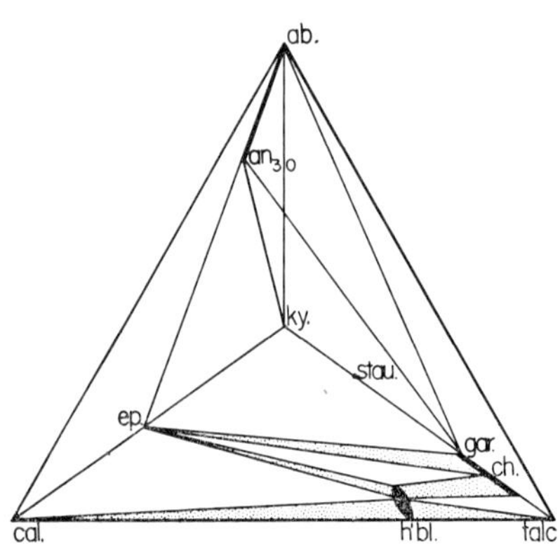

FIG. 80.—Same type of diagram as Fig. 79, but here F contains enough iron to stabilize hornblende. The diagram corresponds to the high epidote and/or low amphibolite facies (see explanation to Fig. 79).

of rocks within the folded mountains of recent and pre-Cambrian age. Many gneisses and gneiss granites belong to this facies, the most famous being, perhaps, the so-called "helsinkites," which are epidote-oligoclase-carrying granodioritic rocks. It is evidently not always easy to decide whether the common gneisses belong to the epidote amphibolite facies or to the amphibolite facies, because the association of quartz, potash feldspar, acidic plagioclase (An_{30}), and micas or hornblendes is stable in either facies. Mica schists, with garnet, kyanite, and/or staurolite; epidote-carrying hornblende schists; tremolite marbles; several soapstones; and numerous other very common schists and gneisses can be placed in the epidote amphibolite facies. Particularly the uppermost part of the facies

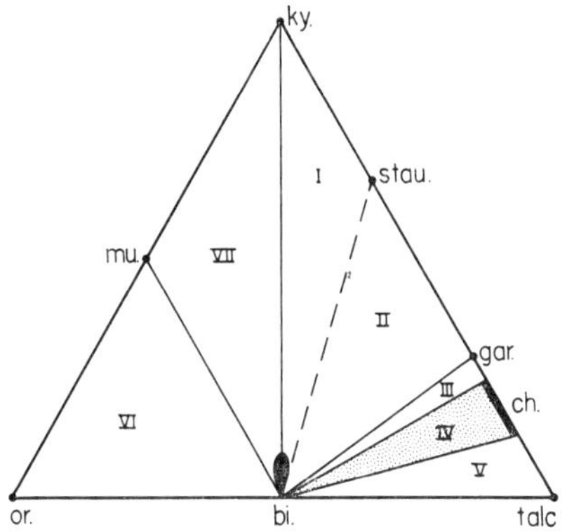

FIG. 81.—Talc-kyanite-potash feldspar triangle diagram of the high epidote amphibolite facies.

in question is, together with the amphibolite facies, the most commonly encountered facies within the vast monotonous Archean gneiss areas all over the globe.

AMPHIBOLITE FACIES

Rocks formed at temperatures (and pressures?) just above those of the epidote amphibolite facies belong to the amphibolite facies. This metamorphic facies is the classic one developed in the famous Orijärvi region in Finland described and discussed by Eskola in 1915. The distinction between the epidote amphibolite facies and the amphibolite facies is by no means clear. Eskola does not allow epidote or zoisite to enter any of the stable mineral assemblages of his originally defined amphibolite facies (1914, 1915, 1939). In his opinion (1939, pp. 355–57), the epidote amphibolite facies is characterized by the stable association of epidote or zoisite and practically pure albite, whereas in the amphibolite facies pure anorthite is stable. However, there is a considerable gap between the temperature at which epidote and pure albite can coexist stably and that at which pure anorthite is stable. Eskola's definitions, therefore, must necessarily leave a certain P,T-gap between the two facies in question, although they were believed to be immediate neighbors in the P,T-scheme. This discrepancy may be canceled by simply using the epidote (zoisite)–$An_{30}(\pm)$ equilibrial association as the lowest boundary for amphibolite facies (see Fig. 27, p. 51). Another indication of the amphibolite ⇄ epidote amphibolite facies transition is the lowermost stability border of diopside in calcite milieu.

The following reaction:

$$\overset{\text{epidote amphibolite facies}}{2SiO_2 + 3CaCO_3 + \underset{\text{tremolite}}{Ca_2Mg_5Si_8O_{22}(OH)_2}} \rightleftarrows \overset{\text{amphibolite facies}}{\underset{\text{diopside}}{5CaMgSi_2O_6} + 3CO_2 + H_2O} \,, \quad (105)$$

referring to the pure Mg-silicate end-members of the mixed crystals, may be the proper indicator of the lowermost border line of the amphibolite facies.

According to Tilley (1927), diopside (associated with vesuvianite and grossularite) occurs together with epidote and rather pure albite within an area of apparent epidote amphibolite facies rocks. It may, therefore, be concluded that diopside is stable in even the lowest part of the epidote amphibolite facies (Turner, 1948). On the other hand, the minerals vesuvianite and grossularite occurring in the rocks described by Tilley indicate, according to common petrographic experience, a much higher temperature than that hinted by the association of albite and epidote. It is therefore not unambiguous to conclude that the association above corresponds to a low epidote amphibolite facies. It can easily be argued that the albite-epidote association is stabilized, not because of low temperature, but because of an excess of calcite, which gives the following reaction:

$$\overset{\text{high temp.}}{CaCO_3 + \underset{\text{anorthite}}{3CaAl_2Si_2O_8} + H_2O} \rightleftarrows \overset{\text{low temp.}}{\underset{\text{epidote}}{2Ca_2Al_3Si_3O_{12}OH} + CO_2} \,. \quad (106)$$

This reaction has a higher equilibrium temperature than does the epidote-anorthite equilibrium where calcite does not occur (reactions [102] and [103], above). Also, the formation of grossularite can take place according to the following reaction:

$$2CaCO_3 + \underset{\text{anorthite}}{CaAl_2Si_2O_8} + SiO_3 \rightleftarrows \underset{\text{grossularite}}{Ca_3Al_2Si_3O_{12}} + 2CO_2 \quad (107)$$

at rather high temperatures (pyroxene hornfels facies, p. 162; see Goldschmidt, 1911). As a necessary consequence, anorthite will be "leached" out of plagioclase, even at high temperatures, if calcite is present. Tilley's mineral assemblage is, therefore, formed at higher temperature than could be expected at first glance from the association albite-epidote. Therefore, it should not be assumed that diopside is stable in the epidote amphibolite facies.

Regarding the pure aluminum silicates, kyanite, as well as andalusite, appears to be stable within the P,T-field of amphibolite facies. Kyanite, most reasonably, is the typical mineral in the highest-pressure parts of the facies, andalusite in the low-pressure (high-temperature?) part of the facies. It seems evidenced by petrographic experience also that sillimanite is stable within the very uppermost P,T-field of the amphibolite facies.

The following reaction:

$$\underset{\text{potash feldspar}}{\overset{\text{granulite facies}}{KAlSi_3O_8}} + \underset{\substack{\text{kyanite}\\ \text{or}\\ \text{sillimanite}}}{Al_2SiO_5} + H_2O\uparrow \rightleftharpoons \underset{\text{muscovite}}{\overset{\text{amphibolite facies}}{KAl_3Si_3O_{10}(OH)_2}} + SiO_2\,, \tag{108}$$

under an H_2O pressure commonly present in rocks, is displaced to the right, so that muscovite is stable over the whole amphibolite facies. Indeed, the equilibrium P,T-curve of reaction (108) may serve as the proper uppermost boundary of the amphibolite facies above which muscovite is unstable and the association sillimanite-potash-feldspar + (H_2O) is stable (granulite facies).

Cordierite, which is hardly stable within the epidote amphibolite facies, is stabilized somewhere in the amphibolite facies, whereas staurolite apparently never forms within the amphibolite facies. Cordierite may develop as follows:

$$\text{garnet} + \text{kyanite} \rightleftharpoons \text{cordierite}\,, \tag{109}$$

$$\text{chlorite} + \text{kyanite} \rightleftharpoons \text{cordierite}\,. \tag{110}$$

The garnet lattice is more inclined to adopt the ferrous ions, whereas the cordierite lattice suits Mg^{2+} better. As a consequence, the equilibrium temperature (pressure) of reaction (109) is changeable, depending upon the Mg^{2+}/Fe^{2+} ratio; so is reaction (110). In rocks high in Fe relative to Mg, garnet and kyanite or sillimanite may exist together stably until high up in the amphibolite facies; very reasonably, cordierite will never form in such rocks. In rocks with an intermediate Mg/Fe ratio, cordierite may exist together with kyanite (sillimanite, andalusite) and garnet, the Mg/Fe ratio in the two ferromagnesian minerals changing gradually with changing P,T-conditions.

Most chlorites seem to be unstable altogether in the amphibolite facies, although it is not yet known whether, for example, exceptionally high H_2O pressure and some unique chemical composition of rocks enable chlorite to develop stably in the amphibolite facies. Not rarely does one find chlorite as apparent syngenetic products, together with rather high-temperature minerals (Bugge, 1943).

The stability conditions of biotite in the amphibolite facies are interesting. Its relationship to garnet and potash feldspar is discussed elsewhere (p. 57). Some other mineral equations showing the boundary conditions of the stability of biotite are given below:

$$3SiO_2 + \underset{\text{biotite}}{K(Mg,Fe)_3AlSi_3O_{10}(OH)_2} \rightleftarrows \underset{\text{potash feldspar}}{KAlSi_3O_8} + \underset{\text{hypersthene}}{3(Mg,Fe)SiO_3} + H_2O \,, \quad (111)$$

$$24SiO_2 + \underset{\text{biotite}}{7K(Mg,Fe)_3AlSi_3O_{10}(OH)_2} \rightleftarrows \underset{\text{potash feldspar}}{7KAlSi_3O_8} + \underset{\text{anthophyllite}}{3(Mg,Fe)_7Si_8O_{22}(OH)_2} + 4H_2O \,. \quad (112)$$

To these two reactions, a third may be added in which hypersthene or anthophyllite is replaced by talc. It seems evident that all three reactions, under "ordinary" Mg/Fe ratios and H_2O pressures found in rocks, are displaced toward the left-hand side in the amphibolite facies. Approaching the granulite facies, reaction (111) tends to go to the right (p. 159). According to the ACF diagrams of amphibolite facies in Eskola's textbook of 1939, one would expect biotite to react with calcite to form hornblende or diopside, together with potash feldspar. Such reactions may take place in some instances (Ramberg, 1944*a*); but, on the other hand, it is common to find biotite or, rather, phlogopite as apparently stable mineral grains in amphibolite facies marbles. Considering reaction (113), we see that excess silica is required to "diopsidize" or "hornblendize" biotite through reaction with calcite (dolomite):

$$6SiO_2 + 3CaCO_3 + \underset{\text{phlogopite}}{KMg_3AlSi_3O_{10}(OH)_2} \rightleftarrows \underset{\text{potash feldspar}}{KAlSi_3O_8} + \underset{\text{diopside}}{3CaMgSi_2O_6} + 3CO_2 \,. \quad (113)$$

Ferromagnesian micas can occur in high-grade marbles, because there is not a sufficient amount of free silica to form diopside or hornblende. Being suspicious because of the abundance of Mg-rich micas and the rarity of Fe-rich micas in marbles, the writer (1944*a*) presumed that the stability of the mica-calcite association was mainly governed by the Mg/Fe ratio. Although this effect plays some controlling role, it is not the crucial matter, because the diopsidic pyroxenes (and hornblendes) in marbles are usually richer in Mg relative to Fe than is the coexisting biotite, showing that the Mg member of the biotite series reacts according to equation (113) more readily than does iron biotite. If the presence of biotite in calcite marbles is explained by the Mg/Fe ratio, one would expect to find most commonly iron-rich biotite as skarn mica, not phlogopite, which is the actual case.

The considerations above show that the placing of calcite or dolomite in ACF diagrams easily causes misunderstanding and confusion. The equilibrium relationships between the carbonates and the several principal silicates need at least a tetrahedron diagram to be visualized. Figures 82 and 83 may serve as examples showing that there is nothing unstable

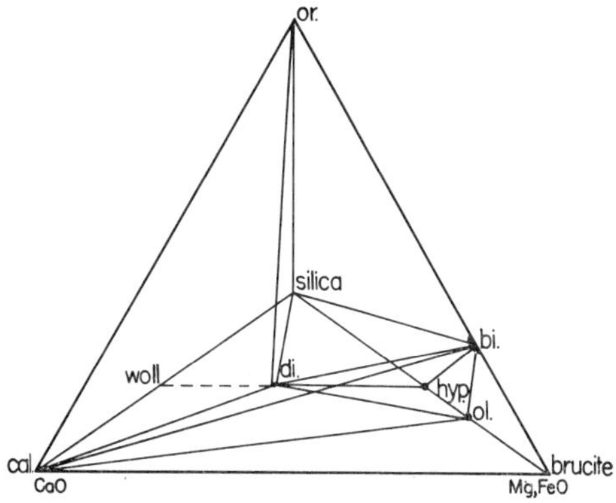

FIG. 82.—Calcite-brucite-quartz-potash feldspar tetrahedron diagram of the granulite facies.

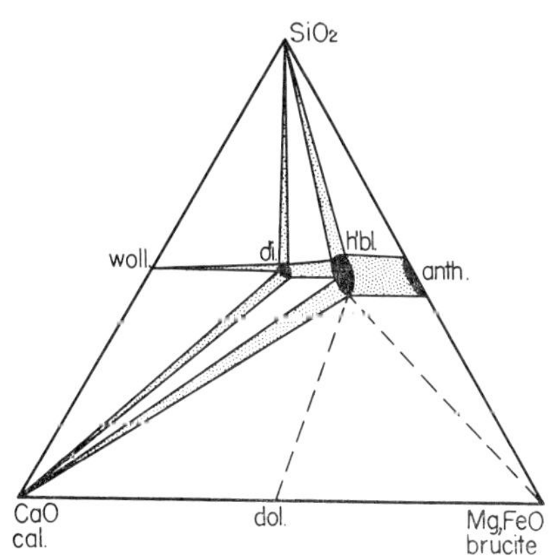

FIG. 83.—Calcite-brucite-SiO_2 diagram of the amphibolite facies

about the coexisting of micas or hornblende and calcite, even if diopside is stable, provided that free silica does not exist in the marble.

Because the calcium-free ferromagnesian micas are so common in amphibolite facies marbles, it seems to be somewhat problematic why other calcium-free minerals, like hypersthene, ferromagnesian garnet, and cordierite, do not occur in skarn associated with calcite (dolomite). On the other hand, the calcium-free olivine is a common skarn mineral, evidently stable together with calcite and dolomite. Let us, in this connection, look at the following reaction:

$$\underset{\text{hypersthene}}{3(Mg,Fe)SiO_3} + CaCO_3 \rightleftarrows \underset{\text{diopside}}{Ca(Mg,Fe)Si_2O_6} + \underset{\text{olivine}}{Mg_2SiO_4} + CO_2\,. \quad (114)$$

This reaction is displaced to the right in amphibolite facies. In distinction to the calcite-biotite reaction (113), the hypersthene-calcite reaction does not require free silica; hence there is no possibility of having hypersthene and calcite in stable associations (except under extremely high CO_2 pressure). One may now ask why talc does not undergo the same reaction but, instead, is a common skarn mineral associated with calcite, at least in the epidote amphibolite facies:

$$\underset{\text{talc}}{3Mg_3Si_4O_{10}(OH)_4} + 5CaCO_3 \rightleftarrows \underset{\text{diopside}}{5CaMgSi_2O_6} + \underset{\text{olivine}}{2Mg_2SiO_4} + 5CO_2 + 2H_2O\,. \quad (115)$$

Because talc is a common skarn mineral in the epidote amphibolite facies and in the amphibolite facies, whereas hypersthene is unknown as a skarn mineral, it may be concluded that reaction (115) goes toward the right at much higher temperatures than does reaction (114).

Almandine-pyrope garnet is not stable with calcite. In an attempt to solve this problem, let us look at the following reaction:

$$\underset{\text{garnet}}{2(Mg,Fe)_3Al_2Si_3O_{12}} + SiO_2 + 2CaCO_3 \rightleftarrows \underset{\text{anorthite}}{2CaAl_2Si_2O_8} + \underset{\text{olivine}}{3(Mg,Fe)_2SiO_4} + 2CO_2\,. \quad (116)$$

It is seen that this reaction requires free silica in order to go to the right. One should, therefore, occasionally expect to find almandine-pyrope garnet in calcite marbles without free silica or other silicates, which might provide the silica necessary for reaction (116). Such findings have not been recorded, and one may, therefore, ask if there are some other reactions which render almandine-pyrope entirely incompatible with calcite (dolomite). The answer seems to lie in the following situation:

$$\underset{\text{garnet}}{3(Mg,Fe)_3Al_2Si_2O_{12}} + 4CaCO_3 \rightleftarrows \underset{\text{diopside}}{4Ca(Mg,Fe)Si_2O_6} + \underset{\text{spinel}}{3MgAl_2O_4} + \underset{\text{olivine}}{(Mg,Fe)_2SiO_4} + 4CO_2\,, \quad (117)$$

where the right-hand side is stable in amphibolite facies (and in granulite facies, p. 156). The diopside-spinel-olivine assemblage is very common in marbles and skarn rocks from these facies. It is interesting to see that a reaction analogous to the formation of spinel and/or olivine cannot take place between biotite and calcite in quartz-free marbles, simply because the potassium content in common biotite-phlogopite would require all the aluminum and silica in the mica in order to build potash feldspar. There would be neither silica nor aluminum left to combine with iron and magnesium to form spinel or olivine (see reaction [113], p. 152). With respect to the experienced absence of cordierite in calcite-dolomite marbles in amphibolite facies, the explanation may be found in the following reactions:

$$\underset{\text{cordierite}}{Mg_2Al_3Si_5AlO_{18}} + CaCO_3 \rightleftarrows \underset{\text{anorthite}}{2CaAl_2Si_2O_8} + \underset{\text{olivine}}{Mg_2SiO_4} + CO_2\,, \qquad (118)$$

$$\underset{\text{cordierite}}{2Mg_2Al_3Si_5AlO_{18}} + 9CaCO_3 + H_2O$$
$$\rightleftarrows \underset{\text{grossularite}}{3Ca_3Al_2Si_3O_{12}} + \underset{\text{spinel}}{MgAl_2O_4} + \underset{\text{olivine}}{Mg_2SiO_4} + \underset{\text{brucite}}{Mg(OH)_2} + 9CO_2\,. \qquad (119)$$

From these reactions, it is seen that several of the common skarn silicates stable in the amphibolite facies can be formed by simple reactions between calcite and cordierite without the help of any other mineral. For this reason, it is obvious that cordierite and calcite are chemically incompatible under the P,T-combinations which commonly have prevailed during rock formation.

There are some doubts as to the minerals in the C corner of the ACF diagram of the amphibolite facies. Wollastonite is recorded from rocks showing several mineral associations placing them in the amphibolite facies (Eskola, 1914; von Eckerman, 1922, 1923; Tilley, 1927; Kuhne, 1932). So far, the classic Orijärvi rocks carry partly wollastonite. There is, thus, no doubt that wollastonite is stable in the amphibolite facies, but there is a question as to whether or not the mineral will form by reaction between silica and calcite:

$$\underset{\text{calcite}}{CaCO_3} + SiO_2 \rightleftarrows \underset{\text{wollastonite}}{CaSiO_3} + CO_2\,. \qquad (120)$$

The temperature of this equilibrium depends significantly upon the partial CO_2 pressure, which again depends upon the permeability of the rocks to CO_2 gas (p. 49). If the permeability to CO_2 is great, wollastonite would form by means of reaction (120) at low temperatures, even below 400° C., a temperature which most reasonably is situated within the amphibolite facies. One should, therefore, now and then expect to find wollastonite formed by reaction (120) in some amphibolite facies marbles. Most commonly, however, quartz and calcite show no tendency to interact in amphibolite facies marbles, indicating either that the temperature has been below 400° C. or that the permeability to CO_2 molecules is so small that a great CO_2 pressure has to be built up before the gas will escape. Concerning the latter alternative, the abundance of diopside, hornblende, scapolite, grossularite, and other reaction skarn minerals, which also require escape of CO_2 gas, proves that the permeability was not small. Furthermore, if it was just a question of permeability and thus rate of reaction, one would expect to find small amounts of wollastonite—indicative of incomplete reactions—commonly associated with quartz and calcite in amphibolite facies marbles. Since that is not the case, *it is logical to conclude that the temperature of amphibolite facies only rarely exceeds 400°–500° C.*

Usually anthophyllite is placed in the F corner of the ACF diagram of amphibolite facies, indicating that neither talc nor rhombic pyroxene is stable in this facies. We have reason to believe, however, that talc is stable in suitable rocks formed in the amphibolite facies, provided that the H_2O pressure is sufficiently high (Bowen and Tuttle, 1949). Actually, under a water pressure of above 5,000 psi, talc would be stable up to 750°–800° C. Evidently the rock pressure during amphibolite facies conditions is at least 5,000 psi, and the temperature should undoubtedly be well below 750°–800° C. The absence of talc is therefore another indication that the water-vapor pressure in rocks is usually much less than the rock pressure.

It is evidenced by field exploration and laboratory tests that rhombic pyroxene is stable in the highest half of the amphibolite facies just below the granulite—or the pyroxene hornfels facies. On the other hand, field evidence has proved that anthophyllite is characteristic of the amphibolite facies, whereas experiments seem to disprove the stability of pure Mg anthophyllite altogether (Bowen and Tuttle, 1949).

In spite of these minor but interesting uncertainties, the facies in question is well explored in a general way, and the association Figures 84 and 85 can be constructed with considerable confidence. If we consider the amphibolite facies to comprise the P,T-field limited by the equilibrium between zoisite and An_{30+}, and by reactions (108) and (111) (p. 151), then rocks belonging to this facies are undoubtedly the most abundant ones in gneissic area. The main part of the old pre-Cambrian shields and vast areas within the Caledonian and Hercynian roots comprise schists and gneisses being formed, recrystallized, or reactivated in the amphibolite facies.

GRANULITE FACIES

At places in deeply eroded pre-Cambrian gneissic areas, some very high-grade metamorphic rocks are encountered—the so-called "granulites" (Saxony, Lapland, India, Ceylon, Uganda, Norway, Antarctica, Greenland, eastern Canada, Ellesmereland). The rocks originally called "granulite" by the German and Scandinavian petrologists are quartzo-feldspathic gneisses characterized by anhydrous minerals (garnet, kyanite, and/or sillimanite, rutile, etc.), besides quartz and feldspars, and often also by a certain structure indicating intense plastic movements. The classic area for such rocks is the Granulitegebirge in Saxony in Germany. The granulites so defined are necessarily of limited chemical composition; otherwise calco-ferromagnesian minerals would have developed. Isograde rocks occurring intermingled with the granulites *sensu stricto* therefore contain different minerals: rhombic pyroxene, diopsidic and augitic pyroxenes, ilmenite-magnetite ore, some characteristic hornblende, etc. The quartzo-feldspathic granulites have given their name to the whole facies within which they develop. From field explorations, as well as from studies

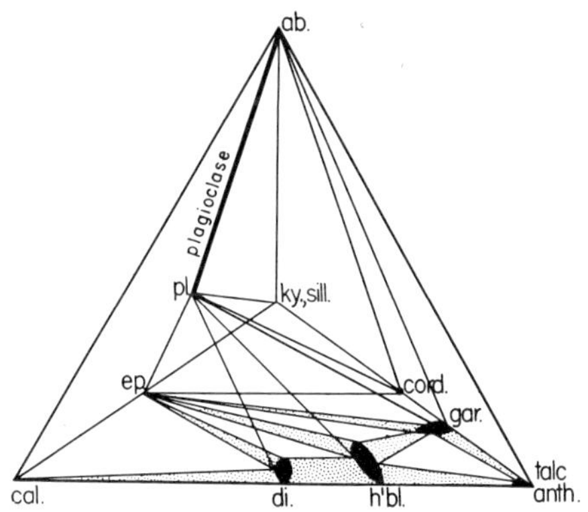

FIG. 84.—Calcite-talc-kyanite-albite tetrahedron diagram of the amphibolite facies. Inside the space: *ep.–di.–h'bl.–gar.–cord.–pl.*, the composition of plagioclase is nearly constant = *pl.* at constant *P,T*. Inside the space above the surface: *pl.–ky.–cord.–gar.–h'bl.–di.*, however, the composition of the stable plagioclases is changeable and depends upon the bulk chemical composition of the assemblages. The base of the diagram represents a common ACF triangle of the amphibolite facies.

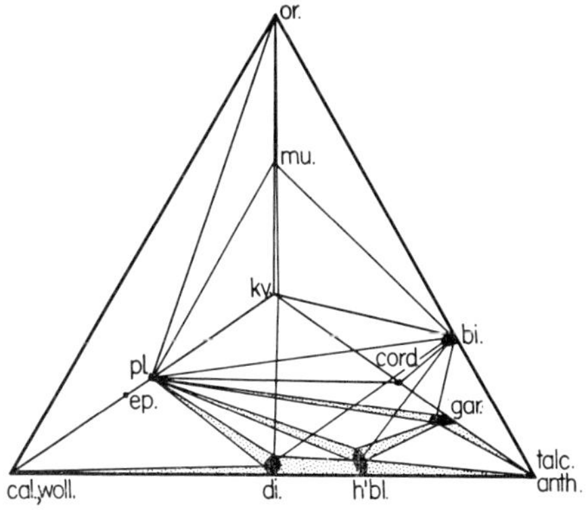

Fig. 85.—Calcite-talc-kyanite-potash feldspar tetrahedron of the amphibolite facies

of the mineral sequences, it is clear that the granulite facies succeeds the amphibolite facies when confining pressure, in particular, but also temperature increase. Many authors are of the opinion that the granulite facies also requires intense one-sided pressure and, in particular, tectonic movements to develop (Schüller, 1948). Such circumstances certainly ease the development of the stabilized mineral association, as would be the case in any facies, but it seems unnecessary to require intense one-sided pressure and rock movement in order to stabilize the mineral associations characterizing the granulite facies.

Although most of the properties of the mineral assemblages characterizing the granulite facies have been developed gradually through the amphibolite facies, there are some changes which take place abruptly around the border line between the two facies. The following mineral reactions are typical:

$$\text{muscovite} \rightarrow \text{sillimanite} + \text{potash feldspar} + \text{water}\,, \tag{121}$$

$$\text{biotite} + \text{sillimanite} \rightarrow \text{garnet} + \text{potash feldspar} + \text{water}\,, \tag{122}$$

$$\text{biotite} \rightarrow \text{hypersthene} + \text{potash feldspar} + \text{water}\,. \tag{123}$$

According to field observation, these three reactions seem to have rather similar equilibrium conditions, all of them occurring rather close to the transition between granulite and amphibolite facies. As the two last reactions depend upon the Mg/Fe ratio (and reaction [122] also upon the Mn percentage in garnet), it seems possible to choose the particular Mg/Fe ratios at which reactions (122) and (123) have an equilibrium temperature identical to that of reaction (121) at constant rock pressure and water-vapor tension.

Reactions (122) and (123) show that biotite may or may not be stable in the granulite facies, depending upon the Mg/Fe ratio and the content of MnO in the rock. It can hardly be doubted that the right-hand sides of the equations are the high-temperature sides. As recrystallizing rocks are permeable to H_2O molecules, high rock pressure would not increase the temperature of equilibrium.

Experience from rocks reveals that pyroxene has a greater Mg/Fe ratio than does the coexisting biotite. That means that increasing Mg/Fe ratios should lower the equilibrium temperature of equation (122), and the general equilibrium diagrams will be as shown in Figure 86. The fact that biotite is present stably in several quartz-carrying granulite facies rocks means that the major part of the equilibrium curves in Figure 86 is above the lowermost border line of the granulite facies. One should not forget, however, that the curves are displaced up and down with changing water-vapor pressure in the rocks.

Still more complicated is the presence of hornblende in the granulite facies. The belief that hornblende is unstable in the granulite facies (Eskola, 1939) is not valid. The matter has been discussed thoroughly on pages 65–70, where it was shown that the stability of hornblende depended crucially upon the bulk composition of the rock because of the susceptibility of hornblende to dissolve all kinds of ions and to change its stoichiometric proportions.

Figures 87 and 88, together with Figure 35 (p. 69), indicate the place

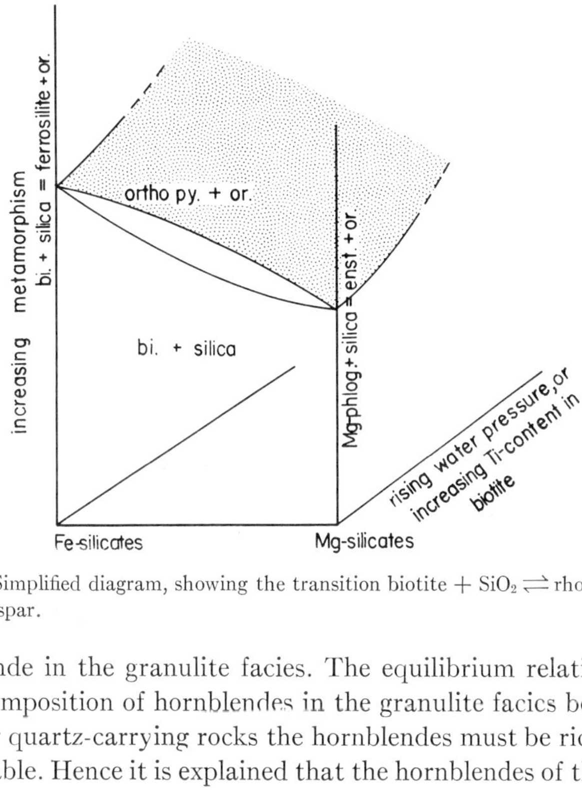

FIG. 86.—Simplified diagram, showing the transition biotite + $SiO_2 \rightleftharpoons$ rhombic pyroxene + potash feldspar.

of hornblende in the granulite facies. The equilibrium relations require that the composition of hornblendes in the granulite facies be restricted: in ordinary quartz-carrying rocks the hornblendes must be rich in Fe and Ti to be stable. Hence it is explained that the hornblendes of the granulite facies are therefore monotonous, even when occurring in different granulite facies rocks and different granulite facies areas. Only in quartz-free rocks may Mg-rich and Ti-free hornblendes occur. In any case, the granulite facies hornblendes have to be rich in Al, in particular in the $(Si,Al)_8$ position in the lattice (see Ramberg, 1949). That is, chemically, the granulite facies hornblendes are close to hastingsites (Foslie, 1945).

The quartz is of a rather characteristic type in several granulite facies rocks. It is often of a bluish milky color, supposedly due to inclusions and minor amounts of dissolved Fe, Mn, etc. The feldspars are also extremely typical of rocks belonging to the granulite facies. Not only do they exhibit a dark greenish to brown or grayish-black color, but also the particular perthitic structures are unshared by feldspars from other facies. A very fine microperthite structure is typical of the orthoclase or microcline, and antiperthite is the typical plagioclase, provided that it is not too basic.

Sphene, which is stable in all other normal facies, does not occur in the granulite facies except in the lime-rich skarn rocks, where sphene is associated with scapolite, diopside, condrodite, etc. In most types of rocks,

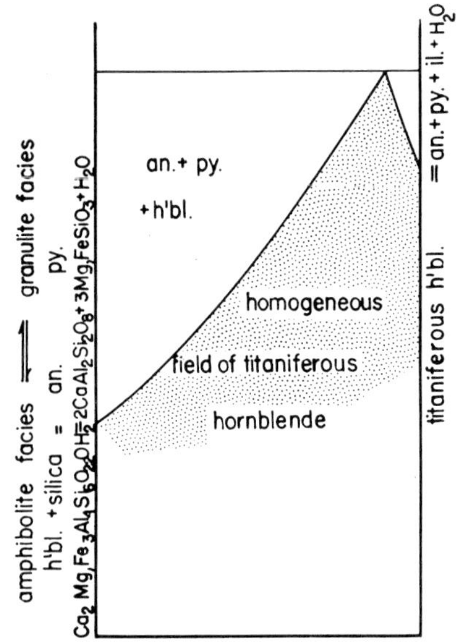

FIG. 87.—Tentative diagram, showing the gradual splitting-up of titaniferous aluminous hornblende in the granulite facies.

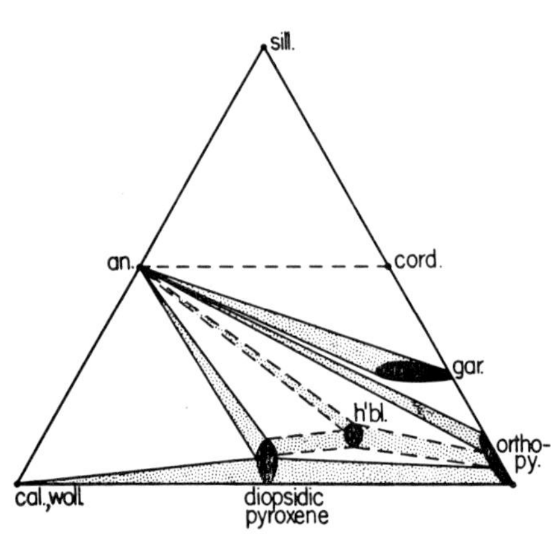

FIG. 88.—ACF diagram of the granulite facies. The presence of the hornblende field depends upon the Mg/Fe and [Ti/(Mg+ Fe)] ratios of the rocks, as explained in the text.

however, sphene is unstable and undergoes the reactions discussed on page 73. That is, either the TiO_2 is liberated in the form of rutile, or the titania goes into ilmenite, biotite, or hornblende in the granulite facies (see Fig. 89). Some mineral associations characterizing the granulite facies are compiled in Figure 90.

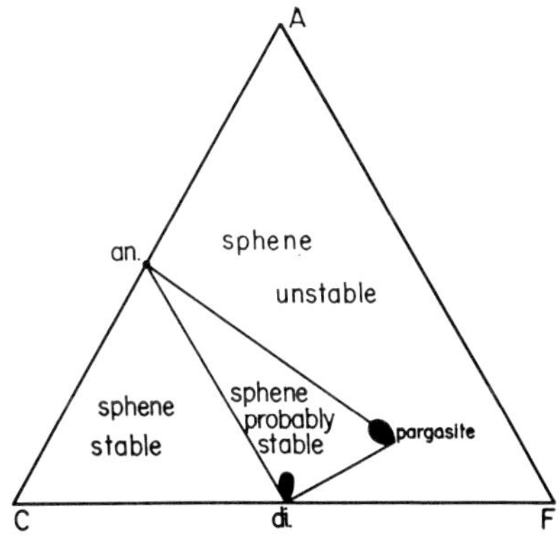

FIG. 89.—The stability fields of sphene in an ACF diagram of the granulite facies

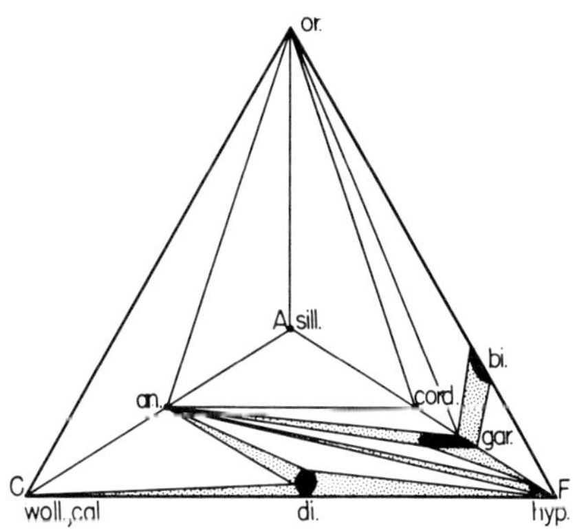

FIG. 90.—Diagram, showing the relationship between potash feldspar and several other minerals in granulite facies. Some lines are omitted.

It is interesting to note that the so-called "charnockite-anorthosite" series of plutonic rocks belongs to the granulite facies (Groves, 1935; Bugge, 1943; Schüller, 1948; Ramberg, 1949). The explanation lies in the conclusion that these plutonic rocks are not igneous but are created through metamorphism and metasomatism under P,T-conditions corresponding to the granulite facies.

The composition of kryptoperthites from granulite facies rocks is often around 50 per cent albite and 50 per cent microcline, indicating that the temperature of the granulite facies is around 550°–650° C.

PYROXENE HORNFELS FACIES

In a similar way as the granulite facies succeeds the amphibolite facies by rising pressure (and a more moderate increase in temperature), so the pyroxene hornfels facies is the successor of the amphibolite facies by rising temperature (and decreasing pressure). The pyroxene hornfels facies is the typical contact metamorphic facies so thoroughly described and interpreted by Goldschmidt in his classical treatise of the *Kontakt Metamorphose der Kristiana Gebiete* in 1911. The previously discussed facies belongs to the regional metamorphism, with the exception of the low-pressure part of the amphibolite facies, which characterizes the outer contact zone around a magmatic rock body. The pyroxene hornfels facies embraces the

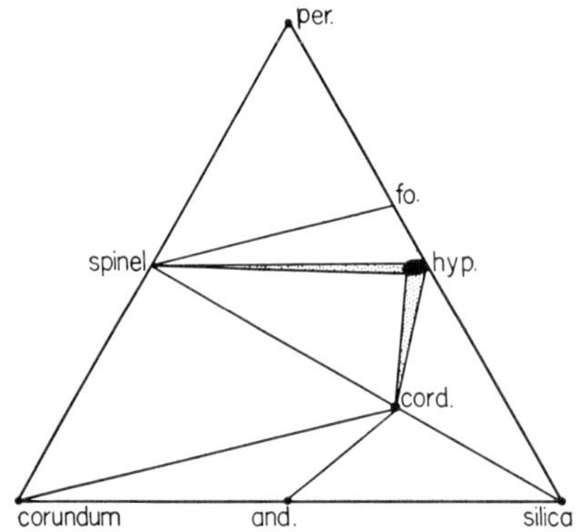

FIG. 91.—Al_2O_3-SiO_2-MgO diagram of the pyroxene hornfels facies

rocks which have been baked in the innermost contact zones surrounding crystallizing igneous rocks. Thus the facies represents rather low pressure but high temperature. The rocks recrystallized within the true pyroxene hornfels facies are usually fine-grained and show no sign of plastic flow. They are, therefore, very different with respect to structure from all kinds of regionally metamorphosed rocks. Based on Goldschmidt's masterly study of the Oslo area, and the compilations of Turner and Eskola, the phase diagrams (Figs. 91 and 92) should hold good for the pyroxene hornfels facies. As the facies is present only around true magmatic bodies, it plays quantitatively a minor role in the evolution of the earth's crust.

SANIDINITE FACIES

Inclusions in lavas are usually baked at higher temperatures than the walls of a subvolcanic magma chamber, the reason being that the temperature of crystallization of the lava is high because of the escape of gases. Such inclusions have been studied from several places (Lacroix, 1893), and the mineral associations are found to be much the same as those formed in laboratories from dry melts. The phase diagrams are seen in Figures 93 and 94. As compared with the hornfels facies, the main differences are found in the following points:

1. The pyroxene in the sanidinite facies are often pigeonites, because the monoclinic hypersthene, which is stable in the sanidinite facies, is partly miscible with diopsidic pyroxene.

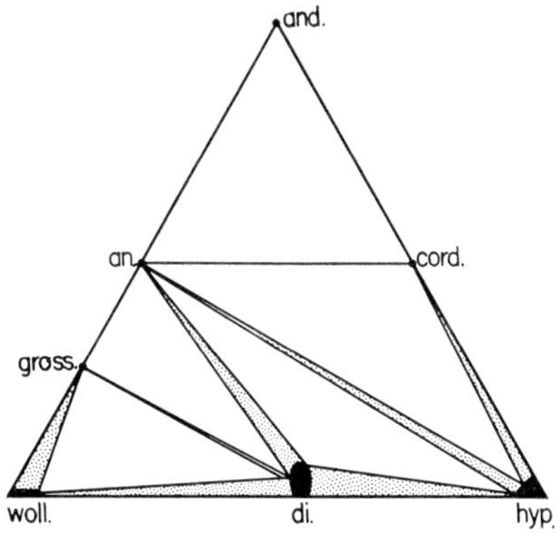

FIG. 92.—ACF diagram of the pyroxene hornfels facies, based on Goldschmidt's data (1911).

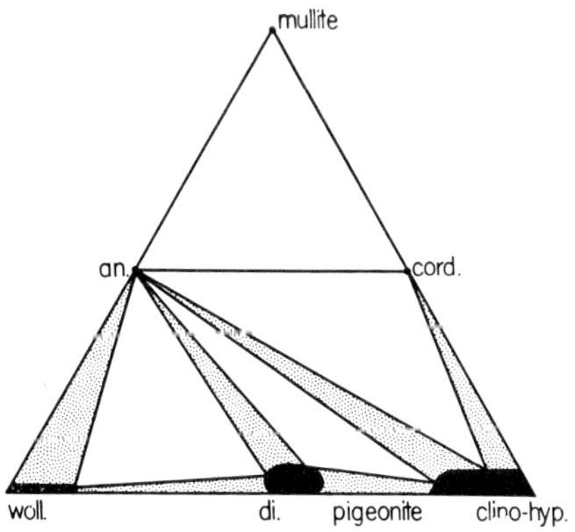

FIG. 93.—ACF diagram of sanidinite facies

2. The grossularite garnet, which is stable in the hornfels facies, has become unstable at the high temperature and low pressure of the sanidinite facies.

3. The plagioclase feldspars in the sanidinite facies most reasonably are of the high-temperature modification (Oftedahl, 1948; Köhler, 1949).

Furthermore, one must certainly expect that the composition of the mixed crystals (cordierite, wollastonite) (Tilley, 1937, 1947) exhibit different margins of variation within the two facies. One should also note that in the sanidinite facies the association ferrous olivine–tridymite (cristobalite) is stable at a greater Mg/Fe ratio than is the case in the hornfels facies.

ECLOGITE FACIES

Eclogite is a peculiar basic rock which consists mainly of a characteristic pyroxene, the omphazite or chloromelanite, and the typical eclogite garnet. This rock has given name to the eclogite facies (Eskola, 1939), which should then embrace the *P*,*T*-field within which eclogites can form stably. In contrast to the situation for most other facies, the bulk chemical composition of the rocks known from eclogite facies is very limited.

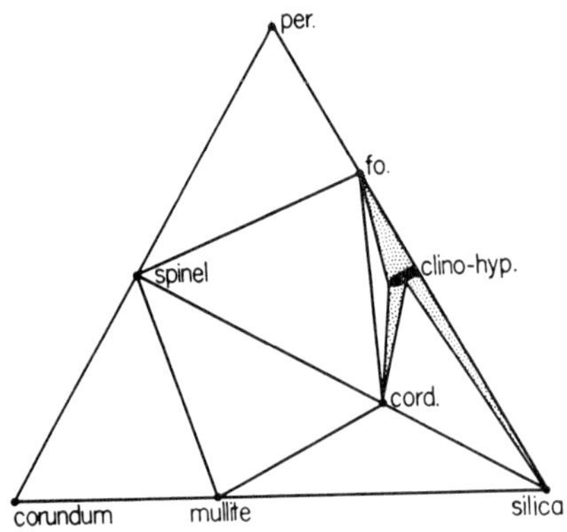

FIG. 94.—Al_2O_3-SiO_2-MgO diagram of sanidinite facies

So far, no feldspathic or acidic rocks are known from this facies. What is the reason for this? Does the quartzo-feldspathic mineral association melt under the *P*,*T*-conditions at which a gabbroic composition will transform into an eclogitic mineral assemblage? Does, perhaps, an acidic quartz-potash feldspar-albitic plagioclase assemblage remain practically unchanged when passing into the eclogite facies, so that acidic rocks are unable to indicate the facies change? Or does the constituent of potash feldspar and acidic plagioclase enter into other minerals in the eclogite facies? The latter possibility is in part true, but it fails to tell the whole story, because, if it was as simple as that, acidic potash and soda-rich rocks should be known from the eclogite facies. Although possessing a granodioritic bulk chemical composition, such rocks should carry some particular nonfeldspathic minerals which take care of the soda and potash in the rock. Actually, there does exist a mineral which (partly) substitutes for the albite in the eclogite facies, namely, jadeite:

$$\underset{\text{albite}}{NaAlSi_3O_8} \rightarrow \underset{\text{jadeite}}{NaAlSi_2O_6} + SiO_2 \qquad (124)$$

and

$$\underset{\text{nepheline}}{NaAlSiO_4} + \underset{\text{albite}}{NaAlSi_3O_8} \rightarrow \underset{\text{jadeite}}{2NaAlSi_2O_6} . \tag{125}$$

It is not settled yet whether one or both of these reactions are displaced to the right in the eclogite facies. It is only known that the common pyroxene in eclogites is able to take up appreciable amounts of the jadeite "molecule." According to Eskola's detailed investigations (1921*b*) at least 35 mol percentage of jadeite does enter the omphacite or chloromelanite. According to Yoder (1950), eclogitic pyroxenes contain from 0 to 8 per cent of Na_2O and 4 per cent in an average of eighteen samples. The maximum concentration of the jadeite component should then be some 40 mol per cent. Since pure jadeite does not occur in eclogites, neither of equations (124) and (125) needs to be displaced toward jadeite under the P,T-conditions represented by the eclogite facies. The amount of diopside and other pyroxene components present in the eclogite pyroxene will change the equilibrium conditions considerably.

According to a recent work by Yoder (1951), thermochemical investigations of Kelly and Kracek show that equation (125) is displaced toward jadeite at room temperature and up to an unknown maximum. But the relative densities show that high pressure favors the formation of jadeite; hence jadeite is a high-pressure, low-temperature mineral relative to the combination albite and nepheline.

Because jadeite is miscible with diopside, an admixture of diopside in reaction (125) will, so to speak, pull the reaction over toward the jadeite side, so that jadeite, as a component in omphacitic pyroxene, is stabilized at relatively higher temperatures (and lower pressure), as indicated by Figure 95.

On the other hand, the admixing of anorthite in albite should act in the opposite sense, that is, to lower the temperature at which jadeite is stabilized, as indicated in Figure 96.

In the eclogites, which are rocks of gabbroid affinity, it does not seem likely that reaction (124) or (125) is the proper one to consider. Equation (126) is more natural:

$$\underset{\text{albite}}{NaAlSi_3O_8} + \underset{\text{olivine}}{(Mg,Fe)_2SiO_4} \rightarrow \underset{\text{jadeite}}{NaAlSi_2O_6} + \underset{\text{rhombic hypersthene}}{2(Mg,Fe)SiO_3} . \tag{126}$$

Most likely this reaction is displaced toward jadeite and pyroxene at low temperature and high pressure. This equilibrium, again, will be controlled by the composition of the olivine and certainly also by the plagioclase composition.

Altogether, we see that the question of the stability of the eclogite pyroxenes is a rather complicated one, and the matter cannot be discussed further without experimental data.

There is no typical potassic silicate to substitute for the potash feldspar in the eclogite facies. In all likelihood, the potash feldspar is, therefore,

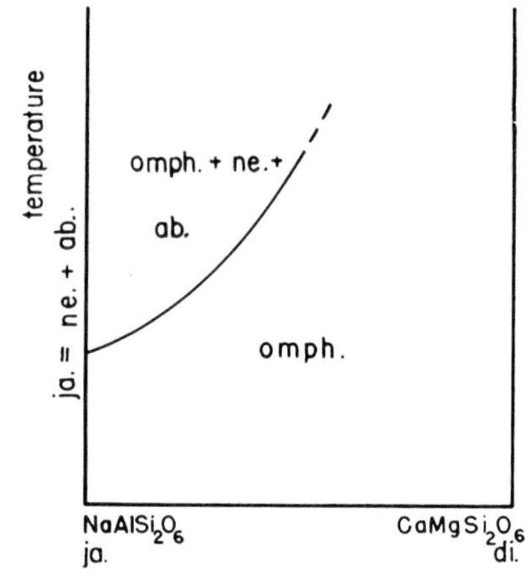

FIG. 95.—Tentative equilibrium diagram of the omphazite-nepheline-albite system. (Based on Yoder's work, 1950 and 1951.)

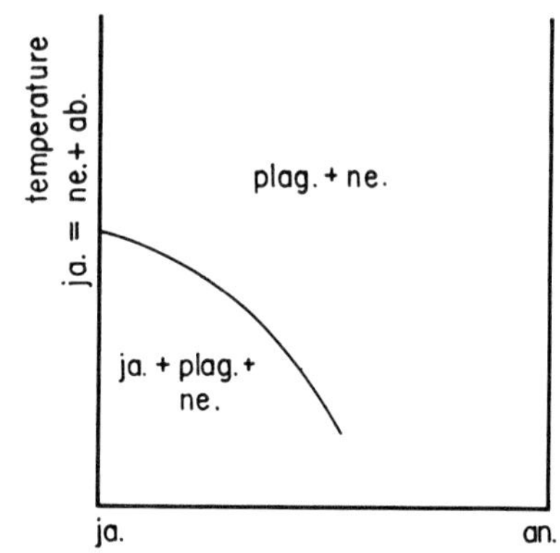

FIG. 96.—Tentative equilibrium diagram of jadeite-plagioclase-nepheline. (Based upon Yoder's work, 1950 and 1951.)

stable in this facies. The writer also strongly suspects that the association albite-quartz is stable, provided that diopside does not exist in the rock (see above). In other words, a rock consisting of potash feldspar, acidic plagioclase, and quartz (and some biotite) is, in all likelihood, a stable one under the physical conditions of the eclogite facies. It is assumed, though, that rather complete miscibility exists between the two alkali feldspar components.

The second significant mineral in eclogites is a garnet consisting of pyrope and almandine, with rather large amounts (up to some 25 mol per cent) of grossularite. In a similar way as albite "dissolves" in the pyroxene in eclogites, anorthite "dissolves" in the garnet.

Several mineral reactions may symbolize this process:

$$\underset{\text{anorthite}}{CaAl_2Si_2O_8} + \underset{\text{olivine}}{(Mg,Fe)_2SiO_4} \rightleftarrows \underset{\text{garnet}}{Ca(Mg,Fe)_2Al_2Si_3O_{12}} \,. \tag{127}$$

Such a garnet does actually correspond to some of the eclogite garnets analyzed (Eskola, 1921*b;* Sahlstein, 1935):

$$\underset{\text{anorthite}}{3CaAl_2Si_2O_8} \rightleftarrows \underset{\text{grossularite}}{Ca_3Al_2Si_3O_{12}} + \underset{\text{kyanite}}{2Al_2SiO_5} + SiO_2 \,, \tag{128}$$

$$\underset{\text{anorthite}}{CaAl_2Si_2O_8} + \underset{\text{wollastonite}}{2CaSiO_3} \rightleftarrows \underset{\text{grossularite}}{Ca_3Al_2Si_3O_{12}} + SiO_2 \,. \tag{129}$$

The two latter reactions are not necessarily displaced toward the garnet side in the eclogite facies. It is therefore not unlikely that anorthite would be stable in the eclogite facies, provided that the rock does not contain ferromagnesian minerals. Or it is possible that zoisite or lawsonite takes the place of anorthite in a manner similar to that in the epidote amphibolite facies (p. 148) and the glaucophane schist facies (below). None of the minerals, garnet, zoisite, or lawsonite, appreciates Na^+ enough to incorporate it together with Ca^{2+} in their lattices. For that reason, the tendency to transform the anorthite molecule into grossularite, zoisite, or lawsonite becomes less and less, the more acidic the plagioclase is (see p. 51). On the other hand, the tendency to "jadeitize" albite decreases with the increasing basicity of the plagioclase. One would therefore expect that intermediate plagioclases would be most likely to be stable in the eclogite facies.

Eclogites are found in different situations in the field: (1) as lenticular masses in gneisses; (2) as bands and lenses in dunites, which again often occur as large inclusions in gneisses; (3) as bands in mica schists and amphibolites; and (4) as inclusion in kimberlite and some basalts, often accompanied by dunitic inclusion.

The position of the eclogite facies relative to the main facies of regional metamorphism is not clear. The previous discussion has shown that the eclogite minerals are gradually stabilized by the displacement of some homogeneous equilibrium among mixed crystals. In fact, we have reasons to believe that eclogites can form from certain basic rocks by rather small deviations of pressure (temperature?) from that corresponding to the amphibolite facies and certain parts of the granulite facies. The eclogite inclusions in the Caledonian gneiss in East Greenland occur in hypersthene garnet gneisses of the granulite facies (Sahlstein, 1935). The eclogites of Norway (Eskola, 1921*b;* Backlund, 1936) occur mainly in amphibolite facies gneiss but are also intimately connected with anorthositic rocks which approach the granulite facies. The later amphibolitization of the eclogitic inclusions in the gneiss of western Norway indicates that the eclogitic bodies are polymetamorphic. It could be that the surrounding gneiss at some time was in the same facies as that which created the eclogitic composition of some of the basic inclusions or remnants. The more reactive quartzo-feldspathic gneiss was able to recrystallize in response to the change of the P,T-conditions to the amphibolite facies, but only occasionally were the large basic inclusions able to recrystallize in the newly stabilized facies. Only when shearing movement took place and sufficient intergranular water vapor was present, could the eclogites recrystallize into the amphibolites now stabilized.

GLAUCOPHANE SCHIST FACIES

Hornblende is the metamorphic mineral which is most flexible with reference to the acceptance of elements in its lattice. Changes in pressure and temperature, as well as changes in chemical milieu, enable hornblende to change widely in composition. Under physical circumstances which are rather similar to those of the epidote amphibolite facies, it appears that the aluminous hornblende can dissolve much sodium in the $(Na,Ca)_2$ positions. This variety of hornblende has been called "glaucophane," and the particular facies within which it is stable is termed the "glaucophane schist facies" (Eskola, 1939). Besides this particular hornblende, lawsonite—$CaAl_2Si_2O_8H_2O$—is typical of the facies. This is a heavy mineral (d = 3.1), which, in connection with the eclogitic variety of garnet present in the glaucophane facies, indicates that high pressure is needed to develop the facies. Also omphacite-like pyroxene may be present in glaucophane schists, showing a similarity to the eclogite facies. On the other hand, epidote, chlorite, and other common medium-grade to low-grade minerals put the glaucophane schists in a facies not much different from the epidote amphibolite facies. It is believed that locally in the epidote amphibolite facies complexes the physical conditions can conform to those necessary to create the typical soda-rich glaucophane and lawsonite. It may also be, however, that the relative concentrations of the chemical elements in the rocks have to be inside certain narrow limits in order that the typical minerals may develop. In any case, the stability fields of glaucophane and lawsonite are so narrow that glaucophane schists are very rare. For a detailed description of the famous Californian glaucophane schists the reader is referred to Taliaferro (1943).

METASOMATISM

INTRODUCTION

METAMORPHIC recrystallization rarely takes place without some changes in the bulk chemical composition of the rock bodies. Solid rocks undergoing recrystallization are almost invariably open systems, allowing chemical elements or compounds to enter or leave the rocks. This fact was recognized long ago by the Scandinavian and French students of metamorphosed rock complexes (Keilhau, 1828–38; Durocher, 1846; Delesse, 1861). The term "metasomatism" (C. Naumann) was devoted to the process which was capable of changing the bulk chemical composition of a solid rock, and the rocks so altered may be called "metasomatites" (Berkey, 1934). That metasomatic alteration of solid rocks is a process which actually has taken—and takes—place in nature is well evidenced by observations described in any comprehensive textbook in metamorphic petrology and ore genesis, but no agreement exists as to what extent and by which mechanism such alterations take place. A few points which justify the establishment of the concept of metasomatism will be mentioned:

1. Ore bodies (sphalerite, galena, etc.) occur frequently in metamorphic limestone, but there is no possibility that such bodies have been injected as an ore melt. Neither can such bodies, in most cases, have precipitated as a sediment together with the limestone. Furthermore, the structure often shows that the ore developed by a volume-by-volume replacement of the limestone.

2. Metamorphosed sediments have been found to change in composition when younger granitic (or quartzo-feldspathic) bodies are approached along the strike. Such changes in sediments can be referred only in exceptional cases to original differences in the sediment layer. Neither can such changes of the sediment be due to a simple mixing between old rock and invading silicate melt.

3. Not infrequently rock bodies are found with a bulk chemical composition inconsistent with that known to form by magmatic differentiation or sedimentation: the magnesium- and aluminum-rich cordierite-anthophyllite rocks, "sedimentary" iron formations (Pettijohn, 1949), a number of ore types which cannot have precipitated out of a concentrated melt, etc.

4. There is, statistically, a relationship between degree of metamor-

phism (metamorphic facies) and average chemical composition of metasediments (Lapadu-Hargues, 1945–49). A similar relationship may also hold for meta-igneous rocks: basaltic lavas (pillow lava), for example, which have recrystallized in green schist facies often have a bulk composition different from unaltered basaltic lavas.

5. Microscopic studies reveal that the one mineral replaces the other by "eating" its way into the lattice of the other. This point alone, however, does not necessarily show anything but a rearrangement of matter within the limited space of rock bodies.

6. The so-called "space problem" is very important: the emplacement of several igneous-looking massifs, dikes, and veins has taken place without dilation or forcing apart of the country rocks (Goodspeed, 1940; Phemister, 1945). It is evident that this could not happen by forceful injection, but it would be in agreement with a slow structure-preserving diffusional introduction of matter.

There are numerous other critical points which could be mentioned, but the reader is referred to the rapidly increasing number of journal articles on the subject. Reference is made in particular to literature on ore genesis, where metasomatism is found to be the most important agency in the evolution and concentration of ore minerals and ore bodies.

There is every gradation between the weak metasomatism which has operated when a rock gains or loses some water or carbon dioxide or other fugitive constituents during recrystallization and the rather complete process which forms an entirely new rock in place of the old, every atom of the old unstable rock being replaced by a new one. Examples of the latter are sphalerite-galena or other ore minerals replacing limestone, and quartz replacing dolomite. The relict structure in the newly formed rock is often an unambiguous proof as to what has happened (Lindgren, 1933, pp. 177–86).

The whole phenomenon of metasomatism is extremely complex and only little understood. In fact, some of the main problems in petrology today—the origin of many deep-seated rock bodies like granodiorites, granites, anorthosites, and a variety of gneisses together with pegmatites and quartz veins—remain unsolved, probably because of lack of understanding of metasomatic processes.

When we are studying metasomatism, we are faced with a phenomenon entirely different from those commonly studied in laboratories; hence we cannot expect to receive aid from physics and chemistry in the form of laws ready to be applied to our complex problem without many modifications. The greatest difficulty lies perhaps in the fact that, on the one hand, metasomatism necessarily implies long-distance transportation of matter through crystalline rock bodies but that, on the other hand, this transportation cannot commonly be similar to the well-understood bodily flow of lavas, ground water, or mineral oil.

The rock developing through metasomatism is significantly determined by the rate of transportation of the several constituents being gained and lost by the altered rock. To determine the velocity of transportation or the power of flow or diffusion of material through such complex systems as rocks seems at present to be scarcely a solvable problem. Other problems we are faced with when studying metasomatism are numerous: Whence do the introduced substances come, and where do the removed substances go? What forces are driving the transportation? What rules are governing the growth and dwindling of minerals in the altered rocks? Etc.

It has been customary to make a primary magma responsible for the substances which a rock gains during metasomatism. The sequence of happening was believed to be simple in principle: a silicate melt, often of immense volume, forced its way through the crystalline crust; by loss of heat to the surroundings, the melt crystallized, and the surrounding rocks were "baked" and hence recrystallized. The solidification of the melt built up a high partial vapor pressure of the fugitive molecules in the rest liquid, and the gaseous molecules streamed into the confining solid rocks, performing there the processes known as "metasomatic." By cooling, the pneumatolytic gases, among which H_2O made up the major fraction, condensed to a watery liquid solution which circulated farther through the rock complexes, performed hydrothermal metasomatism, and precipitated hydrothermal veins.

There is no doubt that such a picture is true in a general way in certain areas. Evidently, the contact metamorphism and the contact metasomatic ore bodies (Fe_2O_3, PbS, AgS, etc.) around definitely magmatic rocks may have originated in a way close to that outlined above (Goldschmidt, 1911). But even the evolution of true magmatic rock series cannot longer be assumed to have proceeded along the superficial scheme: ultrabasic → intermediate → acidic, with a late magmatic-pegmatitic phase and a pneumatolytic-hydrothermal oxide sulfide ore phase. The most severe difficulties are met with, however, if we intend to fit the experience from regionally metamorphosed complexes into this narrow picture frame. The main objections may be three in number: First, the volumes of metamorphosed and metasomatosed rocks are so vast that the relatively small amount of apparent magmatic bodies (homogeneous "granites") could not reasonably have supplied either the heat or the substances required by the observed regional alteration. Second, the solubility of rock-making minerals in common hydrous solutions is so small,[1] their vapor pressure so low, and the flow permeability (p. 179) of the recrystallizing rocks so negligible that one cannot seriously believe the metasomatism to have been performed by a fluid advancing through the rock complexes from some hidden magmatic source. Third, metasomatosed rocks in regional metamorphic areas are not particularly rich in the so-called "fugitive" late-magmatic substances which should concentrate in solutions of magmatic parentage (fluorine, boron, etc.). We may also add that in regionally metamorphic complexes the pegmatites and other veins do not commonly concentrate in the proximity of igneous-looking bodies (Andersen, 1931), although high-grade regional metamorphism often creates granitic bodies as well as pegmatites, so that some space relationship exists between these events.

An alternative general view has, because of field observations together with theoretical physicochemical considerations and data gathered from silicate experiments, forced itself upon an increasing number of students of metamorphic problems. It is not the magmas which are the primary cause of all events in regional metamorphic areas—it is exactly the other way around: increasing temperature, intense stress, and rising pressure met with in deeper sections of folded mountains cause not only a simple recrystallization of the affected rocks but also extensive metasomatism by large-scale rearrangement of the substances within certain volumes of the crust (metamorphic differentiation of global scale; homogenization) "Les roches plutoniques se sont donc formées aux dépends des roches métamorphiques; elles représentent le terme extrême du métamorphisme général; elles sont l'effect et non la cause de ce métamorphisme," wrote Delesse in 1861. The carriers of these rearrangements and displacements of matter are not commonly of magmatic parentage. Only locally does actual refusion take place, so that "palingenic" quartzo-feldspathic melts are created, thus explaining logically the existence of dacitic and rhyolitic volcanism in the course of the formation of folded mountain chains (p. 273). This general view has been advocated for silicate rocks during a period of years by some petrologists, among which Wegmann perhaps has been the pioneer in the modern school (1935). Ore geologists, however, have found it harder to escape from the narrow frame of a juvenile magmatic source for almost any ore body. Recently, Sullivan (1948) has advocated a new line of thinking close to that of many hard-rock geologists.

The relatively high temperature necessary to cause the happenings here roughly outlined may be caused by the exothermic radioactive decay in "sial" and/or by the process of the orogeny itself, which most reasonably

1. For a more detailed discussion of the solubility of minerals see p. 187.

is a heat-liberating event because it is a series of processes aiming toward a re-establishment of a disrupted equilibrium in the outer part of the earth (see also p. 271).

We shall leave the broader problems of metasomatism and restrict ourselves for some time to a discussion of the mechanism of transfer during metasomatism, the forces underlying such a process, and the rules governing the metasomatic alteration of rocks.

THE MECHANISMS OF METASOMATIC TRANSFER OF MATTER THROUGH ROCKS

During metasomatism the altered rock body appears to have been "soaked" thoroughly by the altering medium, so that each mineral grain has been in intimate chemical contact with the altering agency, if not thoroughly penetrated by it. "In metasomatism new minerals develop at countless points in the old rock" (Lindgren, 1933, p. 177). On the other hand, there is no ground for believing that the rock, during the metasomatism, was literally "soaked" or suspended in a true liquid. All experience shows that the common rock during each stage of the alteration was a continuous crystalline aggregate through which the agency of metasomatism had to find its way in order to reach materially each individual mineral grain or each point within the body of such grains. How can matter penetrate consolidated crystalline aggregates like solid rocks? There are two basically different possibilities: (1) There may be a network of interconnected capillary pores penetrating the rock, thus making it a conductor for flowing pore fluids, which carry in solution or colloidal suspension the substances which build up the metasomes[1] developing in the rock. This is a well-known property of partly cemented or weakly recrystallized sediments (sandstone, limestone). (2) The rock may behave like a semipermeable "membrane," permeable only to individual atoms, ions, or molecules but behaving like an insulator against the bodily flowing pore fluids (cf. p. 179). It is suggestive, in this respect, that clay containing large amounts of adsorbed water cannot easily be anhydrized by mechanical squeezing but that the same clay loses its water easily by being kept in a dry atmosphere for some time. In the first case, it is a question of forcing liquid water bodily out of the system of capillary voids in the clay; in the second case, it is individual water molecules which migrate out because the water-vapor pressure is low outside the clay specimen. That is, the clay is easily permeable to individual water molecules but practically not to bodily flowing water, which necessarily must be a polymolecular continuum.

MECHANICAL FLOW[2] OF FLUID SOLUTIONS THROUGH ROCKS

Transportation of matter by mechanical flow is familiar to us from our everyday experience with ground water and mineral-oil motion, with geysers and other springs, and, so far, also with outflow of lavas. The mo-

1. A metasome is a mineral grain formed by metasomatism.

2. The term "mechanical flow" is used in order to stress the fact that the flow of a fluid or plastic solid is a mechanical process in the nomenclature of Brönsted (1938).

tion of ground water through uncemented sandstones and porous limestones may reasonably be believed to represent a direct parallel to parts of the transfer of material through rocks undergoing metasomatism. Although such a bodily flow takes place where open space is available in complexes of rocks, it comes out to be questionable whether such pore flow is the crucial kind of metasomatic transportation except when the transfer takes place along literal fissure openings (some ore veins). The reasons for this conclusion are numerous and will be discussed subsequently. We need, then, a full understanding of the most important aspects of the process of flow of fluids through open channelways or porous media.

W. King Hubbert (1940) has given a comprehensive and rigorous analysis of the motion of ground water, which can also be applied to the flow of a fluid through porous rocks during metasomatism. Hubbert states that a pore fluid flows down the fluid potential gradient (or the mechanical potential gradient) (Brönsted, 1938) just as an electric current runs down an electric potential gradient, or heat flows from high to low temperature, or chemical transfer of matter extends from high to low partial molal free energy. The bodily flow of a fluid or a plastic solid is actually a mechanical process driven by mechanical stresses (mechanical potential gradients). The diffusion of individual molecules, ions, or atoms is a chemical process driven by the same forces which drive any other chemical process or reaction. We shall, therefore, distinguish between flow permeability and diffusion permeability of rocks. We may also speak of rocks as "fluid conductors" (like heat conductors, etc.) and as "diffusion conductors."

In the flow process there are as many fluid potentials as there are individual immiscible fluids. Hence different motion in direction or velocity of the several fluids is possible if the fluid potential gradients are different in magnitude and/or direction or if the resistance to the flow is different for the several immiscible fluids, e.g., if the fluids have different viscosity.

The isothermal laminar flow of an incompressible fluid through a porous rock is given by equation (130), which is equivalent to Darcy's law:

$$\frac{\partial v}{\partial t} = -A \cdot K \cdot \frac{1}{g} \cdot \frac{\partial \Phi}{\partial l}, \qquad (130)$$

where $\partial v/\partial t$ is the discharge in volume per unit time through the cross-section A of the porous rock; K is a coefficient of flow permeability of the rock; g is the acceleration due to gravity; Φ is the mechanical potential per unit mass of the fluid; and l is the length of the flow. In the fluid potential, Φ, are incorporated the mechanical pressure acting on the fluid, the density of the fluid, as well as the position of the fluid in the gravitational field. Equation (130) takes account of the force of gravity and hence holds for flow in any direction in space (Hubbert, 1940).

Considering only the horizontal flow (that is, the flow transverse to the force of gravity) and denoting the horizontal axis X, equation (130) changes into:

$$\frac{\partial v}{\partial t} = -A \cdot K \cdot \frac{1}{d} \frac{\partial P}{\partial x}, \tag{131}$$

where P is the pressure acting on the fluid, and d is its density.

For flow through a single circular horizontal tube, the equation is expressible as follows:

$$\frac{\partial v}{\partial t} = -\frac{\pi D^4}{128 \eta} \frac{\partial P}{\partial x}, \tag{132}$$

where D is the diameter of the tube and η is the viscosity. The flux in the horizontal direction along a plane-parallel fissure is given by the following equation:

$$\frac{\partial v}{\partial t} = -\frac{B D^3}{12 \eta} \frac{\partial P}{\partial x}, \tag{133}$$

where B is the length parallel to the fissure of the cross-section of the flow and D is the thickness of the fissure (Jameson, 1937). Thus the flow per unit square area—the specific discharge—decreases rapidly with decreasing cross-section of the channel.

Equations (132) and (133) are developed under the assumption that the molecular fluid layer lying next to the wall of the tube is fixed to the wall.

Equations (132) and (133) do not hold, therefore, if the cross-section of the channel becomes of molecular dimension, because then bodily flow is possible only if the fluid layer adherent to the tube wall can slip along the solid wall. Such a motion no longer depends on the viscosity of the fluid itself but on the bonding energy between the molecules (atoms, ions) of the fluid and of the porous conductor. In fact, the bodily flow changes into an activated intergranular diffusion, as has been observed by Nissan (1946). Even if the cross-section of the channel is several hundred molecules (1,000 A), the viscosity of the water increases considerably (tenfold) as compared with a large body of water (Henniker, 1949).

In the flow of fluids through porous rocks, gravity plays an important role, always forcing the fluid down to levels of low gravitational energy, if not hindred by a mechanical force acting in opposition to gravity. The isothermal flow in a vertical direction of the incompressible pore fluid will be expressed by the following equation:

$$\frac{\partial v}{\partial t} = -A \cdot K \cdot \left(\frac{\partial P}{\partial y} - d g\right), \tag{134}$$

where g is the acceleration due to gravity and y is the vertical co-ordinate reckoned positive downward. It will be found that this equation is in harmony with the general equation (130) of Hubbert (1940). Equation (134) shows that no flow will take place if the downward-directed pressure gradient affecting the fluid equals the specific gravity ($s = d \cdot g$) of the fluid. Then the conditions of hydrostatic equilibrium are fulfilled. Flow will go upward (in the negative direction of the co-ordinate, y) if $\partial P/\partial y > d \cdot g$, and downward if $\partial P/\partial y < d \cdot g$.

The requirement for mechanical stability—i.e., no bodily motion—in a fluid occupying the pore system of a rock is that the mechanical potential per unit mass of the fluid be constant throughout. Stating this in more common words, one finds that mechanical stability exists if the same pressure acts on the fluid at every point in a given level in the gravitational field, at the same time as the downward-directed pressure gradient equals the specific gravity of the fluid (i.e., the hydrostatic pressure gradient), provided that the temperature is equal throughout.

As we have just discussed, the significant requirements for bodily motion of a pore fluid are two in number, namely: (1) that a driving force (mechanical or flow potential) acts on the fluid and (2) that the resistance to flow is not infinitely large. When these two needs are fulfilled, the flow will take place with a velocity determined by the intensity of the potential and the viscosity of the fluid in combination with the character of the pore system.

In case we are studying ground water, mineral oil, or silicate magmas, the fulfilment of the two requirements for pore fluid flow means that the substances we are concerned with are actually transferred over some distance through the rocks.

In the case of the fluids supposed to be the carrier of mineral-making matter during metasomatism, the situation is very different. The requirements for flow of such fluids are evidently the same as those already considered, but it is by no means thereby insured that effective transportation of mineral-building components takes place from one site to another in the crust. To insure effective transportation in metasomatism, chemical processes must take place within the fluid or between the fluid and its environment, so as to discharge some of the dissolved rock-making solute of the fluid solution. Owing to the very small solubility of rock-making silicates and ore minerals in hydrous solutions of a character reasonably to be present in rocks, there will be only a very small fraction of the transported fluid which actually can consolidate in the form of solid minerals at the proper places.

THE PLACE OF FLUIDS IN METASOMATISM

Let us confine our attention again to ground water and its motion. Why does ground water exist in the outermost parts of the earth? Evidently, because there are some voids between the mineral grains in the loose sediments or porous rocks, or some greater cracks, fissures, breccia zones, etc. The ground water moves because the pores are interconnected, and because fluid potential differences exist or are created occasionally. The ground water carrying porous rock is a network which has not yet collapsed into a compact construction because of its own stiffness under the existing physical and chemical circumstances. The ground water does not usually resist the compaction of the porous rock construction, for the pressure in the pore water at a given depth is generally less than the load of the superincumbent rocks. Hence, as time goes on, the sediment may, if buried sufficiently deep that pressure and temperature are considerable, undergo compaction and cementation, by which the pore water is forced out of the pore network except for the minor quantity which becomes trapped as confined inclusions or adhered to the mineral surfaces in the compacted sediment. This compaction is well understood, and it has been correlated to depth of burial by rather rigorous equations (Pettijohn, 1949). The lack of flowable water below certain depths has been demonstrated by the dryness in deep mines (Lindgren, 1933, p. 34).

Many ore bodies or "veins" of metasomatic origin have developed within only partially compacted or cemented sediments. Here is also the stage of extensive dolomitization and silicification of limestones (Pettijohn, 1949). It is found, furthermore, that typical ore veins, supposedly originated as fracture fillings via hydrothermal action, are confined to the uppermost floors of the crust, that is, within the uppermost zones of regional metamorphism or within the zone of diagenesis and cementation. It is, hence, only logical that ore formation, dolomitization, and silicification have been widely attributed to the flow of hydrous solutions. But silicate metasomatism is most powerful and large-scaled in deep-seated rocks. Within the epidote amphibolite facies, amphibolite facies, and granulite facies is the stage of "granitization," with its sodium and potassium mobility, of magnesium and iron metasomatism, of extensive transfer of aluminum, etc. In striking contrast to the rarity of hydrothermal ore veins in mesozonal and katazonal metamorphic sections, acidic pegmatitic veins and dikes, as well as barren quartz veins, are plentiful within these zones. So far, it is just within the deepest metamorphic zones that the metasomatic dikes of solid pure titanic iron ore are encountered in anorthosites (granulite facies). It seems as if we arrive at incompatible conclusions: (1) The power of metasomatic transfer increases with rising grade of metamorphism and with depth; (2) but, on the other hand, the compaction of rocks also increases with rising grade of metamorphism, thus rendering our rocks "flow insulators."

The reason why the compaction ultimately reaches completion is that pressure differences are thermodynamically unstable, so that open pores or pores filled with a fluid at lower pressure than that in the surrounding solid grains will strongly tend to close by either simple plastic flow or filling by newly formed minerals. In a recrystallizing rock, exposed to even a low grade of regional metamorphism, the power of chemical adjustments is so considerable that the crack-healing process must be practically completed. Therefore, a continuous pore network cannot exist in rocks undergoing regional metamorphism, so that rocks recrystallizing at some depth are, in all likelihood, "flow insulators" except for the temporary fissures and cracks which open up and close in strained rock bodies. But the cracks thus formed have usually a short lifetime because they are thermodynamically and mechanically very unstable. Furthermore, outside extensive joints, thrusts, and breccia zones, it seems that most temporary cracks opened up by strain are not interconnected and therefore that they do not provide avenues for flowing pore solutions.

The pore fluid will be squeezed out of recrystallizing rocks at some depth, with the exception of the fluid inclusions trapped within the mineral grains, the "fluid" strongly adsorbed to the mineral interfaces and mosaic fissures, and that existing in temporary fissures.

Although rocks thus recrystallized at some depth are "flow insulators" because of their compact structure, they are not "diffusion insulators." Thus, for example, water and carbon dioxide usually disappear from the rocks if anhydrous silicates and calc silicates are formed at the expense of hydrous minerals and carbonates. As a concrete example let us consider a chlorite mica schist which is formed from a clay sediment. During compaction and recrystallization the free pore water has disappeared from the altered clay (with the exception of fluid inclusions in the minerals), and the texture is compact. Nevertheless, however, recrystallization of the chlorite mica schist at higher pressure and temperature (granulite facies) necessitates a disappearance of some water, because the newly formed minerals are anhydrous (garnet, sillimanite, feldspar, etc.). The only possible way for water to move out of the rock now is by means of diffusion of individual molecules along the intergranular network and along mosaic fissures. (If zeolites and other open lattices are present, a migration of water inside the lattices is also possible [Tiselius, 1934; Hey, 1935].) On the other hand, basic igneous rocks containing mainly "dry" minerals are often recrystallized to green schist facies or epidote amphibolite facies. The minerals thus stabilized are hydrous and require considerable amounts of water, which in one way or another always finds access to the very points in the rocks where the hydrous metasomes grow.

The adsorbed molecules (ions, atoms) along the intergranular form a surface film that cannot be compared directly with a regular liquid solution; it must be considered an individual phase, very different from a liquid macrobody. Where free supermolecular space opens up (a crack), this will be filled with the most movable particles from the adsorbed intergranular "film," and equilibrium may be established between the intergranular film and the fluid macrophase in the crack. It is most reasonable now that the fluid macrophase thus developed should have a lower total pressure than the pressure in the compact rock (the rock pressure), because otherwise the fluid-filled crack or fissure would persist and would show no tendency to close by either plastic yield or mineral growth (see below). One also recognizes that—provided that the pore fluid is gaseous or supercritical—the possibility that the total pressure of the fluid in the crack is the same as (or greater than) the rock pressure means simply that the rock is at the boiling point at the given depth. Indeed, this does not seem to harmonize with our experience. When the volume of a confined crack decreases because of a slow plastic yield in the surroundings and/or growth of new minerals in the opening, then the pressure of the fluid in the crack will increase and should eventually stop the collapse of the crack. Since, however, this process does not come to a stop in the zones of regional metamorphism, it must mean that, when the crack shrinks, the small rise of pressure in the fluid caused thereby is sufficient to make the constituents of the fluid diffuse back into the intergranular of the country rock. This conclusion is in complete harmony with the great mobility of water molecules through rocks, as indicated by the fact that dehydration and hydration of minerals take place easily during progressive and regressive metamorphism.

At first glance, these considerations seem to be contradicted by the fact that liquid and gaseous inclusions (Sörby, 1858) are very common in minerals in metamorphites. Why are these fluid cavities not closed by plastic yield in the surroundings or growth of mineral fillings, as explained above? The answer seems simple: Such fluid inclusions are found within the body of the mineral grains, and they are isolated from the intergranular channelways along which the fluid molecules can migrate rather easily. As a consequence, the pressure within the fluid-filled cavity in a single crystal is gradually adjusted to be equal to the pressure existing in the body of the host mineral, and, if we assume that the migration of water, carbon dioxide, etc., is very slow in the lattice of the minerals, there are then no further tendencies to close the fluid filling either by plastic collapse or by newly formed mineral fillings. Incidentally, we also see that neither the state nor the composition of fluid inclusions in metamorphic minerals, when heated up to the temperature at which the inclusions are homogenized after the principle of H. C. Sörby, is usually identical with the state and composition of the movable intergranular rock "moisture," the total pressure of which commonly is less than the rock pressure. This movable phase, when present in an intergranular pore of supermolecular size, may be a true vapor or gas, although the trapped fluid inclusions in the minerals under their higher pressure are liquids or dense fluids.

It is interesting to note that fluid inclusions in metamorphic rocks commonly are arranged in some planar structure (Tuttle, 1949*a*). This fact conforms well with the considerations above, and the theory of secretionary growth described on page 216. Narrow, but often extensive, fractures open up in the rock; these are filled first with the rock "moisture," and gradually they are healed, as explained above. But this healing process is not uniform throughout the fissure, so that the result will be that small separate cells of fluid are left along the fissure and these cells will decrease in size until the pressure increased thereby makes the fluid inclusions thermodynamically stable.

For the several reasons discussed above, there is a significant difference between the behavior of a watery solution and a concentrated silicate melt in deeper portions of the crust. A diabase magma, for example, has a very low vapor pressure, even at high temperature, and the diffusion of the individual atoms, ions, or molecules of a diabase through rocks is a very slow process, as compared with the diffusion of water or other gaseous molecules with high escaping tendency. For that reason the melt can stand a pressure equal to, or higher than, that in the surrounding rocks without showing any appreciable tendency to disintegrate and diffuse into the adjacent rocks. A diabase magma, or any other silicate melt, for that matter, can therefore readily force its way through zones of weakness in the rocks by expanding the channelway as the melt proceeds. A hydrous solution, however, cannot commonly force itself bodily through a zone of weakness in a deep-seated rock complex if it is necessary to widen the joint by a forceful injection. It will be easier for the hydrous fluid to disintegrate and diffuse through the intergranular avenues of the adjacent rocks.

We have then arrived at the following picture of the position of a pore solution in metamorphic-metasomatic petrogenesis: The pore solution is only occasionally a fluid three-dimensional macrophase, namely, either when it is trapped as inclusions in mineral grains or when open spaces exist freely for some time because of the sluggishness of the plastic yield and neomineralization of the rocks. Most generally, the pore fluid is identical with the "gaseous" particles which are adsorbed to the mineral surfaces or the mosaic block surfaces.

The free openings are thermodynamically unstable and should close rather rapidly by simple plastic yield or by the secretion effect. Only in a limited way do they provide an avenue for metasomatic transportation. The widespread idea of metamorphic rocks as capable of conveying streaming hydrous solution appears to be inconsistent with our present knowledge of crystalline rocks.

It is highly interesting, though, that actual motion of hydrous solutions is traced by means of unsymmetrically grown crystals in hydrothermal veins (Newhouse, 1941; Engel, 1948).[3] Gordon Smith (1950) has attempted to determine the direction of vein fluid flow by measuring the driving pressure gradient with the aid of the fluid inclusions in minerals along the veins.[4] Such a flow is certainly a very likely process in the course of the time during which the fissure remains mechanically open. On the other hand, no indicators of a flowing pore fluid have ever been traced in the compact body of metasomatosed rocks in which metasomes "grow at countless points."

THE COMPOSITION AND ORIGIN OF PORE FLUIDS OR ROCK MOISTURE

The fluid inclusions encountered in minerals of metasomatic origin usually carry rather considerable amounts of easily soluble salts. Königsberger and Müller in 1906 found that quartz crystals from alpine veins contained hydrous solutions of Na^+, Ca^{2+}, K^+, Li^+, and CO_3^{2-}, Cl^-, SO_4^{2-}, as well as CO_2-gas. Newhouse (1932) found Na^+, Cl^-, and Ca^{2+} in fluid inclusions in galena and sphalerite; Faber (1941) determined the composition of fluid inclusions in pre-Cambrian granite minerals and found the salinity to be 6.6–22 per cent of the fluid. NaCl, KCl, Na_2SO_4, K_2SO_4, and $CaSO_4$ predominated.

Outside these few data, we have but indirect knowledge of the composition of the rock moisture in metamorphic rocks. In an attempt to arrive at this secondhand information, we shall have to consider how a pore fluid originates in a rock and how it gains its charge of rock-making

3. The principle of unsymmetric growth of fracture-filling materials demonstrates that the material building up the minerals (the solute) was in a directed motion. The bodily motion of the solvent is thereby not proved; although there are no doubts that this takes place in open fractures.

4. As stated on p. 181, the pressure in the trapped fluid inclusions in minerals should, in most cases, be different from the pressure in a pore solution which presumably was moving rather freely along the capillary or supercapillary openings. The method of Gordon Smith as a guide in driving potential determination is, therefore, rather questionable.

constituents. The sources for such fluids may be several in number: (1) Fluids may be expelled from crystallizing juvenile and palingenic magmas, as is clearly demonstrated by present-day volcanoes. (2) The fluids may represent aqueous solutions "distilled" from hydrous minerals undergoing transformation into "dry" minerals. (3) Part of the pore fluids in metamorphic rocks may originate from water trapped in sediments during their sedimentation. (4) Ground water may reach down to the zones of metamorphism, thus constituting part of the pore fluid of metasomatism.

The juvenile magmatic rest fluids must contain those of the juvenile magmatic components which show little tendency to precipitate as minerals during the crystallization of the melt. In other words, the composition of the magmatic rest fluid (see definition, p. 4) is determined by the composition of the melt from which the fluid escapes, unmixes, or vaporizes and the distribution of the several possible components among the melt, the minerals, and the rest fluid. As it is almost definite that the rest fluid escaping from the solidifying melt is gaseous or supercritical, the compounds with the highest partial vapor pressure, or which are easily "soluble" in the vaporous phase, will be concentrated in the rest fluid. A very interesting discussion of this matter was initiated by a paper by Graton (1940). Morey and Ingerson (1940) and Fenner (1940) exhausted the reasonable possibilities and arrived at the conclusion that the fluid escaping from a silicate melt had to be gaseous. Scattered investigations from volcanic gases and exhalations have shown that the elements and compounds compiled in Table 15 are concentrated in rest magmatic fluids.

TABLE 15

EXAMPLES OF COMPOSITION OF VOLCANIC GASES AND SUBLIMATES*

Halemaumau Gas (Weight Per Cent)	Katmai Gas (Volume Per Cent)	Fumarole Deposits of Stromboli
H_2O: 45.12	H_2O: 99.65	Fe+Mg in Fe_2O_3 and $MgFe_2O_4$
CO_2: 20.71	CO_2: 0.097	S, Se, and Te
CO: 0.69	CO: present	Li and Tl as sulfates
SO_2: 16.67	H_2+H_2S: 0.024	I
SO_3: 5.51	N_2+A: 0.041	P
N_2: 7.93	CH_4: 0.015	Zn, Sn, Pb, Bi, Cu
S_2: 2.47	H_2S: 0.029	B in boric acid
Cl_2: 0.54	HCl: 0.117	N
A: 0.30	HF: 0.032	As in realgar
H_2: 0.06		K, Rb, Cs in alums
		Na, in sulfates
		Ca, in glauberite

* Reproduced from Rankama and Sahama, 1950.

The composition of hydrothermal solutions of magmatic parentage has often been inferred from field observations under the uncertain presumption that the rocks in question were formed by late magmatic solutions. Quartz veins, pegmatites, ore veins, and even serpentine veins, etc., have been considered unquestionably of late magmatic origin. Therefore, the chemical composition of such rocks has been widely identified with the composition of rest magmatic solutions. We shall discuss this situation presently. For example, granitization in regional metamorphism is often believed to be caused by rest solutions from deeply buried melts. Since granitization is chiefly an alkali metasomatism, it has been concluded that rest solutions after magmas are particularly rich in sodium and potassium. Goldschmidt (1921) and Rosenquist (1942) assume that feldspar is hydrolyzed at the last stage of the magmatic solidification: Alkali feldspar + $H_2O \rightarrow$ muscovite + water glass. The water glass, which can be very concentrated without precipitation, streams out into the surroundings and re-creates alkali feldspar anew. Such a process is very unlikely, because the water-glass solution would be highly unstable in connection with most silicate rocks. By reaction, the fluid will form feldspars and micas immediately, thus being unable to advance to any extent through rock complexes (Eskola, 1939, p. 376). Some of the following reactions will take place:

$$\begin{aligned}&\underset{\text{hornblende}}{Ca_2(Mg,Fe)_3Al_4Si_6O_{22}(OH)_2} + \underset{\text{dissolved potash silicate}}{\tfrac{1}{2}K_2O + XSiO_2} + \tfrac{1}{2}H_2O \\ &\qquad \rightarrow \underset{\text{biotite}}{K(Mg,Fe)_3AlSi_3O_{10}(OH)_2} + \underset{\text{epidote}}{Ca_2Al_3Si_3O_{12}(OH)} + \underset{\text{quartz}}{XSiO_2}\,,\end{aligned} \tag{135}$$

$$\underset{\text{muscovite}}{KAl_2(Si_3Al)O_{10}(OH)_2} + K_2O + 6SiO_2 \rightarrow \underset{\text{potash feldspar}}{3KAlSi_3O_8} + H_2O\,, \tag{136}$$

$$\underset{\text{kyanite}}{1\tfrac{1}{2}Al_2SiO_5} + \tfrac{1}{2}K_2O + 2SiO_2 + H_2O \rightarrow \underset{\text{muscovite}}{KAl_3(Si_3Al)O_{10}(OH)_2}\,, \tag{137}$$

$$\text{hornblende} + Na_2O + SiO_2 \rightarrow \text{soda hornblende}\,, \tag{138}$$

$$\text{pyroxene} + Na_2O + SiO_2 \rightarrow \text{soda pyroxene}\,, \tag{139}$$

$$\begin{aligned}&\underset{\text{garnet}}{3(Mg,Fe)_3Al_2Si_3O_{12}} + 2K_2O + 3SiO_2 + 2H_2O \\ &\qquad \rightarrow \underset{\text{muscovite}}{KAl_2(Si_3Al)_{10}(OH)_2} + \underset{\text{biotite}}{3K(Mg,Fe)_3AlSi_3O_{10}(OH)_2}\,,\end{aligned} \tag{140}$$

$$\begin{aligned}&\underset{\text{pyroxene}}{3(Mg,Fe)SiO_3} + \underset{\text{garnet}}{(Mg,Fe)_3Al_2Si_3O_{12}} + K_2O + SiO_2 + 2H_2O \\ &\qquad \rightarrow \underset{\text{biotite}}{2K(Mg,Fe)_3AlSi_3O_{10}(OH)_2} + \underset{\text{quartz}}{SiO_2}\,.\end{aligned} \tag{141}$$

Now, in granitized gneisses and schists, sillimanite, kyanite, garnet, muscovite, and soda-poor pyroxenes and hornblendes are the common minerals. It is clear, therefore, that strong alkali-silicate solutions are but rare visitors to granitized areas; and field observations from such areas certainly give no evidence for assuming that late magmatic solutions are anything like concentrated alkali water glass.

On the other hand, typical soda minerals like nepheline, analcite, soda pyroxene, and soda hornblende are common minerals in the contact zone around undoubted magmatic bodies of the alkalic kindred (Oslo area, etc.). Solutions rich in soda may therefore be expelled when such magmas crystallize.

Some petrologists (Eskola, 1915) have suggested a magmatic source for the rather widespread magnesium metasomatism encountered in some deep-seated metamorphic complexes. However, to postulate that late magmatic solutions are rich in magnesium is contrary to all experimental knowledge of the physicochemistry of silicate systems. Magnesium concentrates in the earliest minerals to crystallize out of a magma.

The composition of pore fluids of nonmagmatic parentage is not better known. By recognizing the origin of such fluids, however, some conclusions as to their composition may be drawn. Let us consider a marine clay sediment undergoing regional crystallization. Besides the finely divided mineral grains, there are great quantities of water in voids as well as adsorbed tightly to the clay mineral surfaces and interfaces. In the supercapillary voids, the water is an ordinary solution, of a composition, perhaps, close to sea water. The adsorbed film has quite a different composition because some ions (like K^+) are strongly bound to the clay minerals. During the recrystallization, equilibrium is approached within the rock, so that a certain mineral assemblage will develop; likewise, the composition of the true solution and the adsorbed film will approach a mutual equilibrium. As temperature increases, the equilibrium will be displaced, and the composition of the solution will change. Ultimately, the solution will be gaseous or supercritical, and we shall have a situation rather similar to the situation reached when a silicate melt has just solidified: in the voids and cavities a hydrous solution exists which is in equilibrium with the silicate minerals of the rock. Although the fluid solutions created in these two different ways will hardly be identical in composition, some close similarities must be expected. In either case, the solutions must be relatively rich in those elements which show the weakest tendency to enter the lattices of the common minerals or to form specific mineral species. In either case the solid phases are not too different—namely, common silicate minerals[5]—and the major fluid solvent is in

5. Although there certainly are important differences among the several silicates with respect to chemical properties like, for example, type of bonding, etc. (see p. 288), we shall not forget that the silicates as a group of compounds have many chemical properties in common,

both cases water. On account of this, it is to be expected that the same elements which are "fugitive" and mobile in the late magmatic stage will also be relatively easily mobilized during progressive metamorphism (see further discussion, p. 268).

The fluid inclusions in metamorphic minerals may represent approximately the composition of pore fluids which originated in the way discussed here.

There is also a more indirect way of constructing the composition of a pore fluid or the adsorbed intergranular particles in metamorphic rocks. This method is based upon the assumption that a close equilibrium is approached between the intergranular phase and the solid minerals.

In most rocks there are minerals with such different crystal-chemical properties that almost any important rock-making element will be able to enter the lattice of one or more of the minerals present, although some elements are often able to constitute only a negligible fraction of certain minerals. Theoretically, the composition of the solid minerals should represent an indicator of the composition of the adsorbed particles in the intergranular, as well as of the composition of a fluid phase existing in free pores, provided that pressure and temperature are known.

As an example, the content of water in hydrous minerals in a rock (micas, hornblende, etc.) is determined by the water-vapor pressure in the rock, the total pressure acting on the mineral, and the temperature, as well as the particular composition of the hydrous mineral in question.[6] Furthermore, the fluorine-hydroxyl ratio in the micas or other hydrous minerals reflects the concentration ratio fluorine/water in the pore fluid and in the adsorbed interface films. Likewise, the ratio sodium/potassium in feldspars or hornblendes is controlled by the relative concentration of these elements in the pore fluid. The latter relations have been partially investigated by O'Neill (1948).

At the present stage of our knowledge about the crystal chemistry of minerals, we are unfortunately not able to construct the composition and state of a pore fluid which could be in equilibrium with the mineral assemblage of a rock. We hardly know enough to put some limitations to the speculation around the fluids which possibly can carry rock-building substances along to the sites of metasomatic growth.

The fact that common silicates, oxides as well as sulfide minerals, are

which are distinctly different from the properties of, for example, the metals as a group, or of the sulfides, or of the organic crystals.

6. Even if the water content in micas and hornblendes is considered to be stoichiometrically fixed, this does not mean that absolutely no change in water content is permitted. Actually, minor displacements from the stoichiometric ratio are usually possible in minerals, as discussed on p. 20.

practically insoluble in neutral water has been the main challenge to the pore-flow hypothesis. Much work has been done in the search for the proper fluids which could carry sufficient amounts of minerals in solution so that oceans of water would not be necessary to form any modest metasomatic mineral deposits (see, for example, Verhoogen, 1938; Garrels, 1944). With reference to ore minerals, rather concentrated alkali sulfide solutions are capable of dissolving most metal sulfides except the commonly distributed pyrite. Considering silicate minerals, alkaline solutions with sufficient excess of sodium and/or potassium can dissolve many of the common silicates. Hydrofluoric acid solutions are capable of carrying along silica and easily soluble fluorides. But there are other properties than power to dissolve minerals to be required from the fluids which are made responsible for metasomatic transfer. The solutions must be able to build up the very mineral assemblages and sequences which are observed in the metasomatic rocks. If anybody claims that a quartz vein is formed from consolidation of silica carried as silicium fluoride, he should demonstrate the finding of other crucial minerals which necessarily would form in such a fluorine-rich milieu, like fluorspar and fluorine-rich hornblendes or micas. Since such a paragenesis is not present in or around the majority of quartz veins, it is often supposed that easily soluble alkali silicates were carried in solution and that precipitation of silica took place for some reason or other. Then, again, one would expect that soda hornblendes or soda pyroxenes and pure alkali feldspars in connection with lack of muscovite and other aluminum-rich minerals (kyanite, garnet, etc.) would characterize the quartz veins and their closest environment together with the channelways for the solutions (p. 184). Around and within the numerous quartz veins, dikes, and segregations in regionally metamorphosed complexes, the mineral assemblages usually exclude the activity of such alkaline fluids. Indeed, the common distribution of muscovite, sillimanite, kyanite, cordierite, pyrope-almandine garnets, hypersthene, nonalkaline hornblendes, and pyroxenes and the lack of nepheline, analcite, and other typical alkaline minerals, within regionally metamorphic and metasomatosed rocks, testify that solutions with an appreciable excess of potash or soda are rare and cannot by any means be called upon as universal carriers of silicate metasomatism.

As a consequence, the natural solvent which should be able to carry appreciable concentrations of common silicate minerals remains undetected. One may naturally ask: Does such a solvent exist in the earth at all? The answer is not very promising.

The ore geologists are apparently not better off, according to the current discussions (Verhoogen, 1938; Garrels, 1944; and Brown, 1948). Here there are inconsistencies between the paragenetic sequence in sulfide ore, the zoning found along some metasomatic ore veins, and the relative solubilities of the several sulfides in hydrous solutions. There appears to be better harmony between the relative age of fixation of ore minerals in an ore body and the vapor pressure of the mineral (Kordes, 1935) or their hardness (Gilbert, 1924; Bandy, 1940). One may also wonder what reactions will take place between the rather concentrated alkali sulfide solutions necessary to dissolve metal sulfides, and common rock-making gangue minerals. How concentrated can an alkali sulfide solution be without rendering the gangue minerals, like chlorite, hornblendes, etc., unstable, thus putting a limit to the speculations? Garrels has concluded, after exhausting all available possibilities, that ore minerals are carried in colloid suspension (1944). Brown (1948) thinks that they move in a vaporous state.

Mineralized springs should, perhaps, give us information about the composition to be expected from hydrous solutions operating during metasomatism. Such spring and geyser water will certainly depend importantly upon the composition of the rocks through which it propagates. Table 16 gives some analyses of different springs and geysers.

GEOLOGIC FORCES CAUSING FLOW OF HYDROTHERMAL FLUIDS

To squeeze the hydrothermal fluids through rock complexes, we must call upon some geologic reasons for the establishment of the necessary fluid potential differences. Let us consider an interconnected fissure system penetrating a rock complex to some depth. It is easy to imagine that flow currents can originate because of unstable density variations due to temperature gradients, or to unstable pressure differences created by slow collapse of parts of the fissure system, or by the evaporation of a magma.

The slow collapse of a fissure system or a pore system, filled with fluids, is a reasonable cause for flow. This process would, in all likelihood, create an upward motion of the fluid, because the closing of fissures and general compaction of all free voids is more efficient, the greater the depth. The amount of solution originally contained in the free space in the crust is limited, and no long-lasting continuous upward flow can be expected. Such an upward flow would undoubtedly be of significance as a mineral carrier if the fluids were concentrated silicate or ore melts which would consolidate at some level during their motion upward. In the case of hydrous solutions, however, the process seems inefficient to displace significantly other rock-building material than water and components which constitute a major portion of the fluid, as, for example, carbon dioxide. Consider a complex of sediments which are slowly buried, depressed, and folded. Compaction of these rocks will take place extensively before real recrystallization and regional metamorphism start. Therefore, the major portion of the water content is squeezed out of the sediments before they reach the depth at which they may function as a source for metasomatism in the superincumbent beds. At this depth, not only is there hardly more than some 0.5–2 per cent of water left in the rock, but the structure is also

TABLE 16

EXAMPLES OF SPRING AND GEYSER WATERS*
(Composition in Per Cent of Total Solute; Salinity in Parts per Million)

Element and Compound	Chloride Water	Sulfate Water	Carbonate Water	Siliceous Geyser Water
Cl	55.83	6.56	0.03	20.91
Br	0.04	....	....	Traces
I	0.03	....	....	....
SO_4	3.12	57.42	1.31	1.31
CO_3	2.63	6.48	67.56	25.01
PO_4	....	....	Traces	Traces
Na	33.09	13.51	1.13	31.34
K	0.27	0.48	0.26	2.43
Li	....	Traces	....	0.15
Ca	3.72	8.33	2.54	0.17
Mg	1.13	7.03	25.03	0.17
Al	....	....	....	....
Al_2O_3	....	....	0.14	0.17
Fe^{2+}	....	Traces	....	0.13
Fe_2O_3	0.06	....	0.21	....
SiO_2	0.08	0.16	1.79	16.58
NO_3	....	0.02	....	....
NH_4	....	0.01	....	Traces
B_4O_7	....	....	Traces	1.34
AsO_4	....	....	....	0.29
Salinity:	10,589	5,518	7,829.5	1.336

* Reproduced from Clarke (1924). The chloride water is from a Cincinnati artesian well, Ohio, U.S.A.; the sulfate water from Cottage Well, Cheltenham, England; the carbonate water from a spring in Pine Creek Valley, British Columbia; and the siliceous water is from the Excelsior Geyser, Yellowstone Park, U.S.A.

so compact that bodily pore flow can scarcely take place. Disregarding the latter possibility, we may conclude that the remaining 0.5–2 per cent of water may be moved bodily upward. As decreasing pressure has only a limited effect upon the solubility of minerals in such solutions, one must believe that the decreasing temperature met with on the route upward is the main cause of precipitation. In assuming that 2 per cent of the water is literally squeezed out of the rocks at some depth, we find that less than some 0.02 per cent of silicates or ore minerals are thereby extracted from the rocks, because the solubility of minerals in rock-compatible solutions can, under only the most favorable conditions, perhaps reach some 1 per cent (p. 196). Since we are considering flow through the pore system, it is likely that the upward-flowing pore solution passes through a cross-section of approximately the same size as the horizontal area of the rocks from which it is squeezed. We shall assume that the solution completely loses its content of solute when passing through superincumbent rock volumes of the same size as the rocks from which it is squeezed. Then but 0.02 per cent of the constituents of the altered rocks have been gained from the passing solutions.

In this calculation the solubility of the minerals is overestimated, and it is erroneously assumed that the discharge of solute from the solution on its way upward is complete. Actually, only a small fraction of the solutes can precipitate on the route upward. The estimated alteration is consequently unreasonably great, and the significance of such a flow as a mineral transporter becomes even less than estimated. Very likely, though, such upward flow creates considerable hydrolization and carbonatization.

The situation may be somewhat different. One could presume that the pore solution forced out of the compacting and recrystallizing rocks in the depth was led laterally through the closing intergranular pore network toward more or less vertical fissures and breccia zones, these being the main paths for the upward motion of the solution. In that way the elements "leached" out of a great volume of compacting rocks could concentrate in the building-up of fracture fillings and minor replacement bodies of but a small total volume at higher levels. The process would then become more reasonable, as far as the ratio of the amount of built-up metasomatites to the amount of leached-out substance is concerned, but in no case can we avoid the extremely great ratio between volume of solution and volume of newly built-up vein rock. (Attempts to calculate this ratio are found in the next chapter.)

If one, in spite of reasonable assumptions (p. 179), could suppose that the pore system in rock complexes remained open for a long period of time at some depth, the convection flow of pore fluid due to temperature gradients might have been of considerable significance. The higher temperatures at deeper levels would decrease the density of the pore fluid as compared with the superincumbent column of fluid, and an upward-directed flow should occur. Provided that the pore system remains open, this convection current may continue to move upward through a certain column for a long period of time, inasmuch as new pore solutions will be supplied from downward currents in the surroundings. As the temperature is decreasing upward, the conditions for precipitation are also favorable, and metasomatic rock bodies of considerable volume may grow from these solutions.

In an open fissure extending downward in the crust or in an interconnected pore system filled with watery solutions, the hydrostatic pressure at each level will be high enough to keep the solution in a condensed state if the temperature corresponds to the geothermal gradient. At 10 km. depth, the temperature will be some 300° C., the hydrostatic pressure in the free openings will be some 800–900 atm., and the density of the solution about 0.7–0.8 g/cc (Scott, 1948).

Considering a certain level, for example, at 10 km. depth, in a column of water in an open vertical fissure, we may reason as follows: The imaginary isothermal horizontal plane *a–a′* (Fig. 97) is unstable because the density of the fluid underneath that plane is somewhat less than above.

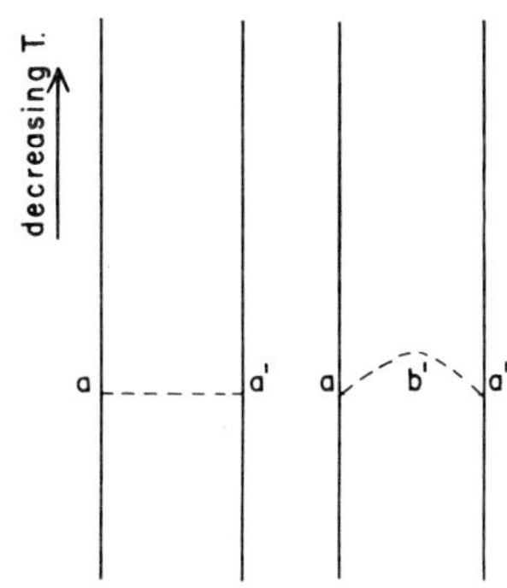

FIG. 97.—The effect of a vertical temperature gradient upon the fluid in a vertical capillary opening (see explanation in text).

(The same is certainly true at any level in the column.) Hence there will be a tendency for the subjacent fluid to move upward, as indicated in Figure 97, *b′*. Because of the somewhat decreased pressure on the upward-moving liquid portion, it will expand slightly, and the temperature within that portion must decrease adiabatically. If it happens that the adiabatically decreased temperature is less than the temperature given by the geothermal gradient at this very level, then no further flow upward will take place. If, on the other hand, the adiabatically decreased temperature in the portion *b′* is higher than the geothermal temperature at this level, upflow will continue. In case the fissure is of some thickness and length, these convection currents may be rather powerful and could provide a possible way of transportation of dissolved material from deeper portions of the crust up to higher levels.

It is a possibility that the upward-moving fluid remains hot enough actually to boil at the higher levels where pressure is low. This situation may cause a subsurface geyser action like the process in a back-boiling apparatus. Much substance could then be precipitated in the upper parts of this subgeyser, as well as a considerable leaching-out of the wall rock in the lowest part.

Although these convection currents of hydrothermal solutions seem to be very reasonable in open fissures of some thickness, it is rather doubtful whether flow in a capillary pore system can be caused by the geothermal gradient. One of the reasons for the flow was that the motion was rapid enough to prevent the temperature of the upward- and downward-flowing fluids to be equalized by heat conduction, but it seems as if the flow permeability of the rocks has to be unreasonably great to prevent such a homogenization of the temperature throughout the pore fluid at a given level.

Another possible way of establishing a mechanical potential difference necessary to move the active fluid through pores and fissures in rocks is found in the "boiling" of magmas (p. 171). Late in the sequence of crystallization of a magma, the rest melt becomes so enriched in fugitive substances that a tremendous pressure can be built up in the magma chamber, and, if space is available, a gaseous phase will develop. What happens with this magmatic gaseous or supercritical phase depends much upon the particular situation. If the melt crystallizes slowly, the gas pressure increases gradually, and, since common rocks are permeable to a gaseous diffusion (note CO_2 and H_2O), it may happen that the gases migrate slowly through the walls of the magma chamber. If there are narrow open cracks and fissures through the walls of the chamber, the liberated fluid phase will stream out "bodily." (The "bodily" flow of a gas means that the flow is driven by the total gas-pressure gradient without discrimination among the several types of molecules in the gas, so that the relative concentrations of the several components remain practically unchanged along the flow.)

A rapid crystallization in a chamber with tight walls may cause an explosion due to the rapidly developing gas pressure, which may reach values sufficient to overcome both the hydrostatic rock pressure and the strength of the chamber walls.

We do not find it likely that the hydrothermal fluid, or the rock "moisture," which circulates during rock metamorphism of hydrothermal character is in direct communication with the true ground-water column. Ground water is confined to the very uppermost part of the crust, rarely extending deeper than some 1,000 meters below the surface. Not only are compact rocks like gneisses and granites remarkably dry in lower levels in deep mines, but even porous sandstones fail to yield water from too great depths (>1,000 meters) (Lindgren, 1933). Therefore, the fluids in rocks undergoing metamorphism are not moved by the common agent for ground-water flow, namely, the lateral change of ground-water level and the consequent change of fluid potential along horizontal planes down in the water column.

THE PORE FLUID AS A MINERAL CARRIER

In order that a flowing solution shall act as an efficient carrier of rock-making minerals, it is necessary not only that the solution contain dissolved minerals but also that it precipitate minerals here and there along its route. Such a discharge of minerals may be due to a simple precipitation of the dissolved minerals or to chemical reactions between the propagating solution and its environments (environments = host rock, solutions of different origin and character, etc.).

Simple precipitation takes place when the solution becomes supersaturated, that is, when the partial molal free energy in the solution of the mineral-building components becomes greater than the molal free energy of the crystalline minerals (p. 40). The reason for these free-energy changes and the corresponding supersaturation and precipitation must be found in changes in temperature and/or in pressures as the solution advances through its channel. Upon reaction with the wall rock, the solution may absorb and discharge solutes concurrently. The controlling factor here is also the free-energy values: the reaction minerals must have less molal free energy than the solute in the fluid solution and the original wall-rock minerals combined. A special case will be one in which all the minerals involved are stoichiometric compounds with no ability to form solid solutions. Only then will the commonly cited mass-action law by Guldberg and Waage control the main paths of the reaction process (p. 41). When mixed crystals are formed or if such crystals react with the solutions, then the situation is more flexible than predicted by the mass-action law *sensu stricto*, inasmuch as small continuous variations of the concentrations in the fluid will reflect themselves as smooth continuous changes of the concentrations in the mixed crystals. As most minerals are solid solutions, the special mass-action law has but restricted application in metasomatism, although this law has usually been supposed to be the most important principle governing replacement (Goldschmidt, 1921; Eskola, 1939; Turner, 1948).

Although supersaturation is a necessary requirement for the flowing fluid to be an effective mineral builder, it is not sufficient in itself. To form the first minute nucleus of the stable mineral is often difficult, and it may happen that the fluid remains supersaturated for some length of time, until it reaches circumstances under which the nuclei form easily.

The nonequilibrium between a pore solution and host rocks may be due to several reasons: (1) changes in temperature, (2) changes in pressure, (3) changes in composition of the host rocks (and their pore fluids), and (4) the solution may primarily be of such a nature that it is supersaturated or in disequilibrium with the rocks it penetrates.

The ease with which nuclei are formed from the supersaturated system (the solution and wall rock combined) depends upon less-known conditions. Most likely the character of the surfaces and interfaces of the wall-rock minerals is very important, for the new minerals often grow on the wall-rock minerals (see metamorphic differentiation, pp. 220–25). The structure of the host rock is also very significant.

When we now proceed to discuss the effective transfer of a flowing solution, we shall recall that there is no relationship between the conditions (the fluid potential) which direct the motion of the fluid and the conditions which govern the precipitation of the solutes or the reaction between the fluid and the host rocks.

If flowing pore solutions are the chief agency of transportation in metasomatism, then the formation of metasomatic minerals and rocks is but a chemical by-product of the mechanical flowage of huge quantities of dilute fluids, the sole aim of this flowage being to iron out mechanical potential differences throughout the pore system. These circumstances are the direct opposite of those of diffusional transfer, at which the transportation and the "precipitation" (consolidation) and so far also the "dissolution" (activation) are departments controlled by the very same power.

Our knowledge of the bodily motion of a fluid shows that all such motion must come to a complete stop when the fluid potential (the mechanical potential) is the same throughout. Therefore, if it is assumed that all long-distance transportation under metasomatism is caused by pore flow, we have to conclude that no metasomatic transfer can take place as long as mechanical stability exists throughout the pore system and the fluid-filled fissures.

Let us consider a fluid flowing through an open fissure. Many vein bodies of ore minerals, quartz, calcite, and pegmatites are supposed to have precipitated from such fluids. Unless our fluid streams out from a hot magma into cracks in the colder country rocks, we have little reason to believe that the fluid is appreciably supersaturated. There is, on the contrary, good reason to assume that the fluid is close to the saturation point at the inlet of the flow. Along the fissure the flow potential must decrease, which means that the pressure decreases as the fluid proceeds, if the flow is in the horizontal direction. Nothing specific can be said about the temperature along the flow nor of the composition of the fluid, besides the fact that the temperature is below the eutecticum of the silicate magma and that the solution is a dilute hydrous fluid which is close to the saturation point. The only necessary requirement in our situation is that the fluid potential or pressure decrease along the horizontal flow, and it becomes a question whether this negative pressure gradient will cause the fluid to precipitate minerals or not. In the case of liquid watery solutions, the effect of pressure upon the solubility is small and, with but few exceptions, such that decreasing pressure increases the solubility (Gibson, 1938). In most cases, therefore, decreasing pressure alone will not cause minerals to precipitate along the flow of a liquid hydrothermal solution. Somewhat different will be the situation if the watery solution contains calcite and an excess of dissolved carbon dioxide which may disappear slowly when pressure decreases. Owing to the fact that an increasing content of CO_2 in solution raises the solubility of $CaCO_3$, calcite may precipitate along the current. When the fluid is gaseous or supercritical, the conditions may be more variable. Usually, decreasing pressure on a gas will move it farther away from the saturation "point," and an isothermally flowing gaseous fluid should show no tendency to precipitate minerals. But the hydrous vapors in rocks may behave more intricately than this, because a watery vapor under pressure (supercritical conditions) behaves like a solvent, being able to dissolve appreciable amounts of silicates. Increasing pressure increases the solubility of the silicate in such a supercritical gas because the density of the gas increases and therefore the power of reaction between water molecules and "dissolved" silicate is more effective. If the circumstances are such, precipitation will automatically take place along a horizontal or upward-directed isothermal flow.

Very informative is the solubility of silica in liquid and gaseous water, which is shown in Figure 98. The data may be used to calculate the efficiency of supercritical water as a carrier of silica in an isothermal current. We shall have to make the theoretical example somewhat artificial and very schematic, in order to be able to employ the few data available. Suppose, for example, that water is flowing upward through a fissure along which the temperature is constant at 400° C. The pressure in the fluid at a certain depth of the fissure is 500 kg/cm^2. According to experiments (Scott, 1948), the density of water is then close to 0.6 g/cc. At a higher level the pressure is 400 kg/cm^2, and the corresponding density is close to 0.5 g/cc. At hydrostatic equilibrium the distance between the two levels is approximately 2,000 meters. As we shall assume that the flow proceeds upward, we must take a smaller vertical distance between the two levels at which pressure is 500 kg/cm^2 and 400 kg/cm^2, respectively. Let us take 1,000 meters instead of 2,000, thus assuming a driving vertical pressure gradient of 50 kg/cm² per 1,000 meters. The concentration of silica in the solution at the 500 kg/cm² pressure level is some 1,200 parts per million; at the 400 kg/cm² level the concentration is some 1,000 parts per million. Hence, owing to the pressure drop, some 200 parts of silica have been precipitated per million parts of solution which have passed through the fissure. It will be seen from Figure 98 that decreasing temperature along the flow will enable less substance to precipitate, because the solubility in the supercritical gas decreases with increasing temperature when pressure is below 600 bars.

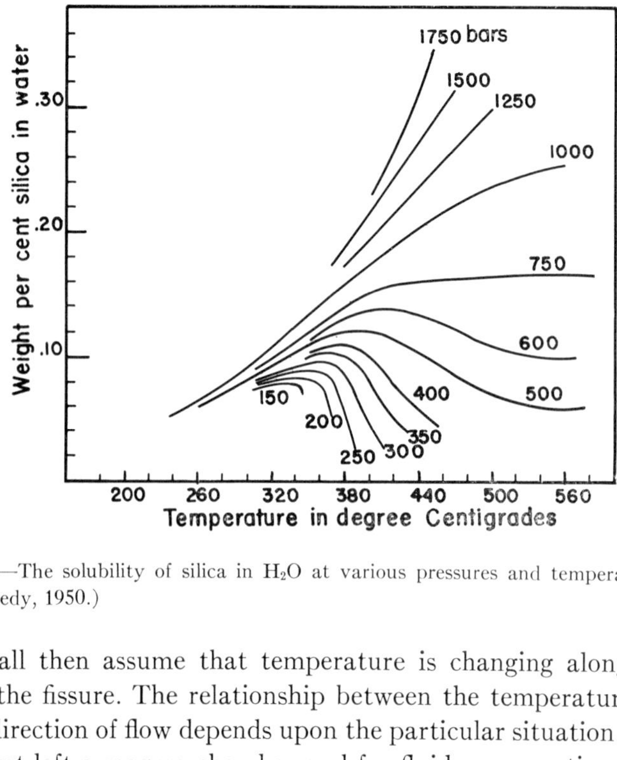

FIG. 98.—The solubility of silica in H_2O at various pressures and temperatures. (After G. C. Kennedy, 1950.)

We shall then assume that temperature is changing along the flow through the fissure. The relationship between the temperature gradient and the direction of flow depends upon the particular situation. For fluids having just left a magma chamber and for fluids propagating upward in the gravitational field, temperature will decrease along the flow. In all other cases the probability for an increase of temperature along the flow will be as great as the probability of a decrease of temperature in the same direction. There is also a great probability that there will be no regular change of temperature along the channel. On the other hand, the solubilities of solids in hydrous solutions usually, but not always, increase with increasing temperature, and one should assume that decreasing temperature along the flow will cause precipitation; only in limited cases should increasing temperature cause precipitation (see above). Let us consider a hydrous solution advancing up through a fissure, and assume that it precipitates minerals because of the geothermal gradient of some 0.03°/meter. Taking the precipitation of calcite from pure watery solution as an example, we may accept that at 75° C. the solubility of calcite is 0.0018 g $CaCO_3$/100 g H_2O, and, at 25°, 0.0014 g $CaCO_3$/100 g H_2O (*Handbook of Physics and Chemistry*, 1947). The depth of 25° is close to 100 meters, and, some 1,600 meters below, the temperature is 75° C. By the upward-flowing solution 0.0004 g of calcite will precipitate along the 1,600-meter-deep fissure for each 100 g of H_2O which passes through. That is, to build up 4 tons of calcite as a thin layer along the fissure, 10^6 tons of water are needed.[7]

As another example, silica may again be chosen. We shall now consider a temperature of 300° at a certain depth and assume that the precipitation takes place because of decreasing temperature in the upward-moving solution. From Figure 98, we can calculate that some 0.8 g of SiO_2 will be precipitated along the route for each kilogram of solution. That means that some 1,200 liters of solution are needed to precipitate 1 kg. of silica.

Reactions between advancing solutions and the wall of the channelway may give rise to precipitation because of chemical incompatibility between the solution and the wall rocks. Such a process may occasionally make the major portion of the solute precipitate out of the solution in the course of but a short column of flow. In this particular case the resulting vein must change suddenly in composition as it extends into the rock (or zone) which causes the precipitation.

Independent of the cause of the precipitation of minerals from a solution advancing through a fissure, the resulting dike or vein must be expected to achieve some properties which would definitely indicate the particular manner of formation. These conditions will be discussed presently.

When minerals precipitate from fluid solutions of some complex nature, the partition functions are generally such that the concentration of the remaining solution will gradually change. As a consequence, the types and compositions of the minerals which precipitate from the solution must change as the process proceeds. Therefore, one must logically require that dikes or veins formed by precipitation from a solution advancing along a fissure change in bulk composition and mineralogical character along the dike. Moreover, one must necessarily assume that such changes along dikes follow definite laws and hence will be repeated among dikes or veins of similar composition all over the world, provided that they are formed in the manner explained above.

7. This calculation is rather artificial and uncertain because of the great effect of the carbon dioxide pressure upon the solubility of calcite (see the recent investigations by Miller, 1952).

Actually, hydrothermal dikes, quartz dikes, pegmatites, alpine veins, etc., are commonly zones so that the composition changes from margin to center, but a regular zone along the dikes is not apparent except perhaps in some ore-vein systems, which then usually exhibit vertical zoning.

As decreasing temperature is supposedly the main reason for precipitation from the flowing vein fluid, one should also commonly expect to find that the mineral assemblages correspond to decreasing temperature along a certain direction parallel to the dike sheet. That is, the mineral facies (p. 136) should change regularly in this direction. Such temperature gradients have not been demonstrated. On the contrary, it has been shown that many silicate mineral veins exhibit the same facies along their visible extension (Ramberg, 1949). That fraction of the solute which precipitates along a certain column of the flow decreases with decreasing temperature difference between inlet and outlet of the column. It is tempting to conclude, then, that, if the presumed temperature gradient is too small to be detected by the mineral assemblages after the facies principle, then the fraction of solute which could be precipitated must also be negligible. Hence the amount of fluid required to build the dike may be impossibly great. One should also require that veins and dikes formed in this way show a dilation geometry (Fig. 123, p. 255) and that they be extensive in depth and length and not only represent a confined fissure filling. Likewise, the zonal construction of the dike should be in harmony with a one-sided transportation of all the mineral constituents parallel to the flow lines of the fluid. This latter statement should apply to each mineral grain (unsymmetric growth) as well as to the whole mineral assemblage (unsymmetric zoning along a dike passing several incompatible rocks).

Let us now treat briefly the effective transfer of rock-building matter by a pore fluid advancing through the bulk body of rocks. In so doing, we must disregard the fact that such a flow is not likely at all because rocks formed at some depth are "flow insulators" (p. 179). We may distinguish between several classes of theoretical possibilities: (1) the fluid may have existed so long in a rock body, or passed so slowly through the body, that the fluid is just saturated with reference to the rock minerals; (2) the fluid may be mainly supersaturated when it enters the rock; or (3) the fluid may be mainly undersaturated. Cases 2 and 3 are reasonably to be expected if the fluid, upon entering a rock, experiences a rapid change in temperature or pressure at the same time or a rapid change in chemical surroundings. In particular, a fluid advancing from a "boiling" magma into the porous wall rock is in strong disequilibrium under the new circumstances. But also a fluid which crosses the border between two chemically incompatible rocks will be in disequilibrium with the new surroundings.

The statement "mainly supersaturated" (2, above) means that the interaction between the pore fluid and rock minerals will tend to precipitate as new minerals more substance than is dissolved of old minerals, the net result being a loss of solute from the fluid, although the kind of solute may change. "Mainly undersaturated" (3, above) means the opposite conditions.

One must assume that the propagation of a fluid through common rocks is a very slow process and that the adjustment between fluid and host minerals takes place rather thoroughly before the fluid can proceed over considerable distances. When we consider the oozing of a supersaturated fluid into a rock, then we find that the discharge of solute should occur just around the inlet of the flow, thereby building a reaction zone parallel to the rock boundary and extending some distance into the altered rock. When the fluid has passed this layer, it has become in equilibrium with the old minerals and can proceed through the rock without reaction unless the temperature changes along the route or the rock is heterogeneous. If the flow continues through the reaction layer, this layer will thicken and gradually eat its way into the original rock. During the interaction between the fluid and the old minerals in the reaction layer, the fluid may most likely dissolve some of the old mineral constituents, which, in turn, could reprecipitate in front of the reaction zone, as shown in Figure 99.

Suppose, for example, that the solution contains dissolved Na^+ ions and the old rock is composed of orthoclase, muscovite, and quartz. We shall assume that the Na^+ concentration is sufficient to transform the orthoclase into albite:

$$Na^+ + \underset{\text{potash feldspar}}{KAlSi_3O_8} \rightarrow \underset{\text{albite}}{NaAlSi_3O_8} + K^+ , \tag{142}$$

thereby dissolving K^+. This increased concentration of K^+ may be sufficient to transform muscovite into potash feldspar:

$$2K^+ + \underset{\text{muscovite}}{KAl_3Si_3O_{10}(OH)_2} + 6SiO_2 \rightarrow \underset{\text{potash feldspar}}{3KAlSi_3O_8} + 2H^+ . \tag{143}$$

This microclinization may take place simultaneously with the albitization, but it may also take place somewhat later, thus building a potash feldspar-rich zone in front of the albitized reaction zone (Fig. 100). It may well be asked whether or not under these reactions the presumed pore openings will soon close in the reaction layer and the continuation of the flow be made impossible.

The likelihood that the solution will be undersaturated when entering a rock is as great as its being supersaturated. An undersaturated fluid will dissolve minerals around the inlet of the current, and an open and porous reaction zone is to be expected. If the pore solution has migrated for some time through a homogeneous rock, it will be saturated and in close equilibrium with the minerals. Such solutions will, therefore, neither precipitate nor dissolve minerals unless the temperature changes along the route. The dissolution process is common in limestones in the crust close

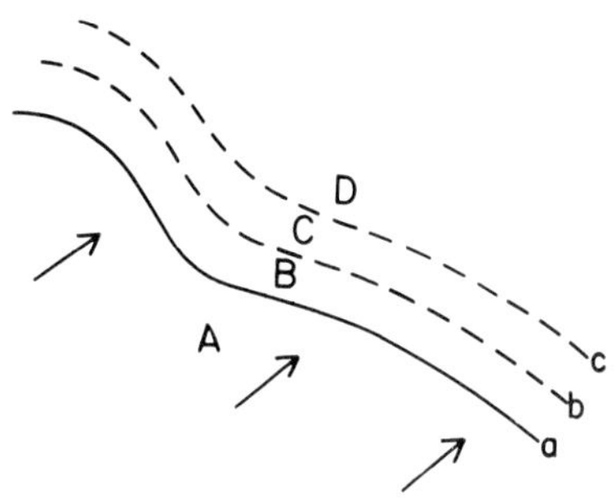

FIG. 99.—Schematic picture of formation of zones when a pore fluid crosses a border between two rocks, *A* and *D*. The arrows show the direction of flow. *B* and *C* are two contrasted zones formed by interaction between the fluid and rock *D* (see explanation in text).

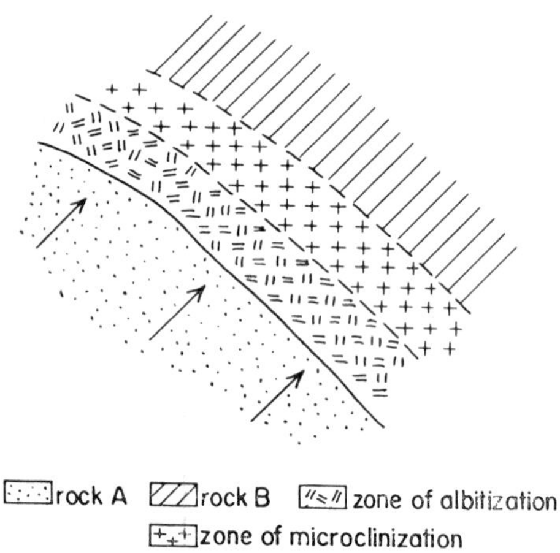

FIG. 100.—Zones of microclinization and of albitization formed by a pore fluid flowing along the arrows.

to the earth's surface, where moving pore solutions constitute an important part of the geologic processes. Often huge dissolved caves are encountered in marbles and limestones formed by the action of circulating ground water. This is in striking contrast to the compactness and lack of free space characterizing regional metamorphosed complexes.

One may tentatively assume that the lack of dissolution cavities in the latter geologic units is not due to the lack of extensive and chemically active pore flow but because the process of leaching out is superimposed by a slow plastic collapse of the cavities before they reach capillary or supercapillary size. If that is the explanation, it seems inconsequent to maintain the concept of flowing pore solutions at all, because any capillary or supercapillary channel for such flow would collapse as soon as it formed. Again we reach the conclusion that rocks are flow insulators under regional metamorphism.

In the case of metasomatism by an advancing interstitial pore fluid, all dissolved elements must move in the same direction along the flow lines. This requirement puts a serious limit to the importance of pore fluid flow.

We have seen throughout this section that the hypothesis of a flowing pore fluid as the chief carrier of rock-making minerals is not very convincing when questioned in some detail.

ROLE OF DIFFUSION IN METASOMATISM

Different kinds of diffusional transfer constitute a crucial part of metasomatism. Within the uppermost portions of the earth's crust, where open fissures are plentiful and some rocks remain porous for an appreciable length of time, circulating fluid solutions may perhaps be the most important long-distance carriers of gained and lost elements during metasomatism. Also here, however, the short-distance communication between the flowing carrier and the surrounding minerals, as well as the even distribution of the substances throughout the compact portions of the rocks, is impossible to understand unless rather powerful diffusion is accepted. Open channels are scarce in deeper portions of the crust and porous rocks practically nonexistent. Yet at these depths of regional metamorphism the evidences for metasomatism and replacement are most striking. To account for these phenomena, one is forced to accept long-distance diffusion through rocks as an important petrogenetic process. Several petrologists have contributed to the illumination of the phenomenon of long-distance diffusion through rocks. C. E. Wegmann is the pioneer (1935), Perrin and Rouboult (1949) are perhaps going much too far with their belief in large-scale lattice diffusion; D. L. Reynolds has, in a series of papers, showed how elegantly petrological problems may be explained by diffusion (1946–47). The present writer has attempted to explain the driving potentials of geologic diffusion and has tried to evaluate the importance of such processes for petrology.

THE LAWS OF DIFFUSION IN ROCKS

Diffusion through systems of the complexity of silicate or ore rocks is not a simple matter comprehended in a few equations. In contrast to the bodily flow, diffusion is a statistic process, entirely depending upon the fact that macrobodies of matter are constituted of a large population of discrete particles, among which a certain amount of kinetic energy is distributed at a given temperature. Under certain circumstances the probability for motion of a particle is greater in one direction than in the opposite, and hence a net diffusional transfer of matter extends in that direction if a great population of particles is considered. The laws of diffusion, in ideal gaseous, liquid, or crystalline mixtures, are relatively simple. The well-known first and second Fick's laws hold for stationary and nonstationary diffusion currents, respectively:

$$\frac{dm}{dt} = -A \cdot D \cdot \frac{dc}{dx}, \qquad (144)$$

$$\frac{dc}{dt} = D \cdot \frac{d^2 c}{dx^2}, \qquad (145)$$

where dm/dt is a quantity of matter transferred in a unit of time through cross-section A when the concentration gradient is dc/dx, and D is the diffusion constant (eq. [144]). Here it is supposed that dc/dx remains constant within the time interval considered and along the entire length of the current. That is, the current is stationary. The more common rule, however, is that the very transfer of dissolved particles (or gas molecules in a gaseous mixture) from high to low concentration will make the concentration gradient decrease. The driving "force" exhausts itself. Under these circumstances, it becomes important to know how rapidly the concentration changes with time at a certain fixed cross-section in the diffusion current. Equation (145) shows how this change of concentration with time, dc/dt, depends upon D and d^2c/dx^2.

There is another important relation to consider when discussing diffusion, the so-called "mean-square displacement":

$$x^2 = 2\,Dt\,, \qquad (146)$$

where x is the average length of the path of the diffusing particles through time t. The importance of this equation may be shown by the following theoretical experiment (Fig. 101). A is a container with saturated NaCl solution, B is pure water, a–a' is a semipermeable wall through which Na^+ and Cl^- can pass, but not water. After a time t, the concentration in the container B is as shown by curve m; after a time t_2, the concentration is given by curve n; etc. Here x is the distance from the wall a–a' to that cross-section in the current at which the concentration at any time is approximately $c_0/2$, when c_0 is the constant concentration at $x = 0$. According to equation (146), x is hence a very convenient estimate of the importance of diffusion in geological processes. Let us assume, for example, that D is some 5 cm^2/day, which is a reasonable value for diffusion of several ions in watery solutions at some 75°–100° C. In 1 million years t is $360 \cdot 10^6$ days, and x is $\sqrt{36 \cdot 10^8} = 6 \cdot 10^4$ cm. = 600 meters; in 100 million years, x = 6,000 meters; etc. Several other estimates of x under different conditions are found on page 80.

The situation in homogeneous, ideal solutions is reasonably simple, and the reason why the diffusion equations here are expressible as direct functions of the concentration gradient is that the fundamental driving energy of diffusion, the partial molal free-energy gradient, bears a simple relation to the concentration (eq. [17], p. 18). When heterogeneous rocks are concerned, the free-energy differences are only partly controlled by concentration differences in space of the different movable particles in a rock.

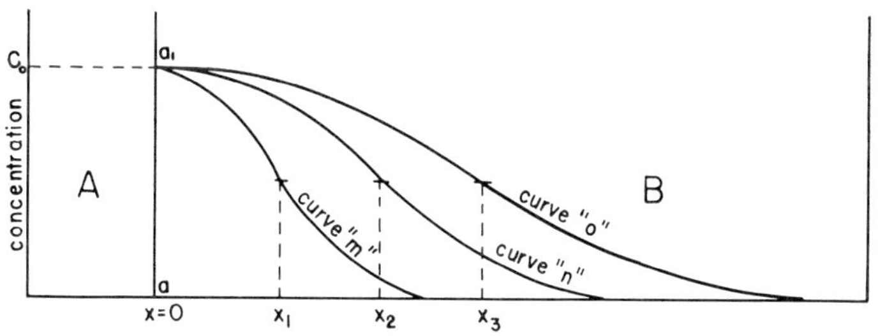

FIG. 101.—Illustration of diffusion through a solution. A is a large reservoir in which the concentration of solute is constant at C_0 at all time. B is a container of solvent which is pure at time $t = 0$ at start of diffusion. a_1–a is a wall which is permeable for solute. Curve m gives the concentration throughout container B at time t_1, after start of run; curve n gives concentration at time t_2; and curve o at time t_3. X_1, X_2, and X_3 give the distances from a_1–a to a vertical section at which the concentration of solute is $C_0/2$ at time t_1, t_2, and t_3, respectively. The mean distances of diffusion, X_1, X_2, and X_3 are given by the following equations: $X_1 = \sqrt{2Dt_1}$; $X_2 = \sqrt{2Dt_2}$; $X_3 = \sqrt{2Dt_3}$, where D is the diffusion coefficient.

Type of mineral lattices, pressure, temperature, position in the gravitational field, chemical milieu, and size of the phases, all together, significantly control the magnitude of the free energies of the several constituents. Fick's laws are thus of but limited value for our consideration. On the other hand, it appears practically impossible to establish a simple equation which would express the process of diffusion of a constituent through even the most simple rock. We shall therefore need to be satisfied with the qualitative statement that isothermal diffusion of a certain kind of particle (atom, ion, molecule) through the several paths in a rock always extends down the partial free-energy gradient of the particle type. Only when thermal gradients or strong electrical potential differences exist in the crust is the driving effect of the free-energy gradients blurred.

It may be said that activation of minerals at places of high free energy, diffusion of the constituents to places of low free energy, and consolidation there constitute the principal reasons for metasomatic rearrangement in the earth's crust.

In the pore fluid flow, a fluid potential could be assigned to each type of individual fluid phases, and each fluid would move along the negative gradient of its corresponding potential. This means that the pore fluid and all its contents will move in one direction, because only in very restricted cases would more than one fluid phase exist in a rock. In the case of diffusion, one should assign a potential—the partial molal free energy, or chemical potential—not to each phase but to each type of individual particle (ion, atom, molecule) within the rock or hydrothermal fluid. In a rock there are as many chemical potentials as there are individual types of constituents, and, since these potentials may vary individually in space, it follows that the diffusion current of the several movable constituents may extend in widely diverging directions. In that respect the diffusion is much more flexible than the bodily flow. For example, potassium in one form or another may migrate in one direction, while sodium moves in the opposite direction and perhaps magnesium or any other constituent moves in a direction inclined to those of potassium and sodium.

The rocks, with their different possible paths for diffusion, are to be regarded as chemical conductors offering different resistances to the several particles. The resistance depends upon the mineralogical character of the rock, the pressure, temperature, structure, presence of catalysts in the intergranular network, and certainly the character of the diffusing particle. The power of diffusion through unit cross-section is determined by the diffusion coefficient, the magnitude of the driving free-energy gradient, and the concentration of the diffusing particle.

Theoretically, a rock will never cease to be a diffusion conductor for some types of particles except at low temperatures. The mineral lattices may certainly often yield practically infinitely great resistance to diffusion, but the intergranular network can hardly stop the passage of atoms, ions, or molecules unless the molecules are extremely large or the adsorption energy is improbably high. It is certainly true that, if there is the slightest possibility of passage of flowing solution, there can be no doubt of conduction of individual particles of atomistic or molecular size. Even when the pore network has diminished to subcapillary dimensions, thus rendering the rock a flow insulator, there is still space enough to let the adsorbed surface particles pass through.

What determines the diffusion coefficient of a rock? For one thing, it is a question of free space for the moving particles. We have already considered parts of that question (pp. 76–88), and we shall point out here the fact that strain in rocks concurrently opens and closes minute cracks of limited extension in the mineral grains as well as along their interfaces. Such cracks will evidently increase the ease of migration, although they are of limited extension and not necessarily interconnected. This is particularly true if we consider a type of particle which is able to migrate not only along the mineral surfaces but in some lattices as well. The rather small and univalent sodium ion is able to diffuse through some mineral lattices very readily, as shown by the ease at which unmixing of alkali feldspar takes place in experiments and by experimental diffusion of sodium through quartz. It would be unlikely if sodium does not possess a rather great mobility along most mineral surfaces (p. 87). Under these circumstances, increasing concentration of minor cracks accelerates the propagation of the sodium ion down its free-energy gradient, because sodium will migrate rapidly along the cracks, and the slow passage from one crack to another will play a less dominant role, the more densely the cracks are spaced. Also for those particles which can migrate only along the surfaces (interfaces) frequent formations of minute cracks will be significant, because, even if some cracks are confined in mineral lattices, most of them will cut across the grains, thus creating a new intergranular channel. As most of these minute cracks are of subcapillary size (p. 207), they will aid pore flow only insignificantly. If the question of free space was the most important in determining ionic (atomic, molecular) mobility in rocks, one would expect small ions to be more mobile than large ions and gaseous molecules like H_2O and CO_2. The size of ions has, in fact, been considered by some authors the very controlling factor of mobility (Lapadu-Hargues, 1945–49; Reynolds, 1946–47; Rouboult and Perrin, 1949). Such a simplification is unsupported by field observations and by crystal-chemical knowledge. Crystal chemistry and colloid science claim that mobility is significantly dependent upon the bonding energy with which a particle is attached to its site. The less this energy, the greater the mobility, other things being equal. This effect will be the most important as far as intergranular and mosaic diffusion is concerned. It is unlikely that, for example, the small Si^{4+} ion, which is so strongly bound within the O-tetrahedra in silicates, will be mobile enough to create a long-distance lattice diffusion. It seems more reasonable that the relatively great mobility of silica in metamorphic rocks is to be explained by intergranular diffusion, in which some adsorbed hydrated silicon ion is the moving particle. The solubility of silica in supercritical water vapor (p. 196) indicates a considerable affinity between Si and H_2O, which reasonably should also be effective in the adsorbed water-rich intergranular "vapor." We have hereby introduced the effect of catalysts on the diffusion coefficient of particles through rocks. The intergranular adsorbed phase will depend, in composition and character, upon the presence of H_2O, CO_2, F, Cl, S, B, etc., and other substances which tend to be concentrated in the intergranular adsorbed phase rather than in the common minerals. These substances generally have relatively high vapor pressure and are weakly bonded in or on the silicate lattices. Considerable differences in composition of the intergranular film may exist among rocks of much the same mineralogical and bulk chemical composition because the intergranular phase is quantitatively negligible as compared with the mineral aggregate. It is thus understandable that apparently similar rocks may, under the same P,T-conditions, exhibit very different diffusion coefficient for a given element (or its compounds).

That increasing temperature increases the mobility of particles along the several pathways in rocks is obvious from our discussion on page 80. One of the interesting facts in this connection is that the increase of the concentration of movable particles and hence the diffusion coefficient increase exponentially with temperature according to the following law (in simple cases):

$$D_T = D_0 \cdot e^{-Q/RT}, \qquad (147)$$

where D_T is diffusion coefficient at T° K., D_0 is the diffusion coefficient at T_0° K., and Q is the molal activation energy of diffusion. As Buerger and others have indicated, this function often shows that it is, practically speaking, a minimum temperature at which diffusion takes place at all. Above this temperature limit, the mobility becomes rather suddenly great. As D_0 and Q are different for the several diffusing particles, the minimum temperature will also be different, and we must expect that the relative mobilities of rock constituents can be reversed and greatly changed by changing temperature or changing grade of metamorphism. A very illuminating example is offered by the surface diffusion of Na and K over a WO surface (Table 7 on p. 87). At low temperature Na has the greater D by far; above 500°, K is more mobile than Na. Such changes with temperature of the relative diffusibility of the several particles in rocks may explain that different kinds of metasomatism operate at different degrees of metamorphism. It is well known, for example, that sodium metasomatism (albitization) is most prominent at low temperatures, whereas potash metasomatism (microclinization) is more intensive at higher temperatures.

We have seen how strain by creation of minor fissures decreased the resistance against diffusion through rocks. Another significant aspect of strain and plastic deformation is its creation of spatial free-energy differences throughout the rock. Short-distance diffusion currents are thereby initiated between points of different stress magnitude (Riecke's principle), and hence it seems reasonable that the ionic mobility of the rock as a whole is intensified as long as the external stress is applied and the plastic yield takes place.

How confining pressure influences the chemical mobility within rocks is not easy to determine, because the effect of pressure upon diffusion velocity in crystalline aggregates has not been investigated with the same enthusiasm as has the effect of temperature. One factor will be the compaction, as already discussed (p. 179). After the vanishing of the capillary and supercapillary openings, however, any further compaction

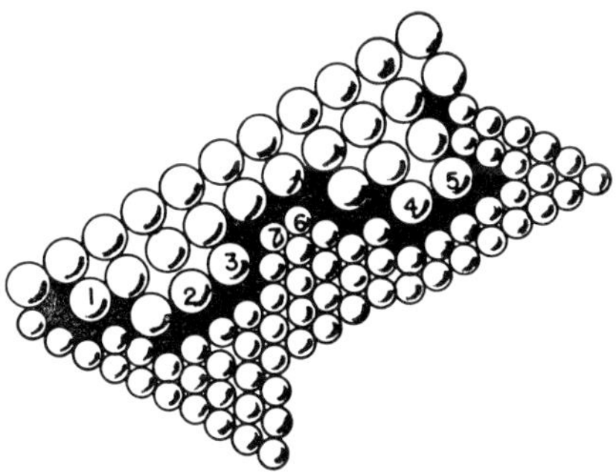

FIG. 102.—Illustration of the contact between two crystals, one consisting of large constituents, the other of smaller constituents. There must exist a rather open and irregular intergranular zone between the two crystals on account of their lattice structure. In this zone, activated particles are easily movable. Some of the constituents of the crystalline lattices along the intergranular border zone must be rather loosely bound because of their unfavorable positions, like particles (*1*), (*3*), (*4*), (*7*), and (*6*), for example. Such particles are likely to migrate rather readily.

of the whole-rock construction is restricted. The remaining openings along the mineral interfaces and mosaic fissures and inside some lattices will not be closed, however effective the recrystallization plasticity, because these atomistic voids are the result of the discontinuous lattice construction of the minerals, and as such they are stable (Fig. 102). Upon further increase of pressure, two effects arise which act in opposite directions in respect to the mobility: (1) The stable openings are somewhat narrowed because of the compression of the rock. This effect will evidently decrease the mobility. (2) Because of the effect of confining pressure upon the free energy and fugacity of solids (pp. 13 ff.), the openings will be increasingly occupied by "vaporized" particles squeezed out from the regular sites of the surrounding lattices, such that the mobility within the rock should increase.

The vaporized or "pressure-activated" particles are not strongly attached to their new sites; otherwise the latter would be nothing but regular stable lattice positions occupied by fixed particles, also at low pressure. The "pressure-activated" particles are, therefore, kept in their uncomfortable positions only if surrounded by close-packed impermeable lattices. If the free positions are interconnected, as they should be along the intergranular network and mosaic fissures, the "pressure-activated" particles should move rather freely.

The equation expressing external pressure, P, and vapor pressure, π, is:

$$\pi_p = \pi_0 \cdot e^{PV/RT} \qquad (V = \text{mol volume in crystal}) . \qquad (148)$$

For silicates, V is of the order of magnitude of 10 cc., and $R = 82$ cc. atm. Assuming $T = 600°$ K and $P = 5{,}000$ atm. (18 km. depth), we find that $\pi_p = \pi_0 \cdot 2.7$. Hence the concentration of mobile particles along the intergranular openings would be some three times as great as under ordinary pressure. At $P = 10{,}000$ atm., $\pi_p = \pi_0 \cdot 7.29$; at $P = 20{,}000$ atm., $\pi_p = \pi_0 \cdot 55$; etc., and the intergranular concentration in the free space would be 7.29, or 55 times that at zero pressure. But there is hardly sufficient free space along the intergranular to consume, say, 55 times as many adsorbed particles as originally existed there. Then our estimate means the following: as soon as a particle leaves its site by jumping along the intergranular or mosaic fissures, the probability that a neighboring particle will move into the newly created free space is 55 times greater at 20,000 atm. than at zero pressure. Assuming that the probability that the surface particles will move laterally is unchanged by pressure, the result is still that the net mobility increases 55 fold. We do not know from experiments how the probability of lateral jumps along the intergranular or mosaic fissures is affected by pressure; but, unless this is considerably minimized, one would reasonably expect the net mobility of at least some constituents in rocks to increase with increasing pressure greater than some 3,000–4,000 atm. At first glance, one would perhaps expect the lateral mobility along the intergranular system to be lowered because of the increased concentration of particles occupying the available space. In this connection we shall note, however, that it is a common rule that surface mobility increases with increasing surface concentration (Barrer, 1941, pp. 337–79).

This contrasts the effect of pressure on pore fluid flow, which, by all logical reasoning, must be expected to be nullified at high pressure because of the increased viscosity of the fluid as well as the complete closing of capillary openings able to conduct such motion.

The theoretical conclusions here arrived at are in agreement with the field testimonies of extensive metasomatism and great recrystallization plasticity of the rocks at the deeper sections of the earth. We shall remember that the viscosity due to lattice slip beyond the yield value of rocks is increased with pressure (Bridgman, 1936–37; Griggs, 1940; Fairbairn, 1949) (see p. 114) and is not much affected by temperature. The ease with which rocks seem to yield at great depths could, therefore, be more simply explained if it was due to a chemical recrystallization flow accelerated by pressure and temperature.

Although our field experience is in the best agreement with the concept of diffusional transfer, as will be discussed subsequently, this mechanism will nevertheless be ruled out as a long-distance carrier if the velocity appears to be too small even for geologic time. Experiments with direct bearing on our problem are practically nonexistent. Diffusion experiments from metals and ordinary laboratory solids, however, are rather plentiful, although they have not been performed under conditions of continuous strain and high pressure which characterize most of our natural metamorphic rocks. A convenient estimate of the geologic importance of the experiments can be achieved if the law $X = \sqrt{2\,Dt}$ (p. 202) is considered. Let us take $t = 100 \cdot 10^6$ years or $3.1 \cdot 10^{15}$ sec., which is a reasonable duration of regional metamorphism during a period of orogeny. The diffusion coefficients of lattice diffusion of different metals are shown in Tables 2 and 3 (pp. 79, 80), together with the mean distances of diffusion in $3.1 \cdot 10^{15}$ sec. We find that lattice diffusion can be of only small-scale importance in explaining phases of recrystallization, narrow reaction veins, and partial recrystallization flow of rocks. Diffusion through lattices seems unable to conduct material more than a few feet through rocks, as shown by experiments conducted by R. Jagitsch (1949). The main path must be intergranular diffusion, as favored by C. E. Wegmann, Backlund, and others long ago.

We have seen that the few data available for surface diffusion are rather encouraging with reference to the distance which the migrating particles can travel in the course of 100 million years. Sodium and potassium will be able to migrate, on the average, some 50 km. along a tungsten oxide surface in the course of that length of time, as seen from Table 3 (p. 80). It is also significant for our problem to know how much material could possibly be transported through the intergranular of rocks through geologic periods.

An estimate of the amount of matter transferred per unit of cross-section across a 100-meter-thick sheet of rock of an "effective porosity" of 0.54 per cent (Garrels, Dreyer, and Howland, 1950) may serve as an example. We shall assume, for convenience, that the rock does not react with Na^+ and that some sodium mineral is precipitated in a fissure at the outlet of the diffusion current. Let the stationary intergranular concentration of Na^+ in the aggregate be, for example, 10^{-2} g/cc of intergranular space at the inlet of the current, and $\frac{1}{2}10^{-2}$ g Na^+/cc of intergranular space at the outlet. The current of Na^+ per square centimeter of cross-section of the rock is, then:

$$\frac{dm}{dt} = -0.54 \cdot 10^{-2} \cdot D \cdot \frac{dc}{dx}, \qquad D = 3.5 \cdot 10^{-3}\ \text{cm}^2/\text{sec}\,,$$

and

$$\frac{dc}{dx} = -5 \cdot 10^{-7}\,;$$

hence

$$\frac{dm}{dt} = 9.5 \cdot 10^{-12}\ \text{g/sec}\,.$$

In $100 \cdot 10^6$ years or $3.1 \cdot 10^{15}$ sec. the amount of material transported per square centimeter of cross-section by the stationary current is $2.9 \cdot 10^4$

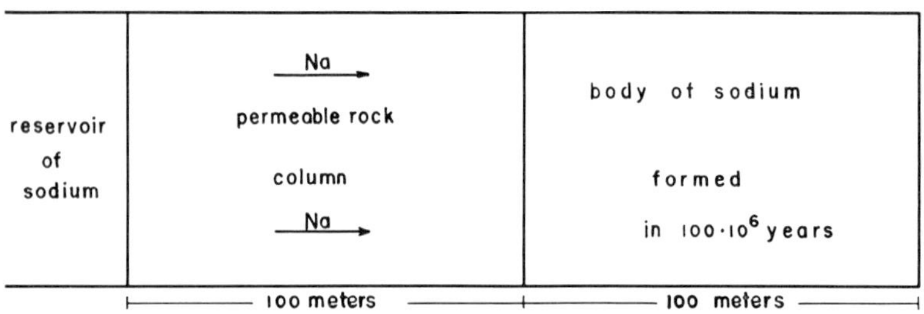

FIG. 103.—Diffusion of sodium from a reservoir through a permeable rock column in the course of 100×10^6 years (see text).

g/cm^2. That is, $2.9 \cdot 10^4$-cm.-thick sheet of metallic sodium would theoretically grow as a crust on the outlet surface of the diffusion current through the 100-meter-thick rock. The amount of actual sodium minerals thus growing would be much greater, because sodium constitutes only a certain fraction of these (Fig. 103).

This is a very schematic example, but, as petrogenetic experimental data are lacking so far, the example above should suffice to illustrate that the process of intergranular diffusion is, in all likelihood, significant in the long transfer of matter in metasomatism.

We have seen that the sole aim of the bodily motion of solutions was to iron out mechanical potential differences within the channel system, thus establishing a mechanical equilibrium. In contrast, the aim of diffusion is to iron out all free-energy differences throughout the space through which diffusing particles can propagate. That is, diffusion is actually aiming at a complete thermodynamic equilibrium throughout the system, but it will be hindered from establishing such an equilibrium for several reasons.

First, the resistance against the process is often practically infinite, so that even the geologic time scale may fail to make the process important. That is particularly true in solid rocks in the outermost parts of the crust, where temperature is low and where thorough shearing and plastic deformation (which greatly accelerate the chemical migration) scarcely take place except in narrow zones. Here, therefore, the paths of the migration are restricted and mainly controlled by open structure. The process is limited to open fluid-filled fissures, mylonitization zones, shear zones, and the true capillary or supercapillary pore system in uncemented and uncompacted sediments. These channels are also conducting flow, and it may perhaps be that the diffusion, as a long-distance carrier, is completely overshadowed by the bodily currents, diffusion being only the communicator between the bodily flowing solutions and the growing or dwindling minerals.

In deeper portions of folded mountain zones, where temperature is high and mechanical deformation takes place continuously and evenly over large volumes of rocks, the resistance against diffusion appears to be low enough to permit considerable rearrangement of matter by this mechanism. But we cannot here speak about the establishment of thermodynamic equilibrium over large volumes, because the resistance to the diffusional transfer is still high. Only a statistic approach to such equilibrium is apparent. The whole phenomenon of granitization and quartzo-feldspathization in the upper floors of the crust (and the formation of anorthosites at deeper levels?) may be nothing but the result of diffusion processes aiming toward a thermodynamic equilibrium.

Over and over again, the approach to thermodynamic equilibrium is partly disturbed by equilibroturbal processes in the sense of van Bemmelen (1949 52). For example, by injection of a magma or by outflow of lava, free-energy differences may be created between the igneous bodies and their environments (chemically incompatible rocks are in contact). Upon later dynamothermal metamorphism, the unstable situation approaches equilibrium by diffusional rearrangement among the incompatible bodies. Similarly, by sedimentation, packages of chemically incompatible rocks are created which may subsequently be exposed to regional metamorphism and substantial rearrangement (example: reaction skarn, Magnusson [1930]).

METAMORPHIC DIFFERENTIATION

THERE is every transition between isochemical recrystallization, in which the rock's bulk chemical composition remains constant, and allochemical metamorphism or metasomatism, in which change of the bulk composition is essential. Some of the transitional steps are found in the phenomenon of metamorphic differentiation, first recognized by Stillwell in 1918 and later treated in some detail by Eskola in several papers (1932). As expressed by Turner (1948, p. 137): "Metamorphic differentiation covers collectively the various processes by which contrasted mineral assemblages develop from an initially uniform parent rock during metamorphism."

Even by simple recrystallization and neomineralization a limited differentiation of chemical composition in space follows, inasmuch as the newly stabilized mineral grains have compositions different from the old minerals which occupied the same space. The presence of a large cordierite, garnet, or feldspar porphyroblast in a metamorphic schist demonstrates that the chemical composition within the space occupied by the porphyroblast has changed radically in the course of recrystallization. In so doing, matter fixed in the lattice of the giant crystal has been collected from throughout the rock itself and often also from introduced material. Occasionally, such porphyroblasts are surrounded by zones impoverished in those elements which are concentrated in the porphyroblast. A bleached iron-poor zone around almandine or staurolite crystals is rather commonly encountered. Such leached-out aureoles around porphyroblasts are by no means always visible in the field; actually, they are exceptions rather than the rule. One must therefore conclude that the growing crystal has been able to collect material from more distant places throughout the rock and that the chemical gradients driving these centripetal migrations have often been too small to manifest themselves distinctly as concentric zones around the porphyroblasts.

The forces underlying the growth of a porphyroblast and the mechanism of this process are complicated in detail but rather easy to explain in principle. By changing P and/or T, a rock has become "supersaturated" to a certain small degree with reference to a mineral, say garnet. Under these circumstances, the probability of forming nuclei large enough to grow further (*positively unstable nuclei*) is small, whereas the very minute nuclei which form more frequently are negatively unstable phases and are likely to disintegrate. A further growth of the few growable nuclei decreases their molal free energy, thus establishing a centripetal free-energy gradient, which causes the crystals to grow with a supply of material from the surroundings. During the whole process, the probability of nuclei formation throughout the rock remains small. Provided that P and T remain constant, the said probability will even decrease with time because the growth of the few porphyroblasts decreases the degree of "supersaturation" throughout the rock.

Commonly, however, metamorphic differentiation means that polymineralic or at least polycrystalline aggregates of varying size separate out in originally more homogeneous rocks. These segregations constitute veinlets, lenses, irregular bodies, or plane-parallel, rhythmetically constructed bands (banded gneisses).

Some kinds of metamorphic differentiation are definitely initiated by mechanical (structural) or chemical heterogeneities within the rock complex. These types are more easily understood than those proceeding without a clear relationship to such initiating heterogeneities (see below). We shall, in the following sections, distinguish between different types of metamorphic differentiation; but, as we find that Eskola's (1939) well-known classification of these phenomena is not unambiguous from a thermodynamical point of view, we shall consider the phenomena differently.

DIFFERENTIATION BY INTERACTION BETWEEN INCOMPATIBLE ROCK UNITS

When two chemically incompatible rocks are in direct contact during recrystallization, substances will migrate in either direction across the border line, and a reaction zone of stable minerals will form. A frequently encountered case is the so-called "reaction skarn" of different lime silicates appearing along the contact of carbonate rocks (Magnusson, 1930) and biotite-rich rocks formed along contacts between amphibolites or green schists and quartzo-feldspathic gneisses. In granulite facies, hypersthene-rich reaction zones are often formed instead of the biotite zone between amphibolites and quartz-rich gneisses (Fig. 104) (Ramberg, 1948*a*). In many cases, one must expect that the mean composition of the reaction layer—or the multiple reaction layers which often form—will be more or less intermediate between the two contacting rocks. A reaction layer of wollastonite, $CaSiO_3$, originating between quartzite and limestone, usually contains less silica than the quartzite and less calc than the marble. Likewise, a biotite-plagioclase reaction zone developing between amphibolite and granitic gneiss may often contain a smaller percentage of potash and silica than the granitic gneiss, and less magnesia, iron oxides, and calc than the amphibolite. Under these circumstances, the effect of the process has been to reduce contrasted composition or to homogenize rather than to differentiate. Usually, however, some elements reach a concentration peak in the reaction zone, and other elements become less concentrated than in either of the original rocks. The concentration peak of certain elements within the reaction layer is denoted "geochemical culmination" by Reynolds (1946). A concentration valley may similarly be denoted a "geochemical depression."

The underlying principle is found in the chemical stability of the several minerals in combination with the mobility of the several constituents through the rocks. If some of the minerals formed in the reaction zone have great "affinity" for a certain element which occurs in the surroundings, this element will tend to concentrate in the zone, being subtracted

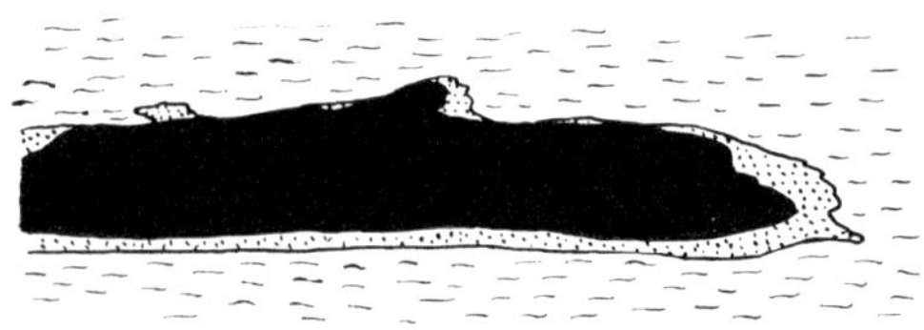

FIG. 104.—Amphibolite remnant in hypersthene carrying quartzo-feldspathic gneiss Note reaction rim of hypersthene along the dark amphibolitic inclusion. Thickness of remnants = 1 foot. West Greenland. After photograph by H. Ramberg.

from the environment by means of diffusion toward the partial free-energy depression in the reaction zone. Elements with weak or no affinity for the newly stabilized reaction minerals will be expelled from the newly formed lattices and are thus forced to diffuse out from the reaction zone.

As an example, let us consider the interaction between a chlorite schist and an arkose or granitic gneiss. Biotite is the main mineral to form: potash feldspar + chlorite $\rightarrow$ biotite $\pm$ muscovite + H_2O. It is directly seen that water has to disappear, but several other less obvious things will also happen. The biotite which forms has considerably greater affinity for Fe as well as for Ti than has the chlorite mineral. Therefore, the reaction zone represents a place of lower molal free energy of titanium and iron than in the surroundings. If these elements are mobile, they are forced to migrate toward the reaction zone of growing biotite and to concentrate therein. The ease with which this migration takes place depends upon several intricate circumstances discussed elsewhere. It is most likely that the basification (Reynolds, 1946, 1947) which now and then advances in front of the granitization (p. 262) is in part explainable as a fixation of some elements with great affinity to biotite, the biotite being created by reaction between the introduced potassium and the original hornblende, pyroxene, garnet, olivine, etc.

The interaction between calcite marble and silicate rocks (reaction skarn) is only rarely able to produce wollastonite, because the necessary temperature is reached only in contact metamorphism and occasionally during the highest degree of regional metamorphism (p. 155). Diopside and hornblende (scapolite, epidote) are most frequently found in regionally metamorphosed skarn rocks. Diopside and hornblende are often formed by reaction between calcite marble and silicate rocks very poor in magnesium (like granitic gneiss, quartzite, arkose, etc.). In such cases the reaction skarn represents a geochemical culmination of Mg and perhaps Fe, which is strongly concentrated in the diopside skarn. Often a hornblende (paragasite)-rich zone develops closest to the silicate rock, diopside-scapolite zones developing closest to the marble. As such hornblendes are usually richer in Fe relative to Mg than diopside is, the hornblende zone represents a culmination of Fe. It is interesting to note that such border interaction necessarily requires transfer of some elements (molecules, ions) in one direction, of other elements in the opposite direction. For that reason, it is evident that the crucial manner of transfer is a chemical diffusion and not a flow of pore fluids.

We have here considered the geochemical culminations of some elements in reaction zones caused by a strong tendency of the reaction minerals to incorporate certain elements in their lattices (Fe^{2+}, Ti^{4+}, in biotite, etc.). Geochemical depressions could arise in reaction zones because of the opposite tendency, namely, to expel some constituents. So, for example, the hypersthenization occurring along the interface between gneiss and amphibolite in granulite facies (Fig. 104, above) is accompanied by the expelling of titanium and water, because the pyroxene lattice is not very susceptible to the incorporation of these constituents. Illuminated by these occurrences, one may ask, What will happen with the expelled constituents? It is a reasonable assumption that the expelled substances can give rise, in turn, to geochemical culminations by fixation of the expelled constituents either closely outside the growing reaction zone or some distance away. The formation of titaniferous iron ores has been explained by this process (Ramberg, 1948*a*).

DIFFERENTIATION DUE TO HETEROGENEOUS PRESSURE

Cracks, joints, and fissures which form in a stressed rock complex represent pressure minima, so that any mineral (or liquid body) growing there has lower free energy than the same mineral (or liquid body) in the compact parts of the complex (p. 13). The opening will first be filled with the most mobile constituents of the host rock, that is, water, carbon dioxide, and other "fugitive" and mobile compounds existing throughout the rock. We concluded (p. 180) that in many cases these substances would first constitute a gaseous phase, filling the openings and fissures, and that the pressure of these gaseous fillings in the beginning would not exceed the total vapor pressure of the rock's minerals with their intergranular adsorbed particles. At lower temperatures (below the critical temperature of water) the fillings will be liquid if the partial water-vapor pressure of the rock minerals is as high as that of a saturated water vapor at the corresponding temperature. Such a hydrous liquid phase would also, in most cases, have a pressure less than the rock pressure.

After the first rather rapid filling of the opening with the most fugitive constituents, there will still be disequilibrium in the rock-fissure system because minerals existing in the fluid-filled fissure have still less molal free energy than the corresponding minerals in the rock. The situation is supersaturated in the fluid-filled fissure. The tendency will necessarily be that minerals slowly disintegrate in the rock and the constituents diffuse toward the low-pressure place, where precipitation must take place. As the several constituents now have very different mobility, the bulk chemical composition as well as the mineralogical composition of the filling will often be radically different from the host rock.

The growth of minerals along the wall of the crack and the plastic yield of the host rock make the fluid-filled volume of the opening decrease, followed by an increase of the pressure of the fluid phase. If the wall was impermeable (for example, because of the crust of newly formed minerals), the pressure in the filling would gradually become equal to the pressure in the rock, and the situation would be mechanically stable. Since the walls are permeable, however, the fugitive constituents will start to diffuse from the fluid filling into the country rock, whereas the elements constituting the minerals in the fissure may continue to migrate into the fissure until the process comes to a stop when the fissure is closed either by plastic yield in the country rock and/or by the newly formed minerals. Although the principle is simple and follows from elementary thermodynamics, the details of the process are but little known. At first glance, for example, one may expect the material filling the crack to be extracted entirely from the closest part of the adjacent rock, thus creating a visible concentric zoning around the secretion, these zones being impoverished in the constituents enriched in the secretion (extraction zones). Such a phenomenon may be seen in the field (Fig. 105), but often external zoning is not macroscopically detectable. In this respect, secretions are similar to many porphyroblasts which do not surround themselves with an easily detectable extraction zone (p. 212), although it is evident that matter once migrated toward the growing porphyroblast.

Let us consider the secretionary growth a little further (Fig. 106). When a small amount of the minerals has been removed from the adjacent rocks and deposited in the fissure, the adjacent rock has shrunk by a cer-

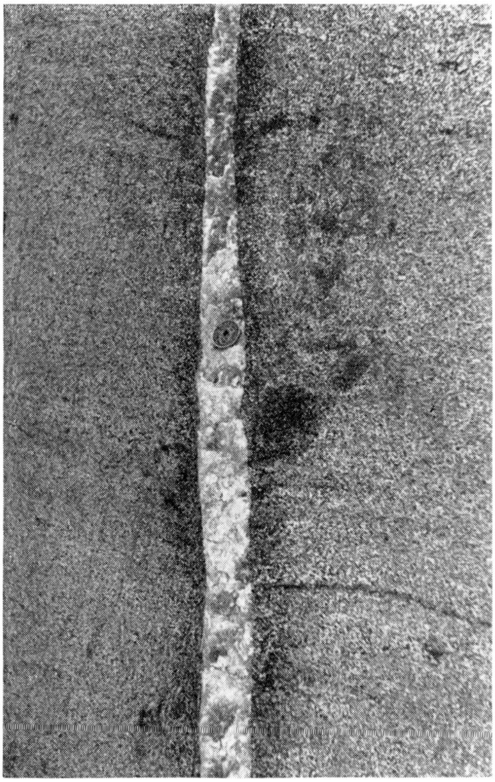

FIG. 105.—Quartzo-feldspathic vein in amphibolite. Hornblende is relatively concentrated along the vein because of extraction of sialic material. The vein is some 2 inches thick (see text). West Greenland. Photograph by M. L. Ramberg.

tain volume fraction and the pressure acting on the minerals in the shrunken part of the rock is less than originally and will remain so until readjusted by plastic flow in the surroundings. For that reason, the tendency to disintegrate minerals in the immediate country rock and reprecipitate minerals in the secretion is not strongly present after the first small "leaching out" has taken place. On the other hand, there is a tendency to displace material from more distant places, where pressure is high, into the weakly leached-out zone, where pressure is low. This process introduces anew mobile constituents into the host rocks adjacent to the crack, and the slow adjustment of pressure by plastic flow in the leached-out zone will intensify the pressure difference between fissure and adjacent rock, so that matter will continue to creep into the fissure. Hence the constituents concentrating in the growing secretion are not necessarily

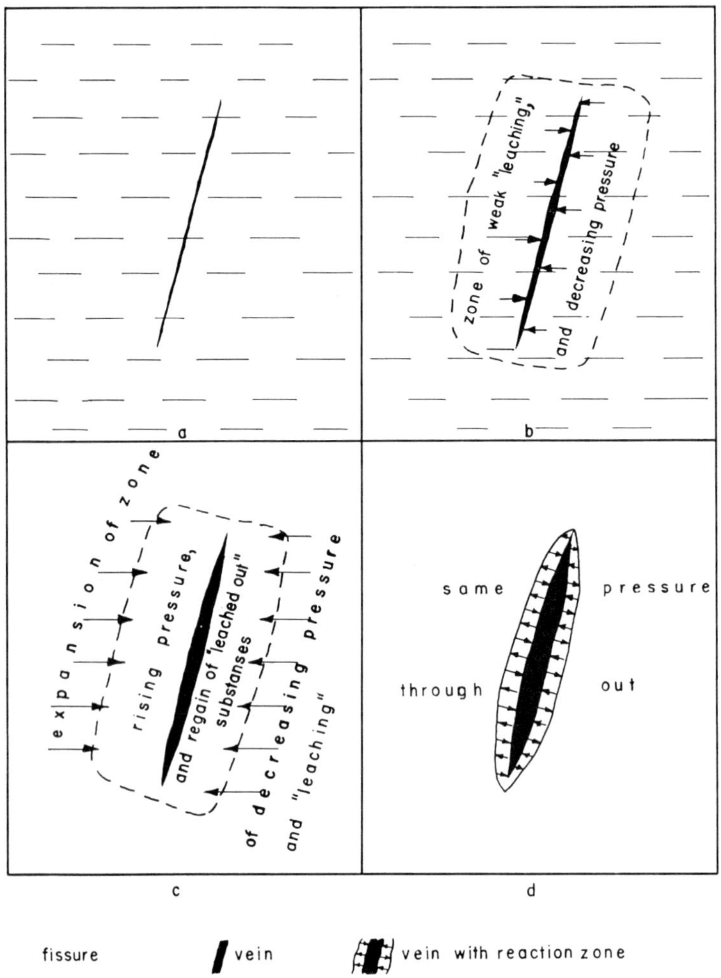

FIG. 106.—Evolution of an "ideal" secretion in a tension crack in a gneiss. Arrows indicate diffusion of material (see explanation in text).

taken from the adjacent rocks but rather from places in the neighborhood which are rich in the most mobile mineral constituents.

When the secretion is filled, the rock pressure will act upon the secretion mineral with its full value, and it is likely that some of the secretion minerals therefore become chemically incompatible with the adjacent country minerals, so that a reaction zone may develop—for example, biotitization of hornblende in amphibolite around secretion pegmatities.

In an extensively interconnected system of joints or fissures crossing through several rock bodies, the mobility along the fissures may greatly disturb any relationship between country rock and fissure fillings formed in the way here explained (Fig. 107).

Cracks and fissures open up most frequently in relatively brittle rocks, particularly when inclosed in more plastic rocks. As the plasticity, to a great extent, is due to recrystallization flow (p. 130), the brittle rocks

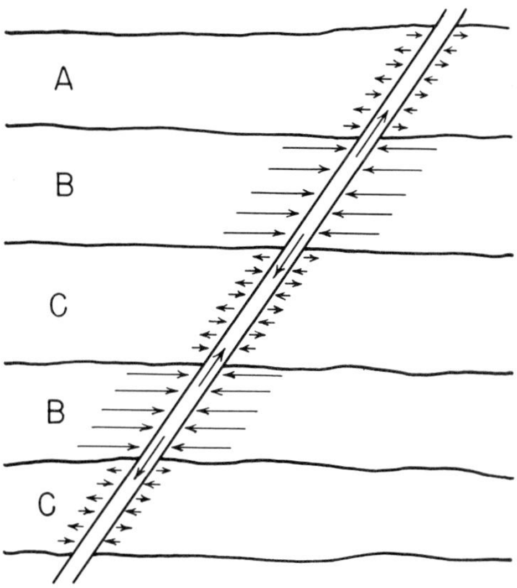

FIG. 107.—Filling and formation of dike in a fissure extending across several contrasted rock layers. The arrows show the major directions of migration in the rock-fissure system. *A* and *C* are rock layers consisting of inactive and immobile minerals, *B* is the layer of active and mobile constituents, for example, granitic minerals. The whole dike will therefore develop as a quartzo-feldspathic vein.

consist mainly of minerals with low chemical mobility (i.e., calco-ferromagnesian silicates). On the other hand, the minerals which concentrate in the cracks in the brittle rocks consist of the most mobile elements (Si, Na, K, Al). For that reason, the secretions effect actually gives rise to a pronounced metamorphic differentiation. It is evident (p. 252) that many quartzo-feldspathic pegmatites have grown in this way or at least under the control of the same fundamental principle.

Open cracks and fissures function also as birthplaces for crystal nuclei of minerals which are newly stabilized by a change of P and T in the country rock. By increasing temperature, for example, biotite will yield microcline, the nuclei of which prefer to form in the cracks before they form within the compact rock. Since potash feldspar is more mobile than the other minerals, feldspar will continue to grow in the crack.

Likewise, some "rare" elements, which are relatively concentrated in the intergranular system in rocks but occur only as traces in the common minerals, may be able to form a specific rare mineral in cracks and other low-pressure zones.

It is evident that stresses exist in recrystallizing complexes without the presence of fissures. Under such circumstances, it is rather reasonable to expect that some large-scale metamorphic differentiation can be caused by the relative mobility of the several constituents of the flowing rock. During recrystallization flow, the most diffusive constituents, like Na, K, Si, Al (O), etc., would tend to concentrate at the low-pressure areas (Fig. 108). This process has been termed "chemical squeezing" (Ramberg, 1944*c*, 1949) in contrast to the mechanical squeezing or filter pressing in the theory of evolution of magmatic rocks.

DIFFERENTIATION CAUSED BY DIFFERENCES IN INTERFACIAL TENSION AND IN ENERGY OF NUCLEI FORMATION

So far, the genetic types of metamorphic differentiation treated have been fairly clear, to the extent that the underlying physicochemical principles are easily understood. There appears to be another genetic class of metamorphic differentiation processes which are not so straightforwardly reconciled with well-understood physicochemical principles. Very frequently clusters, lenses, "schlieren," etc., of certain mineral assemblages develop in a variety of metamorphic schists and gneisses, without any apparent relation to original structural or compositional heterogeneities in the rocks. Examples are numerous from regionally metamorphosed rocks: vein-shaped clusters of quartz and albitic plagioclase are commonly developed in biotite schists. Segregations of quartz, calcite, and/or epidote frequently develop in green schists of various parentage (altered diabase dikes, basalts, gabbros, basic sediments, etc.). Clusters of quartz, plagioclase, diopside, and/or hornblende are frequently encountered in pyroxene granulites rich in hypersthene (unpublished observation from West Greenland). Quartz-kyanite segregations have formed plentifully within muscovite-kyanite schists in Shetland (Read, 1933). Potash feldspar, acid plagioclase, and quartz (and muscovite) develop as coarse-grained segregations within several types of schists and gneisses. At low degrees of metamorphism and diagenesis, monomineralic but polycrystalline concretions are formed frequently in sediments. Examples are calcite concretions in shale, chalcedony (chert) concretions in limestone, pyrite or markasite in black shale, etc.

The facts here mentioned seem to indicate that random mixtures of minerals belonging to a, in the usual meaning of the word, chemically stable assemblage, are not always entirely stable after all. The situation appears more stable if particular groups of minerals within the assemblage separate out of the mixture and concentrate in clusters. The tendency of concretionary differentiation requires that the molal free energy of the differentiated rock be less than the molal free energy of the same minerals randomly intermingled. That is, the molal free energy of, say, albite and

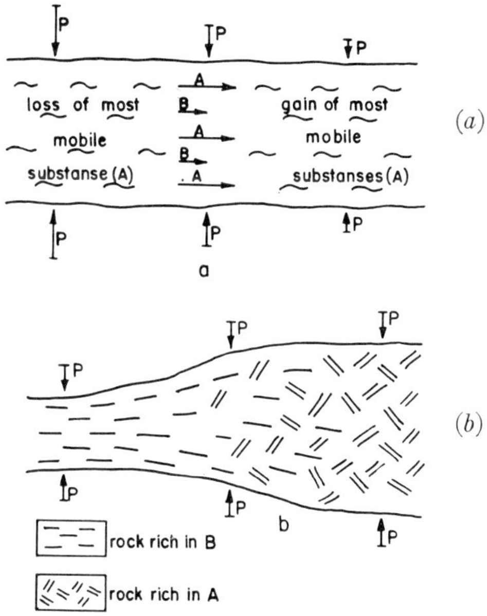

FIG. 108.—Illustration of the concept of chemical squeezing (see text). A pressure gradient acts on a rather homogeneous rock of intermediate composition (a). The chemically most mobile constituents of the rock are thereby transported by diffusion toward the low free-energy site at the low pressure place (b) (see Fig. 122, a and b, p. 254).

quartz when occurring alone should be somewhat less than if these two minerals occurred intermingled with hornblende, pyroxene, muscovite, biotite, etc. According to classic thermodynamics, however, the free energy of a phase (crystal) of appreciable size ($>10^{-4}$ cm.) is definitely determined by the properties of the crystal itself at constant P and T, and not by the surrounding medium. Only when the crystal is very small will its free energy actually depend not only upon its own conditions but also upon the surrounding phases, because the surface energy of the small crystal grain is dependent upon the character of the surrounding phases. It is therefore likely that the free energy of a minute albite grain surrounded by quartz will be somewhat different from the free energy of the same grain inclosed in hornblende or pyroxene. Therefore, in case the mineral or minerals which form the growing cluster (or concretion) originally occur as very small grains or, better, in a colloidal form throughout the rock, then the total free energy of a certain mass of a rock will decrease appreciably by the clustering. Minute grains of calcite intermingled thoroughly with clay minerals in a slate will have greater molal free energy than an aggregate of calcite crystals in which each calcite grain is surrounded by grains of its own species. Finely divided sulfides in black slate are surrounded by carboniferous matter, clay minerals, etc. Owing to surface energy differences, the free energy will be less if the sulfides occur as pure sulfide clusters, and hence they will tend to develop in the course of time. The same is true for the silica gel or finely divided silica occurring in some limestones or dolomites. The segregation of silica in concretions represents a decrease in free energy.

In the several cases mentioned above, there exists a tendency to disintegrate the finely divided calcite, silica, and sulfides, respectively, at places where these constituents are intimately mechanically intermingled with clay minerals, etc., to direct the diffusion currents centripetally toward embryonic clusters, and to consolidate there the migrating activated particles. Pore fluids streaming through the porous sediments will disturb the ideal centripetal diffusion process, the growing concretion would be unsymmetrical, and it may contain constituents introduced from external sources.

Although concretionary growth is rather common in fine-grained sedimentary rocks under diagenesis and weak recrystallization, true metamorphic differentiation after a closely allied principle occurs first in metamorphic rocks in which the grain size would, in all likelihood, be too great to allow varying interfacial tension to influence the free energy of a mineral. Thus an albite grain with a mean diameter of, say, 0.1–1 mm. would have closely the same free energy, independent of the changing chemical-structural character of the inclosing minerals (provided, of course, that the composition of the albite grain, as well as the pressure and temperature, remains constant). Thus the energy gain which is combined with a displacement of the albite grain from a pyroxene-rich milieu to a quartz-rich milieu, for example, may not be sufficient actually to enable such a displacement by disintegration, diffusion, and reconsolidation. However, time is available without rationing for many petrogenetic processes, so that very weak forces may cause large alterations throughout periods of metamorphic recrystallization. It should be noted here that iodine crystals changed from 2–3 mm. diameter to 4–5 mm. in the course of 8 years at 0°–25° C. in an old experiment by Jonstorf (Weiser, 1939, p. 8).

It is more reasonable to assume that the growth of the clusters is intimately related to the stabilizing of new minerals during change of *P* and/or *T* (or introduction of material from external sources). Then the tendency of minerals to co-develop in clusters is determined by the ease with which a certain new mineral starts to grow on, or in, the minerals already present. To start the growth of a crystal nucleus, it is not a sufficient requirement that the mineral, as a macrophase, be stable under the particular circumstances. The milieu must be supersaturated to such an extent that the probability of creating a growable nucleus is not too close to zero. If the pressure and temperature are entirely homogeneous throughout the rock, so that the whole body is in mechanical and thermal equilibrium, then the ease with which a new nucleus is formed depends upon the concentration in space of the necessary constituents (ions), the mobility of these constituents, and the interfacial energy relationship between the newly stabilized mineral and the pre-existing phases. These conditions are so complex and little known that any prediction can hardly be made. As an understanding of the general principles may direct the gathering of data in a fruitful path, however, we shall discuss the matter further. Let us consider a homogeneous rock with no "schlieren," veins, clusters, etc., and let *P* and *T* change so that a new mineral is stabilized (this is an ideal case; usually two or more new minerals are stabilized simultaneously; pp. 42 ff.). The probability of nucleus formation at given *P* and *T* and a given degree of supersaturation is determined by the character of the pre-existing minerals and their interfaces. The new nuclei will, therefore, preferably tend to form in or on one of the pre-existing minerals or in an interface between two particular species of old minerals. In the homogeneous rock the particular favorable places of nuclei formation are evenly distributed throughout, and the resulting rock should also be homogeneous. But if the most favorable place of nuclei formation is an interface between minerals which are rare in the host rock, then only a very few new nuclei will form, and the growth of "glomeroblasts" (polycrystalline monomineralic clusters) around the nuclei thus creates a certain type of metamorphic differentiation. If two or more minerals are stabilized simultaneously by a change of *P* and/or *T*, both minerals may grow together as dimineralic clusters, or one mineral may form glomeroblasts, whereas the other becomes evenly distributed, etc.

We have seen that concretionary growth at low temperature causes heterogenization of sediments. When true metamorphism starts, either such heterogeneous bodies may be still more contrasted (see above), or homogenization may occur, depending upon particular circumstances. The concretions, which at low temperatures were chemically compatible with the inclosing rock's minerals, may become incompatible with the host rock at higher temperatures, so that interaction between concretion and inclosing rock may reduce or intensify the original contrast in composition between rock and concretion. Common examples are calcite concretions in slate, which, at higher temperatures, give lime aluminum silicate nests in mica schists, sillimanite gneisses, etc. Similarly, chert concretions in dolomite or limestone give diopside (wollastonite) nests in marbles.

It may also be that the old concretions are, in effect, easy birthplaces of newly stabilized minerals. These minerals will then preferably grow in the concretion, gradually impoverishing the host rock of some of the elements constituting the newly stabilized mineral.

Metamorphic differentiation after the principle of concretionary growth is explainable if the minerals which statistically tend to co-develop in clusters and concretions exhibit relatively low mutual interfacial energy, because then the molal free energy of a small nucleus of the one mineral grown upon a surface of the other mineral is almost as low as the molal free energy of a macrocrystal of the former mineral. There are reasons to assume that (1) strong chemical affinity between the constituents of two minerals and (2) similarity in crystal structure and type of bonding forces decrease the interfacial energy between two minerals and increase the ease with which they build nuclei on one another (Volmer, 1939).

We shall briefly summarize the main information we possess regarding the types of common minerals which appear to separate by metamorphic differentiation after the concretion principle.

1. Calcite is likely to form concretions in shales as well as in sandstones.

2. Silica (chalcedony) frequently forms concretionary bodies in carbonate rocks, and less commonly in shale. At a slightly higher temperature of recrystallization (green schist facies) quartz develops very abundantly as conformable lenses and small concretionary veinlets in mica schists.

3. Sulfides grow frequently as concretions or replacement bodies in some shales (black shale).

4. Quartz, acid plagioclase, and potash feldspar are undoubtedly the minerals which most commonly go together in forming conformable lenses, veinlets, and irregular clusters in gneisses and crystalline schists of different composition. In general, the salic minerals appear to shun the calcofemic minerals in metamorphic differentiation.

There appears to be a close relationship between the principle discussed above and the fact that, during metasomatism caused by the introduction of matter, certain guest minerals tend to co-develop in some cases, whereas rather pure monomineralic aggregates develop in other cases. The fact that dolomitization tends strongly to create rather pure dolomite marbles with but minor amounts of calcite (Table 18, p. 231) may be due to the interfacial energy conditions between calcite and dolomite. It is likely that a dolomite nucleus does not form so easily in contact with calcite grains as it does in contact with other dolomite grains. Therefore, the dolomitization will most likely proceed in such a manner that a pure dolomite rock develops rather than intermixtures of calcite and dolomite marbles.

The formation of replacement ore proceeds in such a way that rather pure ore bodies are more likely to form than a gradual growth of scattered ore grains throughout the silicate host rock. This fact can actually be explained if the ease with which ore mineral nuclei start to grow on or in common silicates is less than that at which one ore mineral starts growing on (or in) another ore mineral. One must assume that during metasomatism the conditions are but weakly supersaturated with reference to the growing metasomes. Under such circumstances the conditions for ore nuclei formation in the silicate rock may not be fulfilled, whereas the same weak degree of supersaturation is sufficient to cause nuclei to develop and grow in or on the already existing ore minerals.[1]

It is now very interesting that crystal chemistry supports this point of view. It is well known that the covalent bonding forces in the sulfide ore minerals are different in strength and character from the chiefly ionic bonding forces in silicates (see Appendix). For that reason, it is very likely that the energy of nucleus formation of a sulfide mineral on (or in) a silicate is somewhat greater than the energy of nucleus formation of one sulfide on (or in) another. It appears to be a rather strong tendency which hinders sulfide ore minerals from starting to grow in silicate rocks during metasomatism; there may exist a weaker tendency trying to control the types of ore minerals which preferably go together.

A similar explanation appears to apply to the metamorphic separation of calco-ferromagnesian silicates from quartzo-feldspathic silicates. The character of the lattices and the chemical composition are rather contrasted between the olivine-pyroxene-hornblende mineral group, on the one side, and the quartz-alkali-feldspar group, on the other. As discussed in the Appendix, for example, the oxygen in orthosilicates and metasilicates has a strong affinity for iron and magnesium, whereas the oxygen in phyllosilicates and tectosilicates has a stronger affinity for alkali metals (and calcium). For that reason it is rather difficult for the minerals of the olivine-pyroxene-hornblende group to form nuclei on minerals of the feldspar-quartz-muscovite group, or vice versa.

1. This concept is in close agreement with W. Newhouse' view based upon his long experience from ore studies (oral communication, 1951).

DIFFERENTIATION DUE TO SHEARING OF MINERALS WITH CONTRASTED MECHANICAL PROPERTIES

Schmidt (1932) has developed a hypothesis by which he attempts to explain metamorphic differentiation on purely mechanical grounds. The essence is that shearing of a rock causes minerals with great plasticity to concentrate in layers of intense shear motion, whereas less mechanically ductile minerals tend to remain in the less sheared portion of the rock. Schmidt thinks the minerals are moving bodily to their proper places during the mechanical deformation of the rock body.

Schmidt's idea is very interesting, and it may, perhaps, be applicable to metamorphic differentiation in certain cases. It may be fruitful to try a combination of Schmidt's mechanical principle and the chemical sides of the differentiation treated in this book in consistence with the initial viewpoints of Stillwell and Eskola. One would think that shear fissures and minor cracks preferably open up in zones of relatively high velocity of shear motion in the rock complexes. As such fissures function as birthplaces of newly growing crystals (p. 219), the chemically most mobile minerals would tend to concentrate in some zones of intense and rapid shearing.

The metamorphic differentiation is, as we have seen, governed by different principles in which the physicochemical properties of the several minerals and their constituents are crucial, on the one hand, and the geological circumstances (pressure, stress, temperature, time, etc.), on the other hand. It appears perhaps unlikely, therefore, that the same two mineral groups (or two mineral species) which tend to separate in space under some particular geologic condition (for example, intense shearing and recrystallization) also would tend to separate under some different geologic circumstances (for example, under static recrystallization or metasomatism). But it is empirically found that although there are variations in details as to the minerals which tend to cosegregate, it appears to be generally true that minerals which go together in concretionary growth also go together in secretionary growth or in banded growth under shearing. Examples are quartzo-feldspathic dikelets and segregations in gneisses; calcite concretions and veinlets in shale.

THE PRINCIPLES OF METAMORPHIC DIFFERENTIATION AND THE QUESTION OF NUCLEI FORMATION DURING METASOMATISM

Metamorphic differentiation *sensu stricto* means chiefly chemical rearrangement of rock-making matter within limited space during metamorphosis. It is nevertheless evident that principles very similar to those which govern metamorphic differentiation also play their significant role in determining the points where metasomes will start growing when matter is introduced into a rock undergoing metasomatism.

The best-understood cause for growth of a new metasomatic mineral is a simple chemical interaction between some of the old minerals in the host rock and the substances diffusing into the rock. By diffusion of potash into kyanite-carrying sediments, for example, muscovite and/or potash feldspar often grow according to the following equations or similar ones:

$$2H_2O + \underset{\text{kyanite}}{3Al_2SiO_5} + 3SiO_2 + K_2O \rightarrow \underset{\text{muscovite}}{2KAl_3Si_3O_{10}(OH)_2} \tag{149}$$

and

$$\underset{\text{muscovite}}{KAl_3Si_3O_{10}(OH)_2} + K_2O + 6SiO_2 \rightarrow \underset{\text{potash feldspar}}{3KAlSi_3O_8} + H_2O\,. \tag{150}$$

Albite may form similarly if soda is introduced:

$$\underset{\text{kyanite}}{Al_2SiO_5} + 5SiO_2 + Na_2O \rightarrow \underset{\text{albite}}{2NaAlSi_3O_8}\,, \tag{151}$$

or

$$\underset{\text{muscovite}}{KAl_3Si_3O_{10}(OH)_2} + Na_2O + 6SiO_2 \rightarrow \underset{\text{potassic albite (perthite)}}{Na_2KAl_3Si_9O_{24}} + H_2O\,. \tag{152}$$

Reactions of this type were believed by Goldschmidt (1921) to explain the albitization and microclinization of aluminous sediments. Undoubtedly, a large number of metasomatic transformations can be explained by similar reactions. On the other hand, there are other metasomatic alterations which require basically different explanations. In these cases there are no classic chemical interactions between unstable host minerals and incoming substances. Very often the host rock and the metasomes have no elements in common such as would be the case when, for example, sulfides replace or grow in carbonatites. There can clearly be no question of any of the introduced elements reacting with any of the old elements. Yet ore geologists know that fixation of sulfides in carbonatites and in silicate host rocks so far takes place. It is even well evidenced that such a process often, if not always, takes place as a volume-by-volume event. That is, calcite (or some silicates) have to disintegrate and leave the stage in order to make space for the growing metasomes. The case is analogous when silica replaces marble in a volume-by-volume fashion (Fig. 109, p. 233). In this case, though, it may be argued that silica and calcite have oxygen in common, so that silicon atoms or ions may be supposed to migrate into the carbonate rock and react with oxygen, thus liberating carbon, which is forced to move out:

$$Si^{4+} + CaCO_3 \rightarrow SiO_2 + CaO + C^{4+}. \tag{153}$$

However, there are neither theoretical nor field grounds to think that this is what happens. Undoubtedly, Ca^{2+} and CO_3^{2-} or $CaO + CO_2$ are migrating away, although some oxygen may be transferred from CO_2 to Si so that C^{4+} is liberated. In these cases and in numerous similar cases, it is most likely that the energy of nucleation is significant: *the incoming activated particles consolidate into nuclei in those rocks and on those mineral interfaces where the certain crystal nuclei form easily at but weak supersaturation.* Once formed, the further growth of the metasomes should take place rather easily, because matter will migrate freely toward the free-energy minima which the newly formed nucleus represents. It should be understood that, as in most other cases of metasomatism, it is not necessary that the elementary constituents of the metasomes be supplied from a common source or even carried in the same diffusion current. Actually, some constituents of the metasomes may migrate to the nuclei from one direction, other constituents from quite a different direction.

As explained elsewhere in the chapter on chemical kinetics of metamorphism, the volume-by-volume growth of, for example, silica in limestone or sphalerite in dolomite, etc., may be explained by an elaborated Riecke's principle. The metasomes execute a certain force of crystallization, thereby increasing the stress in the closest surroundings, and hence raising the chemical activity or the free energy of the inclosing minerals, so that they are forced to disintegrate and migrate away from around the growing metasomes.

The fixation of the constituents migrating through a rock is certainly also eased by shear zones, cracks, and other openings or low-pressure places which also control metamorphic differentiation. A rock complex may, therefore, remain "fresh" and unaltered even if metasomatites have grown in zones of weakness to which the substances have been supplied by migration through the bulk of the rock complex.

CONTACT METASOMATISM

MAGMAS contain some fugitive substances which cannot enter completely the solid phases in the course of crystallization. Among such substances are H_2O, CO_2, B, S, F, Cl, or some intercombinations of these constituents or fugitive molecular compounds between these and metals: FeF_3, SnF_4, SiF_4, FeF_2 (Table 15, p. 183). These mobile compounds must leave the magma chamber under or after crystallization, and they may perform metasomatism in the contact aureole. Some rocks in the aureole function as "absorption" media for these escaping constituents: aluminous hornfels are susceptible to tourmalinization by reaction with the boron escaping from the magma (Flett, 1909) and to albitization (adinolization, Milch, 1917) by reaction with soda being liberated by the crystallizing melt. Limestones are particularly efficient adsorption media, as was clearly demonstrated by Goldschmidt (1911, pp. 212–20). He thinks that the major factor is implied in the formation of fluorspar by interaction between fugitive fluorides and calcite:

$$Me_yF_x + \tfrac{1}{2}xCaCO_3 \rightarrow \tfrac{1}{2}xCaF_2 + Me_yO_{\frac{1}{2}x} + \tfrac{1}{2}xCO_2\uparrow \ . \qquad (154)$$

Since metal oxides do not always form, but commonly sulfides (galena, sphalerite), one may assume that sulfur in one form or another is carried with the escaping gases so that the more stable heavy-metal sulfides form instead of the less stable oxides.

METASOMATISM IN PRACTICALLY UNMETAMORPHOSED AND LOW-GRADE SEDIMENTS

THE sedimentary petrologists have been led to the conclusion that a number of dolomites, dolomitic limestones, chert formations, iron-bearing sediments, phosphate rocks, etc., have achieved their present chemical composition by metasomatism. Referring to occurrences of this kind, Pettijohn writes (1949, p. 309): "That such metasomatic replacements occur is proved beyond doubt. The time or date of such replacement, however, is less clear, as is also the extent of such replacement. The question of whether all dolomites, cherts, iron-bearing sediments, and phosphate rocks are results of such replacements is not settled by any means." So we see that the specialists in the field of sedimentary petrology are even seriously considering the possibility that all the sediments of the kind mentioned are due to some sort of metasomatism.

DOLOMITIZATION

Most sedimentary carbonatites are formed by accumulation of shells of calcaceous algae, mollusks, corals, and foraminifera. Only occasionally is the ocean supersaturated with calcite so that inorganic chemical precipitation takes place. The shells which accumulate in the limestones are mainly built of calcite and/or aragonite, with lesser amounts of magnesium carbonate, although some algae may carry up to 25 per cent $MgCO_3$ in their shells.[1] The carbonate sediments are made up of a mixture of shells with an average magnesium carbonate concentration much less than 25 per cent. Table 17 gives some data showing the percentage of $MgCO_3$ in some rock-forming shells.

There are no known possibilities for forming a primary sediment with the some 50 per cent of $MgCO_3$ needed to give a dolomite. It is important to recognize that the most abundant carbonatites are the rather pure calcite limestones, on the one hand, and dolomites, on the other, with intermediate mixtures falling far behind (Table 18).

There are abundant proofs of replacement of calcite by dolomite. Generally, it is found that in dolomites clastic structures and fossils are much less preserved than in calcite limestones (Pettijohn, 1949, p. 314). Dolomite often forms euhedral outlines against calcite and detrital quartz or

1. An interesting study of the Mg content in calcite shells has been recently completed by K. Chaves at the University of Chicago.

chert showing epigenetic growth. Furthermore, dolomite replaces fossils and oölites, and often dolomite metacrysts inclose detrital quartz grains. Certainly, the latter structural relationships do not prove the introduction of magnesium (or $MgCO_3$) into the whole limestone in question, but they suffice to show beyond doubt that $MgCO_3$ was rearranged extensively within the layer. With reference to wholesale introduction, it is more important to note that, statistically, dolomites are more abundant, the older the sedimentary column (see Table 19, compiled by R. Daly). To sum up,

TABLE 17

	Per Cent $MgCO_3$
Algae	17.4
Mollusks	1.0 (?)
Foraminifera	8.2
Coral	11.1

TABLE 18*

Per Cent Dolomite	No. Samples
0– 10	48
11– 50	8
50– 89	5
90–100	97
	158

* Data after Pettijohn, 1949.

TABLE 19

Age	Ca/Mg Ratio	No. of Analyses
Pre-Devonian	3.35	392
Devonian	6.29	106
Carboniferous	12.45	238
Cretaceous	56.32	77
Tertiary	53.09	26
Quaternary	35.0	26
Total		865

the evidences for metasomatic dolomitization are: (1) dolomite is not precipitated as pure layers in the present oceans or lakes; (2) the older the sedimentary column, the more frequent the dolomite limestone; (3) structural investigations reveal that dolomite replaces calcite and quartz frequently and was the last mineral to be fixed during the diagenesis.

There is a great question as to the source of the introduced magnesium. There are several possibilities, among which sea water has frequently been suggested. There is no background for assuming that magmatic solutions have been operating. It seems clear that dolomitization has largely taken place during cementation, consolidation, and compaction of the sediments. Under these conditions the rocks are very porous, with wide channelways for chemical communication even at low temperature. Under such circumstances, it is only reasonable to assume that large-scale rearrangement may take place by diffusion of dissolved material in the pore solution, as well as by flow of these solutions. Dolomitization may be a sort of low-temperature metamorphic (diagenetic) differentiation affecting the carbonatite sediment complex. $MgCO_3$ may dissolve where dolomite is intimately mixed with calcite and from shell containing up to 25 per cent $MgCO_3$, but it will precipitate again as dolomite in purer form. In this manner irregular volumes of rather pure dolomite will gradually grow in calcite limestone, and layers of dolomite may develop alternating with calcite limestone. Such a diagenetic differentiation could explain why the end-members of the dolomite-calcite rocks are so much more frequent than intermediate mixtures. But, in addition to this internal diagenetic differentiation, there must be extensive introduction of magnesium, because the primary content of magnesium in the shells is much too low to account for the large quantity of dolomites which occurs.

SILICA METASOMATISM DURING DIAGENESIS

In the porous sediments, silica appears to be rather mobile, even at the low temperature of diagenesis. Chert, flint, chalcedony, and opal often form secretions, concretions, and irregular replacement bodies in limestones and dolomites or less frequently in shale or even in sandstone. There seems to be convincing evidence that these cryptocrystalline or partly amorphous occurrences are emplaced by colloidal silica solutions. Apparently, the temperature has been too low to permit the rearrangement of the unstable amorphous and cryptocrystalline chert into the stable low quartz. Even the metastable cristobalite may be found as a constituent of chert. In limestones, dolomites, and rarely in shales (black shales), chert is present as flints or nodules of chalcedony. These are irregular and lenticular bodies which may coalesce into continuous layers with pinch and swell structure. There are several structures which show that the flints or nodules of chert are regularly formed by epigenetic replacement. Bedding planes cross undisturbed through the nodules (Fig. 109) (White, 1947); chert replaces oölites, fossils, and the one side of stylolites. It is also believed by some authors (Correns, 1939) that chert is precipitated syngenetically with the limestone or shale. Submarine volcanic action may enrich the ocean with silica, which may precipitate in the form of chert layers. Chert is also commonly believed to originate by diagenetic rearrangement within sedimentary columns containing diatoms, radiolarians, and silicoflagellates which carry opal shells. It is experimentally demonstrated that chalcedony is much more soluble (has a greater free energy) than common quartz, and it is likely that opal is still more soluble in watery solutions than chalcedony is. Dissolution of the silica algae shells and redeposition as chert or flint are reasonable processes. Although the chert problem is as little understood as is dolomitization, the great abundance of chert demonstrates that silica is very mobile during conditions of compaction, cementation, and consolidation of sediments. Some principle of diagenetic differentiation is evidently operating here also. The chert tends to shun the carbonates and clay minerals in such a way that precipitation and dissolution give rise to rather pure segregations of chert. The spacing between the chert layers and chert nodules in carbonatites may, among several other factors, depend upon the velocity of migration.

Quartz veins and "dikes" are common in weakly recrystallized sedi-

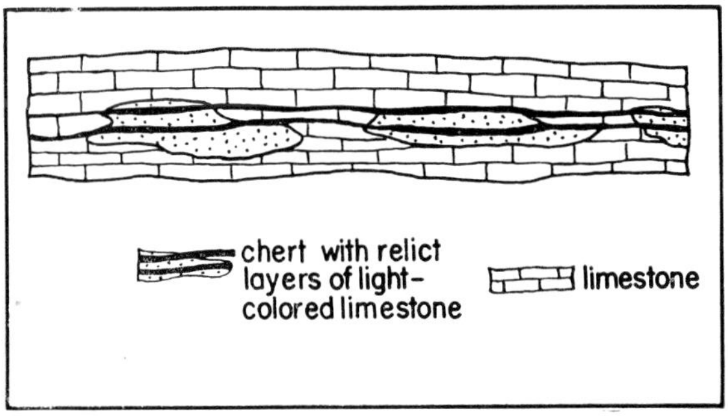

FIG. 109.—Replacement bodies of chert in limestones. (After White, 1947)

ments like some slates, graywackes, and sandstones. Not infrequently these veins are carbonate-bearing, or low-temperature silicates like chlorite are found in the veins. Such veins are evidently of metasomatic origin, although it is rather impossible to determine to what extent the vein quartz is remobilized within the sediments or to what extent introduced from distant sources. The veins may be cross-cutting or partly conformable. Breddin (1930) has given examples of an interesting relationship between the occurrence of quartz veins and the intensity of developed schistosity in the sediments. The veins apparently grow up wherever cracks or intense shearing appeared (see Fig. 110). The veins may often represent fissure fillings, as shown by Breddin, but often, also, the veins are volume-by-volume replacement bodies—they are nondilation veins (Figs. 111 and 112).

LOW-TEMPERATURE METASTOMATIC ORES

There are numerous types of metastomatic ores which have formed at the low temperature (and pressure) characteristic of diagenesis. Such formations should logically be placed in this chapter. However, as this book is not particularly concerned with ore formation, only a few examples will be mentioned.

Galena and sphalerite ore bodies occur in world-wide distribution in limestones, dolomites, cherts, and calcareous shales—all rocks which have

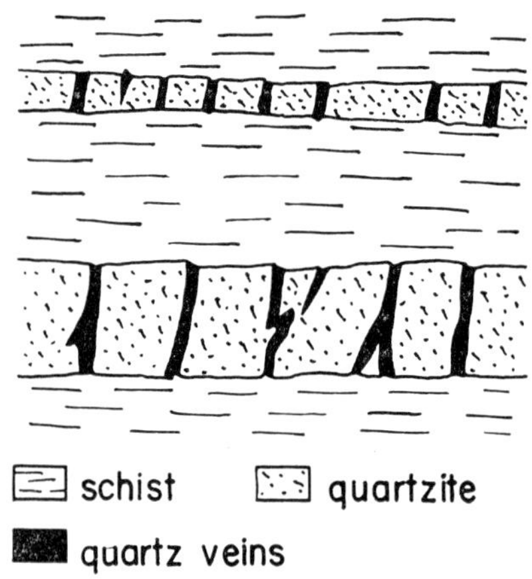

FIG. 110.—Quartz veins in quartzite bands in low-temperature micaceous schist. (After Breddin, 1930.)

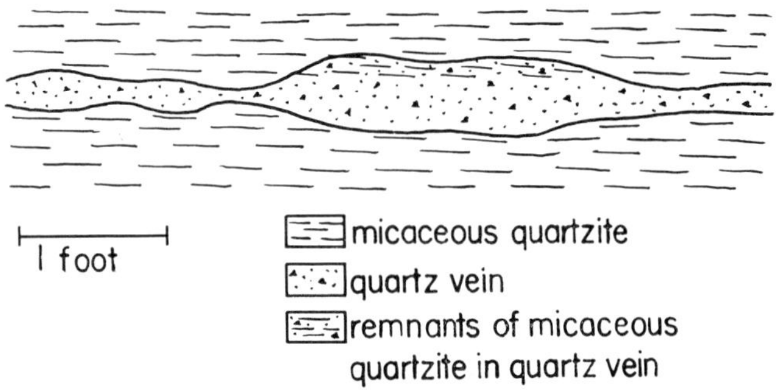

FIG. 111.—Vein quartz replacing micaceous quartzite, Great Smokies National Park, U.S.A. Note that the schistose micaceous quartzite is only partly replaced in the upper part of the swelling quartz vein.

been altered at low temperature (100 °C. ±) and, according to Lindgren (1933), without the aid of igneous agency. The most famous occurrence of this kind may be the Mississippi Valley ores in the United States and occurrences in Belgium and upper Silesia in Europe. Several hypotheses are competing in attempts to explain the origin of these ore types. One reasonable explanation—also in part favored by Lindgren—is that the ore bodies are concentrated through some sort of diagenetic differentiation, the elements of the ore originally having been minutely disseminated throughout the sediments. What could be the reason for such a "secondary" concentration and purification? It is believed by this author that the clue is found in the interfacial energy between sulfide calcites or sulfide silicates, on the one hand, and between the several sulfide mineral species, on the other hand (p. 220). The very minute grains or nuclei of sulfides scattered throughout the original calcite or clay matrix are chemically less stable than sulfide grains in a sulfide matrix. Therefore, the metasomatic processes proceed in such a way that larger bodies of sulfides

FIG. 112.—Typical replacement quartz vein in amphibolite. The dark mineral in the quartz is hornblende of the same type as that in the amphibolite. A thin zone of acidic plagioclase is found along the border. Thickness of vein, 75 cm. Godthaab Fiord, West Greenland. Photograph by H. Ramberg, 1951.

originate slowly, by disintegration of the sulfides scattered throughout the sediments, migration toward more stable sites, and reconstruction of the minerals there. The stable sites often correspond to faults, joints, and open cavities or porous beds. The driving tendency of such a sequence of happenings is rather well understood in a general way (p. 220). It is interesting to note that dolomitization and chert formation have commonly taken place simultaneously with the introduction of sulfide ores. Although the ores of this kind often are connected with faults and joints in the rocks, there is little evidence of channelways which could have led the ore-carrying agency from deep-seated hidden magmas (Lindgren, 1933). For descriptions and discussions of origin see Report VII of the International Geological Congress (1948), No. 18 (London).

Sedimentary iron formations (siderite, iron silicates of chloritic type, and hematite) may, in part, be of replacement origin in the sense that secondary enrichment has operated during diagenesis of poor iron-containing sediments.

Some sedimentary phosphate occurrences may, to a great extent, have concentrated into valuable ore bodies, owing to allied low-temperature metasomatic rearrangement (Lindgren, 1933; Pettijohn, 1949).

Another example of ore formed by low-temperature metasomatism under diagenetic conditions may be provided by the well-known "carnotite deposits" of the Colorado plateau in the United States. These vanadium-uranium ores are believed to have been emplaced with the aid of ground water in porous sandstones (Fischer, 1950).

METASOMATISM IN REGIONALLY METAMORPHOSED COMPLEXES

The contact aureoles around undoubted magmas occupy but a minor fraction of the accessible part of the earth's crust; consolidated unmetamorphosed sediments cover a considerably larger area; but by far the major portion is occupied by regionally metamorphosed crystalline schists and gneisses and by deep-seated plutons, the origin of which is uncertain. These rocks are born in the intermediate and deeper levels of folded mountains of different geologic ages. The more that is added to our knowledge about the pre-Cambrian shield areas, the more established becomes the concept that any portion of these complexes has undergone one or more periods of orogeny. The ancient stable shields are composed of the deeper roots of several folded mountains which are welded together, as it were. Postorogenetic diabases often densely penetrate these complexes, and now and then typical postorogenetic alkaline magmatic rocks penetrate the structure, but the major portion of the structure is synorogenetic or late orogenetic.

Under these circumstances, recrystallization and plastic deformation of the rocks have been so thorough that the criteria of metasomatism will be very different from those applied in the diagenetic sediments, where the chemical processes have been less complete. Clastic grains and original bedding planes are but rarely present, so that inclusion of clastic grains in metacrysts or preservation of original bedding planes in metasomatic segregations may be used only occasionally as an indicator of replacement. Actually, each mineral grain in the altered rock has formed by neomineralization (recrystallization) during some stage of the evolution of the whole rock, the bulk of the minerals have recrystallized practically simultaneously, and it is not easy by any means to construct rigid age relationships by the help of microscopic investigation. Even if the age relations of the minerals were unraveled by means of microscopic investigations (porphyroblasts of cordierite, staurolite, garnet, or feldspar, etc., may clearly be demonstrated to grow by incorporating some characteristic feature of the ground mass), this need not be a significant criterion of the sequence of rock metasomatism in such a way that the apparent ages of the minerals indicate the time of introduction of the corresponding substances. The youngest mineral may, for example, only be the last mineral to grow by internal neomineralization in response to changing P and T. We shall therefore need to find different criteria for large-scale metasomatism during regional metamorphism. For various independent reasons, it is generally agreed among the students of gneisses and crystalline schists and the ores situated in such milieu that metasomatism plays an overwhelming role in creating such rocks. The evidence will be discussed subsequently.

GRANITIZATION OR QUARTZO-FELDSPATHIZATION AND ALLIED METASOMATISM

A great number of dissimilar rocks, if not any kind of rock, show a tendency to achieve a mineralogical and chemical composition approaching a granite or a quartz-diorite or any intermixture of these types by the action of regional metamorphism of intermediate to high degree. There is abundant evidence that migmatitic gneisses, veined gneisses, or rocks rich in remnant structures of sedimentary parentage have formed by a slow, gradual change of the composition in the solid state. The evidence is more indirect, but yet equally convincing, that numerous massive granitic bodies, pegmatite, and aplite dikes or veins (situated in schists and gneisses) were formed in a closely similar way.

Regional and dynamothermal metamorphism occurs invariably connected with orogenesis. In a general way the degree of metamorphism increases with depth as well as with decreasing distance from the central axis of the folded complexes. Therefore, elevation of the folded mountain and erosion will uncover increasing degrees of regional metamorphism from the margin to the central zones of the eroded orogeny. In detail, many deviations from this general feature are present. Local areas of low-facies metamorphism may be found in the high-grade rocks (for example, along post- or late orogenetic breccia zones), and high-grade rocks may be situated in the upper and marginal ordinarily low-grade complexes because of local injection of hot magmas or overthrust high-grade rocks.

In the marginal zones of the folded mountains, the old crystalline substratum is readily distinguished from the weakly altered younger overlying sedimentary and volcanic sequences. The old basement is here, perhaps, mylonitized and subjected to low-temperature—or so-called "secondary"—alteration along such mylonitization zones. Otherwise, the structure and mineral composition of these old basement rocks remain practically intact during the process of orogenesis. Within the central zones of deeply eroded mountains, such as the Hercynian, Caledonian, and the several pre-Cambrian orogenies, however, it is difficult, if not impossible, to distinguish between the old fundament and the superincumbent sediments and volcanites. The whole complex is welded together, plastically moved, metamorphosed, and metasomatosed so thoroughly that there are no means of distinguishing the basement from the younger strata. The younger strata have achieved coarse-grained gneissic structure and have changed their composition unrecognizably. The older rocks have been equally intensively altered, and, with little exception, everything is conformable. Besides the crystalline schists and gneisses, more massive and homogeneous bodies of quartzo-feldspathic rocks appear. These have commonly, but not invariably, conformable contact relationships to the gneissic surroundings. Schistosity and crystalloblastic flow structure are often easily recognized within these masses. Pegmatitic veins and quartz veins are abundantly present, evidently developed through a long interval of time in the course of the orogenic evolution, because some are highly "mylonitized" and plastically deformed, others clean-cut and late- or posttectonic. Such are, in a general way, the circumstances under which granitization has contributed most extensively to rock formation.

What is the evidence for assuming that many of the quartzo-feldspathic rocks occurring in such deeply eroded mountain roots are formed by metasomatism and not by mere injection of granitic melts? We shall mention some of them.

a) Gradual changes along strike.—One of the most striking evidences follows from the observation that feldspar and quartz metablasts occur more and more frequently in some metasediments or metavolcanites as massive granitic or granodioritic bodies are approached. The old sediments gradually achieve a more or less granitic or granodioritic composition. An example of such an area was described in detail and interpreted in a masterly manner by Goldschmidt in 1921. The fabric of such rocks shows that the minerals which give the metasediments or meta-igneous rocks a granitic composition have grown by metasomatism and replacement in a solid matrix (see point *e* below).

b) The large-scale architecture of gneissic areas.—Some large-scale structures in folded mountain roots are conclusive. Extensive continuous or discontinuous thin layers or folded sheets of marble, quartzite, or some other characteristic sedimentary deposits like rusty graphite and pyrite-rich remnants of black shale, can be traced tens or hundreds of miles along the strike. These thin sheets are often imbedded in granitic or quartz-dioritic gneiss, which may be the only other rock type within thousands of square miles (Adams, 1929; Ramberg, 1944*d*–49). For reasons easy to perceive, such lithologic architecture is hard to explain unless the inclosing gneiss or gneissic granite represents the recrystallized or metasomatosed strata among which the still well-preserved sedimentary layers once were precipitated. In some cases the gneisses may represent a rather thoroughly altered and reactivated substratum, upon which the sediments were once deposited. One may well ask: Where are the thick layers of graywacke, the shales, the thick sandstone beds, etc., which we know necessarily once occupied the geosynclinal column (Pettijohn, 1949)? In an attempt to answer this question, it becomes rather evident that much of the pre-Cambrian, Caledonian, Hercynian, and Alpine gneisses have originated through replacement and recrystallization of such geosynclinal columns of sediments and lavas. Another major portion of the gneisses of different age are most likely just reactivated and thoroughly altered old basements.

c) Common conformity in gneiss-schist complexes.—The structural relationship between schists, gneisses, and acidic massives in regionally metamorphosed areas is chiefly conformable, although cross-cutting pegmatite, aplite, and quartz veins and dikes are rather common, and locally the contact between a schist and a more homogeneous granitic body (batholith) crosses the schistosity. Despite such observations, however, the experienced students of gneissic complexes are impressed by the generally existing conformity among the several rock units. (This is in striking contrast to the true magmatic rocks encountered in, for example, the alkaline petrographic provinces of Montana, New Hampshire, Julianehaab, Oslo, East Africa, etc.)

The prevailing conformity must mean that the various rock units had very similar fluidity. We know that rocks like quartzites, amphibolites, mica schists, marbles, and several gneisses were true solid aggregates during the whole evolution of the complexes (see point *e* below). It is, therefore, only reasonable to assume that the differences in fluidity of these rocks were not too extreme. Because the field situations also show that the difference in fluidity of the rocks mentioned above and the more igneous-looking gneisses and quartzo-feldspathic massives could not have been too large, one could assume either (1) that the latter rocks also developed without passing through a magmatic state or (2) that the physical circumstances (high pressure) made the fluidity of the magmas approaching that of the solids. In the latter case it will be just as easy to introduce a granitic body in the form of a solid aggregate as in the liquid magmatic state, and nothing is gained, after all, by sticking to the idea that the plutonic granitic massives are emplaced as melts. Actually, an unnecessary impasse is thereby introduced, namely, the requirement that the temperature of the mass must be high and that the fabric must be of a pyrogenetic character. What we actually find is that the temperature of evolution of orogenetic granitic bodies and migmatic gneisses is usually well below the solidus (see below) and that the fabric is a typical metamorphic crystalloblastic one (Drescher-Kaden, 1948).

Let us consider the situation somewhat further. Diabase dikes are frequent in gneissic regions. Some of the dikes which are exposed today represent an erosion surface very deep down in the channel through which the magma once flowed (Wegmann, 1938). Hence the basaltic magma of these dikes crystallized under the same pressure as that at which gneisses and some granitic bodies are generally formed. Yet there is no evidence in the diabase minerals or their fabric to indicate that the diabase melt under these circumstances was viscous like a crystalline aggregate. If it was, how could enormous quantities of basaltic melt flow through the channels toward the surface of the earth? But a "dry" acidic magma is known to be much more viscous than a basaltic melt, and high pressure may, perhaps, make the viscosity of such a melt approach that of a solid aggregate. However, in order to avoid too big a gap between the crystallization temperature of the melt and the actual temperature at which most plutonic granites are known to have formed (see below), one will at least need to saturate the melt with water under the prevailing pressure. Although only occasionally may such a procedure drop the crystallization temperature sufficiently, the water content will always lower the viscosity considerably, so that there would hardly be an appreciable difference in viscosity between such a hydrous granitic melt and a diabase magma. It does not seem likely that such a fluid melt would create the conformable structure which so commonly exists between crystalline schist, gneisses, and rather massive granitic bodies.

On the other hand, the veined gneisses, with their quartzo-feldspathic veinlets and lenses and the numerous cross-cutting and conformable pegmatitic and aplitic veins and dikes in gneissic complexes, show that the mobility of quartzo-feldspathic substances must be relatively great under the conditions of gneiss formation. Actually, to form so-called *lit par lit* gneisses as well as pegmatitic bodies by means of a magma will require a very fluid melt indeed, because it had to penetrate the crystalline rocks in order to feed the pegmatites. How can we reconcile this last conclusion with the high viscosity of the quartzo-feldspathic matter indicated from other field observations (see above)? The natural solution would be that the veinlets, the pegmatites, as well as the larger quartzo-feldspathic masses, developed in the solid state by activation, diffusion, and consolidation of ions, atoms, or molecules in the rock complexes. In other words, the emplacement of these bodies has nothing to do with bodily moving liquids, and the viscosity of the melts is therefore irrelevant to our problem. The constituent particles of quartz, potash feldspar, and acidic plagioclase are the most mobile ones among the common rock constituents; therefore, rocks of "granitic" affinity develop most frequently by this process. The metasomatic growth of a quartzo-feldspathic body may create conformable as well as cross-cutting relations to the adjacent host rocks, depending upon the structural situation in which the fixation takes place.

For several reasons, it is most likely that the petroblastic body will end up with a conformable structure against the surroundings. First, in the pre-metasomatic complex, there are usually alternating layers of contrasted compositions, these layers being parallel to the schistosity. It is likely now that the granitization selectively attacks some layers and thereby creates conformable structure. Shearing often takes place parallel to the main schistosity, and, as zones of shear function as birthplaces of metasomes, conformable structures will also often arise for this reason. A paper in preparation by W. H. Newhouse, A. Hagner, and G. DeVore shows by numerous data that this effect is much more significant than hitherto assumed. Third, it is usually thus that plastic flow takes place in the granite, as well as in the country rock, during the metasomatic growth of the granite. Because both rocks are crystalline, the difference between the magnitudes of their viscosity is not very great, and conformable structure will often arise. To be more precise, however, the solid granites flow more easily by recrystallization flow than do more basic gneisses and amphibolites, so that disconformable border structure is occasionally created, as in the case of moving salt domes in more rigid surroundings (Balk, 1949). "Diapire granites," in the sense of Wegmann, arise.

d) The evidence of the fabric of quartzo-feldspathic plutons.—There commonly exists no evidence in the fabric of granitic plutons or "magmatic"-looking gneisses that such rocks have moved bodily as a melt or that they have solidified from a melt. Actually, the gneisses and quartzo-feldspathic plutons have a crystalloblastic fabric showing all the characteristics exhibited by most other metamorphites and metasomatites. They show all evidences of recrystallization, recrystallization plastic flow, and extensive mineral replacement. Drescher-Kaden (1948) has made a very thorough study of the fabrics of granitic rocks, gneisses, and some pegmatites and points out all the characteristics of these fabrics which exclude primary crystallization from a melt. It seems apparent, on grounds of the fabric of granitic plutons, that either the large majority have recrystallized after their birth out of a melt or they originated entirely along a nonigneous path.

e) The P,T-conditions.—Regional metamorphism embraces a wide P,T-field, with P ranging from a few hundred atmospheres to some 5,000–10,000 atm. and with T ranging from less than 100° C. to as high as some 700° C. The very lowest grade of metamorphism does not contain "granitized" rocks, migmatic gneisses, veined gneisses, "augen" gneisses, and pegmatite bodies. As a rule, the kind of occurrences here mentioned start to develop in the epidote amphibolite facies (p. 145)—particularly around the border line between this facies and the amphibolite facies, as defined earlier in this book. The temperature of the start of these phenomena may then correspond to some 300°–350° C. or even lower. (Only rarely do pegmatitic veinlets develop in rocks of the green schist facies and low epidote amphibolite facies. Examples are encountered in the low-grade ferrugineous sediments in Wisconsin [U.S.A.], where occasionally pegmatitic veins develop with a low-temperature nonperthitic potash feldspar, quartz, and the same iron mica [stilpnomelan] which occurs in the country rocks. Low-grade recrystallization of arkoses is also likely to create small veinlets of quartzo-feldspathic composition.) A few cases of wholesale granitization at low-grade metamorphism are known. Misch describes convincing examples from the Himalayas (1949); Elliot describes an equally clear example from South Africa (1947). The mineral facies of these two examples appears to be the low epidote amphibolite facies. Outside these few examples, feldspathization, granitization, and pegmatitization are rare in schists of these low grades. Silica is very mobile, however, and so is soda, and consequently veins, dikes, and different segregations of quartz ± albitic feldspar are abundantly encountered in a variety of low-grade rocks. In greenstones, epidote, chlorite, albite, and carbonates commonly joint the quartz in the segregations and veinlets. So-called "alpine veins," where quartz, albite, and adularia are the major minerals, are formed in low-grade rocks. Recently, Bearth (1948) has shown that albitization of synorogenic age is very widespread and significant in the Alps and their pre-orogenetic gneisses.

Returning to the migmatites and other quartzo-feldspathic gneisses, granitic bodies, and pegmatitic veins, we find that these types of rocks occur abundantly throughout all degrees of regional metamorphism ranging from the epidote amphibolite facies through the granulite facies; that is, from some 300° to some 700° C. The general behavior of such complexes of rocks shows that the quartzo-feldspathic components usually have been the most mobile portions of the complex, and the latest to consolidate. Based on superficial knowledge of the field situations, it is easy to conclude, then, that most of the emplacements of the quartzo-feldspathic components have taken place via the molten state, simply because the eutecticum of common rock-making silicate systems is potash feldspar, acidic plagioclase, and quartz in a certain proportion. Either "juvenile" quartzo-feldspathic melts have invaded the deeper roots of folded mountains on a big scale, or refusion (anatexis-palingenesis) has affected the most easily melting mineral assemblages, namely, quartz, potash feldspar, and acid plagioclase in nearly eutectic ratio. The granitic "ichor" in the sense of J. J. Sederholm is created. The hypothesis of partial anatexis and palingenesis is simple and attractive, but closer studies reveal that it is difficult to reconcile with the observations in field and laboratory. The lowest temperature which can create a silicate melt in natural rocks is the eutectic temperature of the corresponding system. To be specific, it is the temperature of the granitic or granodioritic eutecticum, which, in the case of dry melts, is around 800°–900° C. This high temperature is not reached until the metamorphic grade has advanced well into the granulite facies or, most likely, beyond that facies. It seems, therefore, perfectly safe to conclude that dry silicate melts do not originate at any of those stages of regional metamorphism which are known to us. At a maximum content of water and other fugitive substances, the eutectic granitic melt may originate somewhere between 550° and 600° C. These temperatures are supposed to correspond to the highest part of the amphibolite facies (p. 156). How, then, can we account for the mobility of quartzo-feldspathic material in the lower parts of the amphibolite facies and in the epidote amphibolite facies? There seems to be no way to make a hydrous silicate melt responsible for this low-temperature mobility. The melts with excess of potash and/or soda (water-glass solutions) which do have a very low eutecticum are ruled out because of the stability conditions of the rock-making minerals (p. 184). The anatectic melt which starts to form under progressive metamorphism should, theoretically, have a eutectic granitic composition. Actually, the mobilized or introduced material is often monomineralic; for example, albite or microcline "augen" grow in gneisses, or quartz lenses develop in schists. And, alas, the core of most pegmatites, which should be the last part to consolidate on the liquidus theory and hence should have a complex eutectic composition, is almost invariably monomineralic quartz. See the description by Cameron, Jahns, McNair, and Page (1949).

Provided that sufficient amounts of water and other fugitives were present, the temperature and pressure of the high amphibolite facies and granulite facies should be favorable for the generation of quartzo-feldspathic silicate melts. What is the chance of having sufficient amounts of gaseous substances present in the metamorphic rock complexes under these conditions? Owing to the compaction and recrystallization which the rocks have undergone, most of the water has been "squeezed" out of the rocks and the hydrous minerals prior to reaching the stage of high facies metamorphism. The water which still remains usually constitutes less than 1.0 per cent of the rocks, most frequently less than 0.5 per cent. To a great extent, this water is rather tightly bound in biotite, muscovite, hornblende, epidote, etc., and only a minor fraction of it would be available for an anatectic melt. It becomes unlikely, therefore, that appreciable quantities of hydrous low-melting silicate magmas can generate in this way, and the theoretical minimum temperature at which hydrous silicate melts of petrogenetic significance are fluid in the laboratory, i.e., 550°–600° C., is considerably below the limit at which true refusion would take place in natural rock complexes. Again we reach the conclusion that granitization, migmatitization, and pegmatitization to a great extent take place without the co-operation of a silicate melt.

The relative temperatures of the introduced quartzo-feldspathic veins, pegmatitic dikes, and massives and the country rock are of considerable interest. When the typical magmatic diabase dikes invade gneisses or schists, the mineral assemblage of the dikes corresponds to the high temperature of crystallization of the basic magma, which certainly is independent of the degree of regional metamorphism which the surrounding rocks happen to belong to. Only the immediate external hornfels contact may reflect the crystallization temperature of the diabase magma. The same is true also in the case of syenitic or granitic magmas which invade older rocks: a definite external aureole of high-temperature hornfelses is created. Conditions are radically different for the granodioritic or granitic massives and pegmatitic dikes, "sills," and veins which are encountered in regionally metamorphosed areas. Only in a few cases, in such areas, are granitic massives surrounded by a typical and well-defined thermal contact zone representing a temperature of magmatic magnitude. The general rule is that not only do the surrounding rocks belong to the same regional metamorphic facies as that prevailing in the area but also the granitic bodies themselves reflect this mineral facies (Misch, 1949; Ramberg, 1949). There are several possible ways to explain such a state of affairs. First, one may assume that most granitic bodies in orogenic zones are syn- and premetamorphic, so that the regional mineral facies has been passively impressed on the "granites." One may also, however, take the view that the granite is the egg and the metamorphism the hen (p. 273). This was pointed out by V. M. Goldschmidt in his excellent article about the laws of rock metamorphism (1912). He emphasized the discrepancy between the relatively low temperature indicated by the mineral assemblages in the schists adjacent to granites in the Caledonian orogeny and in the Alps, and the high temperature reflected in the mineral assemblages in the external contacts around the true magmatic granites and syenites in the Oslo province. Our advanced knowledge of the relative stability conditions of the rock-forming minerals has revealed that the temperature impressed upon schists and gneisses adjacent to many quartzo-feldspathic bodies in orogenic zones is close to that which prevailed during the regional metamorphism of the area. Moreover, the temperature indicated by the minerals in the bodies themselves is—as far as investigations go—the same as that of the prevailing regional metamorphic facies. In epidote amphibolite facies areas, granites or quartz-diorites of helsinkitic affinity (epidote-granite) are common; in granulite facies areas hypersthene-carrying charnockites and enderbites prevail (Ramberg, 1949). Unless one takes the view that all such bodies are postgenetically metamorphosed, such a state of affairs cannot be easily reconciled with the hypothesis of their emplacement in a liquid-magmatic state. It would be more in harmony with the observations to assume that such bodies partly grow by metasomatism or partly are emplaced bodily in the solid crystalline state.

Only rarely do quartzo-feldspathic bodies in orogenic belts fail to show a preferred structure. This is often parallel to the prevailing structure in the schistose or gneissic surroundings and may hence be interpreted as a so-called "secondary" structure, indicating that the regional metamorphic reworking also affected the quartzo-feldspathic massives. For such bodies, then, it would be only natural to find the mineral facies identical to the prevailing facies in the surroundings. There are also bodies which show so-called "primary" flow structure against the gneissose surroundings. Such bodies are not reworked by the dynamothermal metamorphism which created the gneissic environment, and, if they are emplaced in a liquid state, their mineral facies should correspond to the crystallization temperature of the corresponding melt. If these bodies have been emplaced in a solid state, like a salt dome, their mineral facies could correspond to any temperature below the liquidus of the corresponding melt, and most likely the body and surroundings would belong closely to the same mineral facies.

With reference to the pegmatites, so commonly encountered in gneissic complexes, it is more disturbing for the magmatic theory that these late, often cross cutting, "primary" rocks, as far as research goes, indicate the same temperature as the regional metamorphism in the area in which they are situated (Ramberg, 1949). There seems to be no way to avoid the conclusion that these rocks are a result of the metamorphism rather than the metamorphism being a result of them (see also p. 248).

f) Statistic approach.—Students familiar with the kind of areas which contain rocks now under discussion know in a general way that the amount of quartzo-feldspathic rocks (gneisses, massives, and veins) increases as a higher grade of metamorphism is reached. Lapadu-Hargues (1945–49) made a statistical investigation of the relationship between the composition of schists and gneisses and the facies in which they belong.

His results were in harmony with what one would expect from common field experience from several regionally metamorphosed complexes. Increasing amounts of K and Na in particular and lesser amounts of Ca, Mg, and Fe are found when the facies changes from green schists to amphibolite (Fig. 113).

The present writer compared gneisses from the granulite facies and amphibolite to epidote amphibolite facies in western Greenland. There

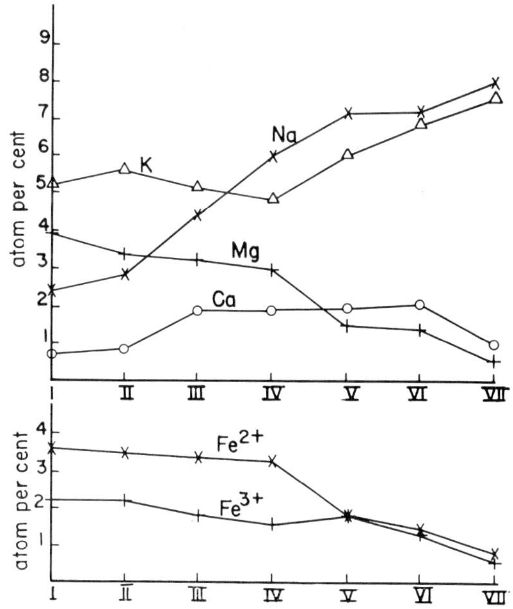

FIG. 113.—Variation diagram, showing the change of atom per cent of several metals in schists and gneisses. *I, II, III, IV, V, VI*, and *VII* represent increasing degrees of metamorphism. (After Lapardu-Hargues, 1945.)

was a definite difference in composition, as shown by the two average analyses in Table 20 (p. 264), although there were the best reasons to believe that the compositions of the two complexes were substantially the same prior to the metamorphism. The change in composition has hence taken place during the metamorphism. It is believed that chemical squeezing (p. 220) has "pushed" out K, Na, Si, O, and H_2O from the deep-seated granulite facies rocks, so that these rocks have functioned as the source of the upward-moving granitic elements.

g) The space problem.—Students of batholiths have speculated on how such vast bodies could be emplaced in the earth's crust, because often there seems to be no sign of mechanical disturbance of the adjacent schist complexes. It looks as if a huge volume of the crust has been cut out and removed, the batholith being emplaced in the outcut. Daly has tried to give an explanation by his "overhead stoping" theory, in which blocks are falling down from the roof and being "dissolved" in the magma chamber or sinking down into it. This may work in some cases, but it does not explain how some structures of the old pre-existing rock complexes are preserved *without geometrical rotation or displacement* in the newly born batholith. Such, for example, is the situation in parts of the huge Coast Range batholith in British Columbia, according to Phemister's (1945) interesting description.

We are led to the conclusion, then, that the quartzo-feldspathic components in gneisses, migmatites, and pegmatites are emplaced by migration of potassium, silica, sodium, etc., through the solid rocks. Quartz, potash feldspar, and albitic plagioclase are particularly mobile, not only in the form of migrating ions or molecules through rocks, but also as plastically flowing aggregates (p. 128). Hence the relative youth of quartzo-feldspathic matter in complexes of crystalline schists and gneisses is explained. Granitic bodies of homogeneous character and considerable size may grow in a similar way, or occasionally they may simply be emplaced bodily by recrystallization flow.

It is interesting to recognize that this view is also arrived at by students who have their major experience from active volcanic areas. The deep-seated plutons (acidic) are formed by metasomatism and metamorphism; when these processes reach their ultimate stage, magmas are created by refusion, and hence, volcanoes arise (van Bemmelen, 1949–52).

Figures 114, 115, 116, and 117 are representative photographs of structures in gneissic complexes.

PEGMATITIZATION IN REGIONAL METAMORPHISM

The growth of pegmatites in regionally metamorphosed areas is just a phase of "granitization" and is mentioned above under this heading. On the other hand, the pegmatites are such interesting and striking rock bodies that it may be proper to consider them in some more detail.

REPLACEMENT PEGMATITES (NONDILATION PEGMATITES)

Often the geometry of cross-cutting pegmatites and their adjacent rocks shows that the former grew as a nondilation replacement body in the latter (Figs. 118–19). The proof is implied in the continuation in the same plane of typical gneissic layers on either side of cross-cutting pegmatites (Goodspeed, 1940). Most likely, the replacement started along a narrow joint in the gneiss or amphibolite and from this joint spread out in the country rock. It is interesting to note that the borders often are macro-

FIG. 114.—Photo of mica schist, in which quartzo-feldspathic porphyroblasts have grown. The dark rock along top of picture is an amphibolite dike. West Greenland, 1948. Photograph by H. Ramberg.

FIG. 115.—Typical "augen gneiss" formed by growth of feldspar porphyroblasts in a green schist. West of Trondheim, Norway. Photograph by H. Ramberg, 1941.

FIG. 116.—Typical veined and banded gneiss. Note the pinch and swell structure of the pegmatite veins. Godthaab Fiord, West Greenland. Photograph by H. Ramberg, 1951.

FIG. 117.—Biotite schist with amphibolitic layers (*upper part of picture*) being replaced by quartzo-feldspathic gneiss (*lower part*). Note remnant stringers of biotite continuing into the newly formed gneiss. At point *X* the situation is particularly clear. West Greenland, 1948. Photograph by M. L. Ramberg.

FIG. 118.—Typical example of nondilation replacement pegmatite. The narrow vein in the middle of the picture crosses an amphibolite layer without dilation. West Greenland, 1949. Photograph by H. Ramberg.

FIG. 119.—Example of nondilation replacement pegmatite. The dike crosses a dark layer in the gneiss without dilation. Note the sharp and fine-grained borders. West Greenland, 1949. Photograph by H. Ramberg.

scopically sharp and that these replacement dikes often are distinctly zoned (Fig. 119).

The evidence of nondilation of pegmatitic emplacement is of a different character in the case of conformable replacement pegmatites. Many pegmatite veins in gneisses are surrounded by an external rim of hornblende or biotite. The external basic rim usually just represents a local concentration of the calco-ferromagnesian minerals in the country gneiss. To a certain extent, at least, such pegmatites have grown in the matrix of the gneiss by assimilating the quartzo-feldspathic matter and pushing away the calco-ferromagnesian minerals (Fig. 120).

Often pegmatites grow in shear zones in gneisses and schists (Fig. 121). It is difficult to prove that such bodies are replacement pegmatites, but it seems highly probable in several cases. It seems to be so that the metasomatic growth of the pegmatites in shear zones may go on simultaneously with the shearing, so that the pegmatites often achieve a crystalloblastic deformation structure.

DILATION PEGMATITES (SECRETION PEGMATITES)

We have seen how a pegmatite can start to grow on a joint and expand as a replacement body from that joint. It is also possible that the pegmatite grows in a joint which either opens up "freely" or is pushed apart by the growing pegmatite because of its force of crystallization. The common filling of pegmatite material in the opening between boudins must in most cases be a consolidation of the pegmatite in freely opening space (Figs. 122*a* and *b*) (see secretion effect, p. 218). Figures 123*a* and *b* also show how the pegmatite gradually grew simultaneously as the fissure opened up (see also Credner, 1875).

CONCRETION PEGMATITES

Some pegmatites have grown by pushing apart the country gneiss or schist in a plastic manner (Fig. 124), in order to provide room for themselves. Such a structure can best be explained by assuming that the pegmatite grew as a crystalline body in crystalline surroundings much like a porphyroblast. No silicate melt could push away the surrounding solid schists plastically.

RECRYSTALLIZATION PEGMATITES

It is not uncommon that local recrystallization along shear zones or in certain parts of a heterogeneous schist or gneiss creates a local coarse-grained texture. A pegmatitic body is thereby developed practically without any introduction of material. Andersen (1931) considered that some of the cordierite-anthophyllite pegmatites encountered in cordierite gneisses had such an origin. A recent discussion by Clabaugh (1950) re-

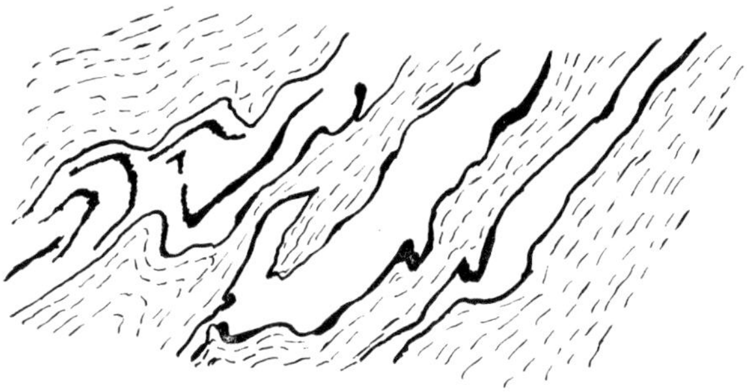

FIG. 120.—Pegmatite veins grown as concretionary bodies in gneiss. Note the concentration of dark minerals along the borders. This is due to a "filtering-off" of the dark minerals in the gneiss as the growing pegmatite expands out into the gneiss. West of Trondheim, Norway. Traced after photograph by H. Ramberg, 1941.

FIG. 121.—Pegmatite vein, approximately 3 inches thick, grown in a cross-cutting shear zone in banded gneiss, West Greenland. Photograph by H. Ramberg, 1951.

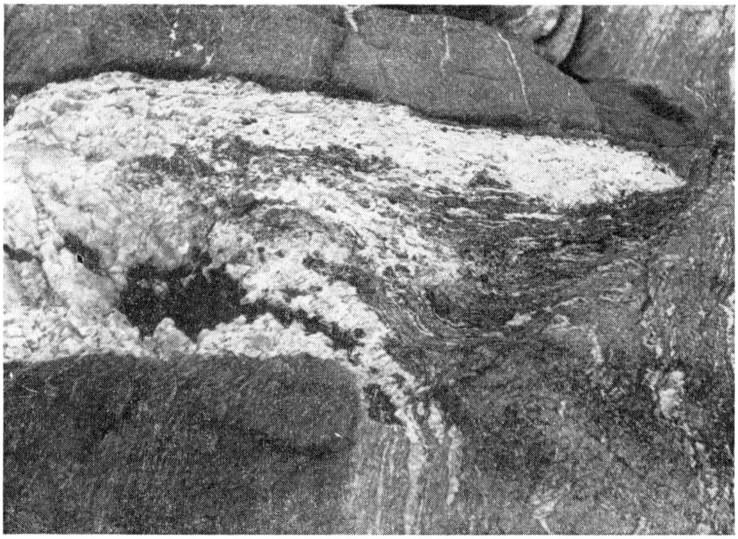

FIG. 122*a*.—Part of pegmatite situated between two boudins of amphibolite in mica schist. The schist is on the right-hand side of the photograph, the ends of the two boudins of amphibolite in the upper and lower parts, respectively, of the picture. The pegmatite material has migrated from the schist into the cracks between the boudins. The black mineral in the lower part of the pegmatite is tourmaline, which most likely is taken from the schist which also contains tourmaline. West Greenland, 1947. Photograph by H. Ramberg.

FIG. 122*b*.—Pegmatite situated between boudins of amphibolite in schist. The pegmatite has formed in the same way as that shown in Fig. 122*a*. This is a typical example of "chemical squeezing" of the most diffusive material into low-pressure places. West Greenland, 1949. Photograph by M. L. Ramberg.

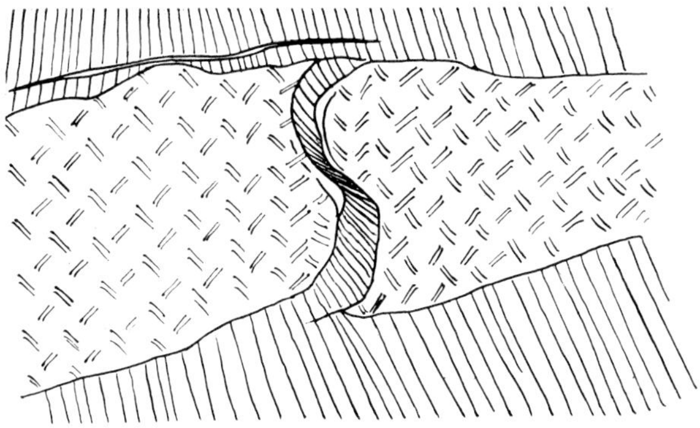

FIG. 123*a*.—Pegmatite grown in a gradually opening fissure in gneiss. Traced after photo. The pegmatite is virtually a plane-parallel "dike"; the apparent irregular shape is a delusion caused by the irregularity of the erosion surface. The composition of the pegmatite is quartzo-feldspathic, and its structure is rather coarse grained. The country rock is a common "gray" gneiss. The "dike" is 1 meter thick. Note the sheared and folded gneiss "bridge" across the pegmatite. The pegmatite does not show macroscopic signs of shearing or cataclasis. West Greenland, 1948. Photograph by H. Ramberg.

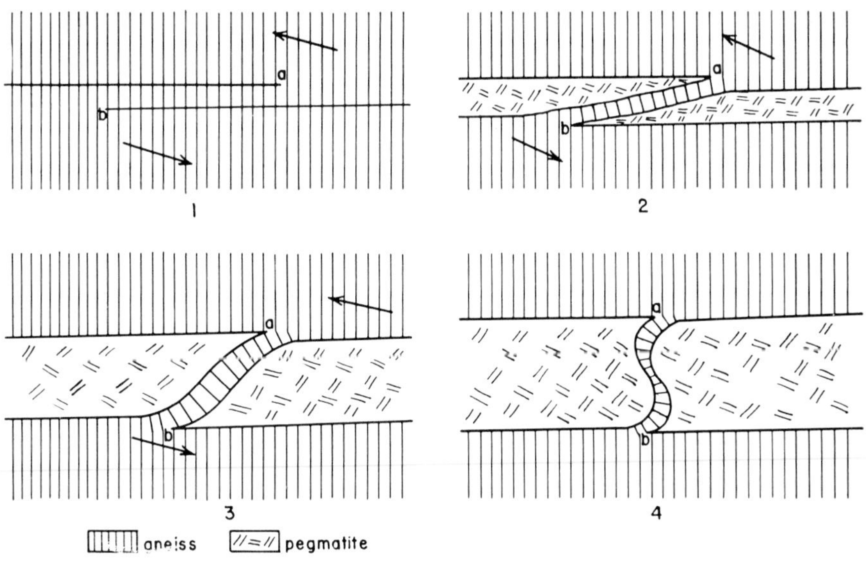

FIG. 123*b*.—Key to Fig. 123*a*. The arrow indicates the movements during formation of pegmatite. *1*, *2*, *3*, and *4* are steps in the evolution of the pegmatite, which grew under a continuous supply of material by diffusion and under continuous recrystallization.

veals that the interesting corundum-sillimanite-carrying pegmatites in Montana represent nothing but recrystallized and metasomatosed aluminum-rich layers in sediments. The present writer found that several of the cordierite-garnet-carrying pegmatites encountered in granulite facies complexes in Greenland owe their existence to a similar process, although commonly some of the processes mentioned above aided in the petroblastic evolution of the pegmatites.

PTYGMATIC VEINS

In consistence with the extensive chemical (and mechanical) mobility of the quartzo-feldspathic mineral during dynamothermal metamorphism, the ptygmatic veins (Figs. 125, 126) form by folding and plastic deformation of pegmatitized rock complexes. Thorough recrystallization of the

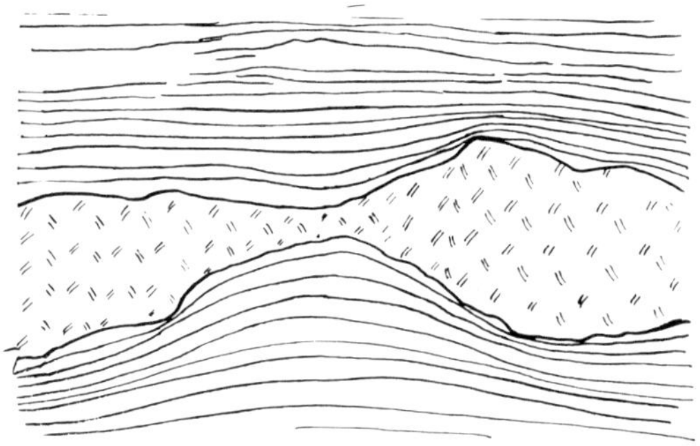

FIG. 124.—Pegmatite with "pinch and swell" structure in gneiss. Note the structure of the gneiss around the pegmatite. The thickest part of the pegmatite is 1 meter. Faringehavn, West Greenland, 1948. Traced after photograph by H. Ramberg.

pegmatitic veins during the drag folding and other kinds of deformation obliterates any drastic sign of cataclastic deformation in the folded veins. However, undulatory quartz is very common in all kinds of pegmatitic veins in gneisses and schists. Amphibolitic bands in gneiss may also be folded in a similar "ptygmatic" manner (Fig. 127).

There can be no doubt that the fundamental cause of the origin of quartzo-feldspathic pegmatites in gneissic complexes is the same as that which creates granites—namely, the fact that these mineral combinations are those which are most stable in the upper portions of the earth's crust (p. 261). It is believed that pegmatites grow in sites where the free energy of the granodioritic minerals is at a local minimum mainly because of low-pressure sites caused by heterogeneous structures (joints, shear zone, pressure shadow, etc.). For that reason, pegmatites grow in a complex undergoing regional metamorphism more readily than the bulk of the rocks is granitized by formation and growth of countless scattered quartz

FIG. 125.—Ptygmatic folds of pegmatite veinlets in mica schist. Signal Hill, District of Kenora, Ontario, Canada. Photograph by F. J. Pettijohn. Thanks are due Dr. Pettijohn for permission to publish this photo.

FIG. 126.—Ptygmatically folded pegmatite in veined gneiss, West Greenland. Note the structure in the gneiss on the convex side of the folds and note also the plasto-mylonitic texture of the pegmatite. These features are explainable if the pegmatite originally started to grow as a more or less straight cross-cutting "dike" and then was folded and deformed together with the gneiss in a more or less plastic manner, the gneiss perhaps being continuously "granitized" and the pegmatite continuing to grow during the plastic deformation.

and feldspar grains. One may say that the whole rock complex is just on the "saturation point" with reference to quartz, potash feldspar, and acidic plagioclase but that, in the local sites of low free energy of growth, the conditions are somewhat supersaturated. Hence pegmatitic bodies grow slowly with a sluggish supply of material through the different channels of migration.

Figure 128 shows an example of thin remnant stringers of country rock in a conformable pegmatite.

FIG. 127.—Folded amphibolite band in quartzo-feldspathic gneiss. The thickness of the band varies from some 10 to 50 cm. Pegmatitic veinlets fill the space between the boudins. Note the similarity between this figure and the pegmatite in Fig. 126. Traced after notebook sketch. Ameralik Fiord, West Greenland, 1951.

FIG. 128.—Conformable replacement pegmatite of quartz and feldspar in a biotite-hornblende-bearing gneiss. Note the remnant of thin stringers along the internal border of the pegmatite and even some in the more central portions of it (X in photo). Godthaab Fiord, West Greenland. Photograph by H. Ramberg, 1951.

THE CAUSES OF "GRANITIZATION"

We have discussed the field and petrologic evidence for concluding that quartzo-feldspathic rock bodies grow slowly in the solid state during regional metamorphism. We have also discussed the kinetics of this process as far as possible on the basis of present-day understanding. But the potential energy underlying this complex of processes may still be less clear.

The migration of quartzo-feldspathic matter in a granitization zone may extend in almost any direction, depending upon the particular conditions which control the driving free-energy gradients of the movable material, as discussed elsewhere. There seems to be no doubt, though, that the average direction of motion of granitic matter is upward from deeper portions of the crust to shallower levels. It is plain mechanics that a granitic magma body or a crystalline granite of relatively low density will slowly move upward through a somewhat heavier crust possessing some plasticity. To evaluate the efficiency of such a motion, let us apply Stokes's law on a sphere of granite 10 km. across, with a density of 2.65 g/cc, inclosed in the crust with an average density of, say, 2.85 and the commonly accepted viscosity of 10^{22} poises. The granite sphere will then move upward with a velocity,

$$v = \frac{2}{9} \frac{g \cdot r^2 \cdot (d' - d)}{\eta},$$

where g is the acceleration due to gravity, r is the radius of the sphere, d' and d are the densities of the crust and the granite, respectively, and η is the viscosity of the crust. This equation tells us that the granite sphere moves with a velocity equal to 10^{-9} cm/sec. In 1 million years, or $3 \cdot 10^{13}$ sec., the granite body has moved 300 meters; in 100 million years it has moved 30,000 meters.

These values are surprisingly large, and they may even be still larger because the viscosity of the crust may be considerably less than 10^{22} poises. This figure is determined on the basis of the rate of postglacial uplift of Fennoscandia (Haskell, 1935, 1936, 1937). The rigidity of the uppermost part of the crust is therefore incorporated in this figure, and it is rather likely that the viscosity of the rocks in the zones of regional metamorphism is appreciably less than 10^{22} poises. It is hence evident that bodily plastic movements of granitic bodies through the solid crust take place with sufficient velocity to account for the well-known anticline granites or domed gneisses, diapire granites, and many batholiths. In the course of a period of orogeny, such plastic upflow of solid granites will proceed many kilometers, even if the bodies are much less than 10 km. across (see Fig. 129).

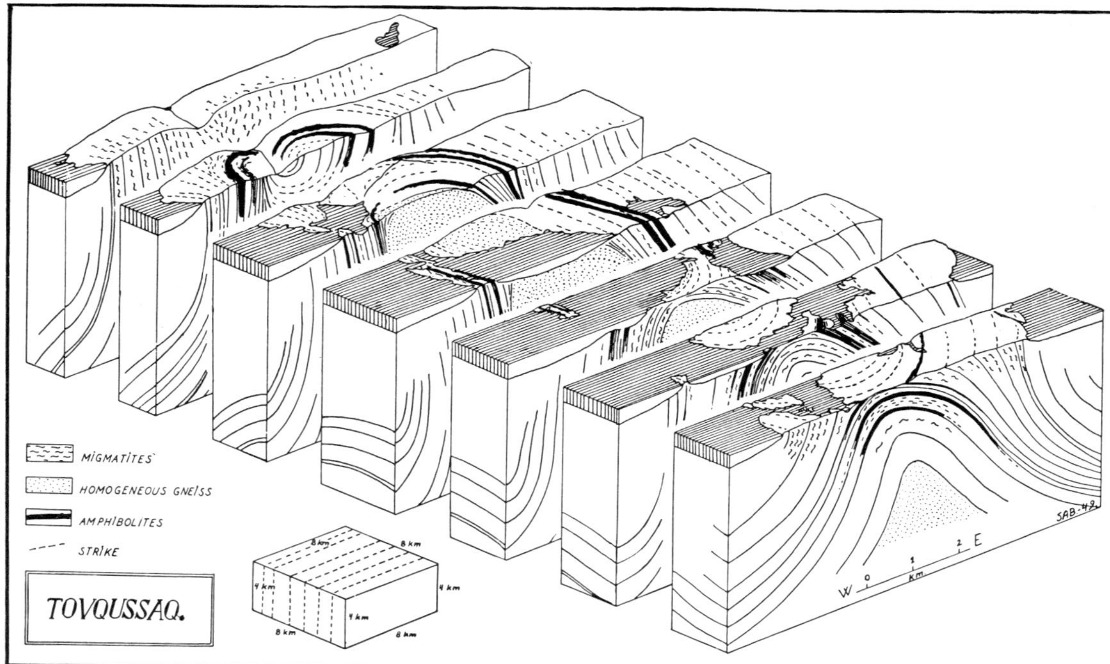

FIG. 129.—Block diagram of the Tovqussaq diapire "granite" in West Greenland, after A. Bertelsen (1950). This striking structure was detected by the writer and M. L. Ramberg in 1948 and was mapped by A. Bertelsen in 1949 on the writer's proposal.

This type of mechanical upflow in a bodily manner has little direct connection with the diffusional transfer which plays such an important role in regional granitization. But the fact that the granitic minerals—quartz, alkali feldspar, and micas—have low densities makes these minerals also thermodynamically most stable in the uppermost part of the earth's crust, as discussed on pages 37 ff. For that reason the granite minerals or minerals containing the same elements, such as quartz, feldspar, and micas (Si, Na, K, Cs, Rb, Al, O, H), will tend to disintegrate when existing at deeper levels in the earth, and, following the demands of thermodynamics, these elements must migrate upward and will not come to rest until they have consolidated in the quartz and alkali feldspar lattices within the zone of granitization. The minor elements which have strong affinity for tectosilicates and phyllosilicates may possibly, by the same process, tend to concentrate in the zones where granodiorites are stable. We have seen, in the chapter on metamorphic differentiation, that for certain reasons particular mineral groups attempted to separate out in a recrystallizing rock. Certain minerals tend to aggregate together. This effect must also be of importance on a large scale, so that minerals which find it particularly convenient from a crystal energy standpoint to consolidate in a quartz-feldspar milieu will preferably be found in quartzo-feldspathic rocks. As Rankama (1946) has expressed it: some elements are granitophile, others granitophobe.

The geothermal gradient would most likely act in the same direction with regard to the slow building-up of granitic gneisses, pegmatites, etc., in orogenic zones. At higher temperatures in deep zones, minerals tend to "evaporate" ("internal evaporation," [Frenkel, 1946]), and the movable constituents will creep toward shallower depths under continuous consolidation. Because the most mobile constituents are those which build granites, quartz and feldspar will preferably be displaced upward by this process also. To compensate for the upward diffusion current, the whole "crust" sinks bodily, so to speak, through the swarm of slowly rising atoms, ions, and molecules. In part, the calco-ferromagnesian elements are also migrating chemically downward. In any case, the net result will be that granodioritic matter is rising and calco-ferromagnesian matter is sinking in the evolving folded zones, thereby minimizing the thermodynamic instability which causes the orogenic evolution. In other words, we may take the view that the energy potential gradient which drives the orogeny is the same as that which drives the granitization and other kinds of large-scale metasomatism. Is it possible that the folding, thrusting, and plastic flow of the rocks, as well as their recrystallization and metasomatism, together with their local refusion, are all visible proofs of the exhaustion of the giant energy potential gradient which stands behind the whole phenomenon of orogeny?

BASIFICATION

In intensively altered gneissic complexes, granodioritic gneisses are by far the most prominent rocks. Among them, "inclusions," layers, bands, remnants, "schlieren," etc., of basic rocks containing different hornblendes and plagioclase ± garnet ± pyroxene ± olivine ± cordierite are often the only nongranitic components. To a certain extent, such rocks are formed by a kind of metasomatism either by subtraction of "granitic" or "salic" elements from the original less basic rocks or by introduction and fixation of "basic" or "mafic" elements: Mg, Ca, Fe, Mn, Ti, etc. There is evidence also that some ultrabasic rocks rich in olivine and pyroxene have originated in a similar way.

There is, by no means, any likelihood that the major portion of the amphibolites in gneissic complexes have originated through basification of originally less basic rocks. There are reasons for these rocks commonly being recrystallized basic igneous rocks, like basalt layers and diabase dikes.

We stated above that "granitization" is quantitatively a much more important process than "basification." That holds true within the outermost layer of the earth, in the so-called "granitic shell," where basification plays an interesting but minor role. But the granitic layer is supposed to be but some 15 km. thick (Bullen, 1947), and one may ask, What happens below this layer, in the deeper portion of folded mountain ranges? Is that the place where rather basic rocks like anorthosites are created by large-scale basification in somewhat the same way that huge volumes of granitic gneisses are created by granitization at a higher level? That is what thermodynamics claims, because the calco-ferromagnesian silicates are more stable than the granodioritic minerals below a certain level in the earth's crust.

Reynolds (1946) was the first to point out the phenomena of basification as an integral part of granitization. When gneisses, metagranites, and pegmatites form by replacement of old rocks of less granitic composition, then Mg, Fe, Ca, and perhaps some other "basic" elements must have escaped, provided that the granitization proceeded as a volume-by-volume replacement. Although granitization by no means always takes place as a volume-by-volume job (some swelling is evident in several instances), it seems evident that Mg, Ca, and Fe^{2+} (Ti, Fe^{3+}, etc.) often have to leave the rock volumes which are affected by granitization. These elements may become fixed again outside the granitization front or in remnant inclusions inside the metasomatic granite. Or, which seems more likely, the basic elements migrate downward against the advancing "granitic" elements.

Within the zones of metamorphism where quartzo-feldspathization takes place, basification appears to be a rather local and small-scale phenomenon, in agreement with the reasonable conclusion that calco-ferromagnesian elements are less mobile in ordinary rocks than are granitic elements. Along growing pegmatites, calco-ferromagnesian minerals may concentrate (Fig. 120), and inclusions or remnants in metasomatic granites may be basified. The Mg-Al metasomatism creating cordierite-anthophyllite rocks may also represent a facies of basification (Bugge, 1943). Several authors have reported field observations indicating that dunitic bodies have grown by Mg metasomatism of dolomites or limestones. Relationships between small dunitic bodies and amphibolites in gneisses from West Greenland are interpreted as if the ultrabasites formed by extraction of Ca, Al, Na, and Si from common amphibolites (unpublished observations by the writer). The latter are but small-scale processes. It has been known by field petrologists for a long time that dunites certainly never crystallized from a silicate melt and that the classic pyrogenetic view fails as an attempt to explain their genesis. It is encouraging to see that experimenters have reached similar conclusions (Bowen and Tuttle, 1949).

Lapardu-Hargues (1945) has demonstrated that statistically low-grade rocks are richer in Mg, Ca, and Fe than are high-grade rocks, which are richer in K, Na, Al, Si, etc. He concludes that Mg, Ca, and Fe have migrated farthest and have been fixed in the low-grade rocks. His data as compared with recent sediments may also be interpreted in a different way, because his low-grade rocks appear to represent a fairly good average composition of recent sediment, so that these low-grade rocks are not significantly altered with reference to bulk chemical composition. The data thus demonstrate that Na, K, and Ca (Al) have been added to the high-grade schists, rendering them more granitic or granodioritic.

One would expect that the "basic" elements (Fe, Mg, Ca) would concentrate below the crustal sphere of granitization. But this assumption is not supported by field observations, which show that the deeper down in the crust erosion has dug, the more quartzo-feldspathic are the rocks. This could mean, however, that erosion is usually not able to uncover the basic rocks which should exist in deeper parts of the earth's crust. On rather rare occasions, though, it seems as if the deep zones of wholesale basification are partly uncovered by erosion, namely, in areas of extensive granulite facies rocks. In West Greenland, for example, there are vast areas of granulite facies gneisses, and it appears that the rocks in the granulite facies areas are, on the average, somewhat more basic than the main gneiss in surrounding lower facies areas (Table 20). It is believed that K, Na, Si, O, and H_2O are "squeezed" out chemically from the once deep-seated granulite facies areas and that these upward-migrating elements have concentrated themselves in higher levels. This means either that "basic" elements migrated downward or that the upper granitized portions underwent swelling at the same time as the lower levels shrank. In either case, the effect is basification of the granulite facies area.

TABLE 20

AVERAGE COMPOSITION OF GNEISSES IN WEST GREENLAND*

Oxides	Epidote Amphibolite and Amphibolite Facies Gneiss	Granulite Facies Gneiss
SiO_2	70.13	64.37
TiO_2	0.32	0.59
Al_2O_3	15.83	16.26
Fe_2O_3	0.83	1.93
FeO	1.78	4.40
MnO	0.03	0.09
MgO	0.69	2.21
CaO	3.02	5.69
Na_2O	4.23	3.50
K_2O	2.36	0.63
P_2O_5	0.13	0.32
H_2O^+	0.44	0.15
H_2O^-	0.03	0.12
Density	2.67 g/cc	2.78 g/cc

* Data from Ramberg, 1951.

It is highly suggestive now that anorthosites practically always are situated in granulite facies areas. Because of the difficulties involved in having anorthosites crystallize as primary magmatic bodies and the numerous indications of replacement and metasomatism participating (Newhouse, oral communication) in the evolution of anorthosites, it may be worth while to consider the idea of deep-seated basification as a clue to the evolution of these puzzling rocks.

SKARN FORMATION IN REGIONAL METAMORPHISM

Originally, metasomatism was believed to be confined to limestones: calcite and dolomite are more soluble and chemically reactive than the rigid and chemically inert silicates. Sulfide ore and iron oxide ore were frequently found to replace marbles (as, for example, described by G. von Rath and B. Lotti in 1870 from Elba, and by Goldschmidt in 1911 from Kristiania). Together with the ore minerals, there occurred the so-called "skarn" minerals, which are mainly iron-rich dark lime silicates—e.g., andradite, vesuvianite, hedenbergite, etc. Although it was later demonstrated that silicate rocks are also easy subjects for replacement and metasomatism, the replacement of limestones results in such an abundance of interesting reaction minerals that such occurrences naturally attract special attention in areas of "common" monotonous gneisses and crystalline schists. Metasomatism of calcite and dolomite marbles under regional metamorphism results in the typical diopside, scapolite, hornblende, phlogopite, etc., rocks which are so characteristic that they leave no doubt as to their real origin.

"Impure" marbles will yield a large number of lime-carrying silicates just by internal chemical reactions (Laitakari, 1921; von Eckermann, 1922–23; Magnusson, 1930; Ramberg, 1944*d*). Introduction of elements into the marbles from outside sources is also very common. This process creates the same types of minerals as those formed by mere recrystallization. It is therefore very difficult to discriminate between these two states in the specific cases. Field observations regarding the skarn rock in structural relations to the surroundings usually give evidence in one way or another. Advanced alterations of marbles often creates amphibolite with relicts of skarn diopside and scapolite. Such diopside may remain stable even after complete granitization has taken care of the old marble, and a diopside-carrying granitic gneiss may, for example, be the final result. Ore minerals, as is well known, preferably replace marbles and dolomites.

ORE FORMED BY REGIONAL METASOMATISM

There are several types of ore bodies which develop under regional metamorphosis and owe their existence to metasomatic concentration of ore-making constituents originally scattered more evenly throughout the altered igneous and sedimentary rocks. Concentration of calcopyrite and pyrite bodies during dynamothermal alterations of basic lavas and gabbros (green schists) are typical (Lindgren, 1933). Magnetite-hematite ores in metasediments and titanic iron ores concentrated in granulite facies are other examples. Some cobalt ore is believed to have formed in a similar way (Rosenquist, 1949). Ore in skarn rocks formed through dynamothermal metamorphism without the participation of melts ("reaction skarn," Magnusson, 1930) are concentrations created by regional metamorphic differentiation. The isotopic ratios of lead in lead-zinc ore are commonly such that one may conclude that many such ores simply represent a concentration of rock lead which was once dilutely scattered throughout the surroundings (Holmes, 1947). The field and physicochemical evidence for nonmagmatic origin of many ore bodies is increasing in strength and amount. It is rather significant that students of volcanoes and lavas do not consider ore to be an effect of igneous activity. The volcanologist, Verhoogen, wrote in 1946: "The igneous intrusions associated with certain mineral deposits are not necessarily the origin of these deposits—they might in some cases simply be a product of process through which the deposit was formed." It is demonstrated that many ore bodies found in the geologic units here discussed actually represent concentrations of ore minerals in physically, structurally, or chemically favorable places (Ramberg, 1948*a;* Sullivan, 1948; Newhouse, 1940). Just as potassium, sodium, silicon, aluminum, lithium, etc., are relatively mobile during metamorphic activation, so also are some of the ore-building elements. Typical examples of ore formed by regional metasomatism are some pyrite and calcopyrite occurrences encountered in, or close to, metagabbros or metabasalts. The recrystallization of the basic magmatic rocks in epidote amphibolite facies enabled pyrite and other ore sulfides to move around and separate out as concentrated bodies. The places where this precipitation occurred are determined by several less understood factors. That structure, openings, shear zones, etc., are very important, just as in the case of pegmatitic growth, is demonstrated beyond doubt by Newhouse and his co-workers. Gravity apparently also plays a role, for the metagabbros in the Caledonian folded chain are often situated above the pyrite bodies, which are believed to have been segregated out of the gabbro during the recrystallization. As discussed elsewhere (p. 37), this effect of gravity is a natural consequence of the stability of minerals in the gravitational field. During the recrystallization of the gabbro which contains a small concentration of sulfides throughout, the whole body and it closest surroundings are saturated with the sulfides. Owing to the effect of gravity, the system will be faintly supersaturated with reference to sulfides in its lowest portions. For that reason, the tendency will be to move the sulfides down to those portions by activation, diffusion, and consolidation. Owing to the similarity of surface physical properties, the several sulfides will tend to cluster together and try to escape from being intermingled with silicates.

Another example of ore formed in a similar way is the ilmenite-magnetite ore encountered in anorthosites and in other granulite facies rocks. It has been demonstrated how the silicates, which are stable in the granulite facies, are unable to incorporate appreciable amounts of titanium in their lattices (Ramberg, 1948*a*).The appearance of titanic iron ore in this facies is therefore, in all likelihood, connected with the exsolution of Ti from the lattices of biotite, hornblende, and sphene when amphibolite facies rocks are rearranged within the granulite facies.

VARIOUS QUANTITATIVELY LESS IMPORTANT KINDS OF METASOMATISM IN REGIONAL METAMORPHISM

SPILITIZATION

Some diabases and basalts which contain albitic plagioclase instead of the basic plagioclase typical of the ordinary basic rocks are called "spilites" (see the discussion and references in Gilluly, 1935). Usually such rocks have an ophitic relict fabric and contain pyroxene and/or uralitic hornblende and often chlorite, biotite, and serpentine, together with the critical albitic feldspar (Eskola, 1939, p. 380). The low-temperature minerals, like uralite, chlorite, and serpentine, show that such rocks have been altered after their solidification from the basic melt, and the sodic plagioclase has therefore, in all likelihood, originated by some kind of soda metasomatism. It is particularly interesting that Eskola, Vuoristo, and Rankama (1937) have shown experimentally that such an albitization of basic plagioclase may take place in connection with Na_2CO_3 solutions.

Now the spilites with ophitic relict fabric have evidently not undergone dynamothermal metamorphism, and they should perhaps not be classed as regionally metamorphosed rocks. It is actually believed that these types of spilites are autometasomatic: i.e., that rest solutions left after the main solidification of the basaltic melt are responsible for the spilitic composition (Bailey and Grabham, 1909). According to Dewey and Flett (1911), the gain of soda is related to the common submarine nature of the spilites, which often are connected with pillow lavas.

However, most spilites show mineral assemblages corresponding to the epidote amphibolite and green schist facies (p. 145), and it is logical to assume with Turner (1948) that spilitization is generally connected with the regional metamorphic recrystallization of common basic lavas or dikes in these low facies. One may assume that the unstable anorthite component under certain circumstances is withdrawn from the spilitized greenstone at the same time as soda and silica are gained. After all, albitization is a very common type of metasomatism in low-grade schists and gneisses, and the numerous epidote, silica, and calcite veinlets in greenstones prove that Ca and Si (in one form or another) are also rather mobile in low-grade basic rocks.

BORON METASOMATISM—TOURMALINIZATION

In the chapter on the mechanism of metasomatism (p. 201), we concluded that the mobility or diffusivity of the substances through rocks depended chiefly upon the bonding or adsorption energy between the mineral lattices and the moving particle. In harmony with this concept, one would think that boron in one form or another would be very mobile in rocks, because the common silicates show little affinity to boron. Boron is mainly found in a few specific minerals, among which tourmaline is the most common. Boron metasomatism in regionally metamorphosed schists is, therefore, easily understood, although the question of the source of the boron may be more confusing. The small, but varying, amount of tourmaline found in most mica schists may, to a great extent, be formed by recrystallization of the original argillaceous sediments, which, according to Goldschmidt and Peters (1932), contain rather significant amounts (some 0.1 per cent) of B_2O_3.

Turner points out, however, that so straightforward an explanation cannot be accepted with respect to the boron-carrying complex of southern New Zealand. These rocks are quartzose or albitic graywackes and green schist of volcanic parentage. Since the amount of tourmaline increases when the granitization becomes intensified and the amount of pegmatites increases in the schists, Turner believes that the boron has been derived from the same granitic magma which caused the granitization. It may certainly also be believed that the comparatively mobile boron constitutes a relatively large fraction of the moving granitizing particles, even if the granitizing agency is not of magmatic parentage. Large-scale metamorphic differentiation could also explain the concentration and fixation of tourmaline in certain rocks. In this case, recrystallizing boron-carrying sediments may function as the source of boron just as easily as could a granitic melt. It is known that some pegmatites grown as secretions by metamorphic differentiation are able to concentrate tourmaline from the country schist (Fig. 122, p. 254).

CHLORINE—CARBON DIOXIDE METASOMATISM—SCAPOLITIZATION

From the presumption that metasomatism is identical with the outflow of fugitive substances from a solidifying melt, it was commonly assumed that minerals containing some fugutive constituents like B, F, and Cl (CO_2, S) are of metasomatic parentage, whereas minerals lacking such elements were "primary," igneous, clastic, or metamorphic. It has therefore often been taken for granted that tourmaline in schists was of metasomatic-pneumatolytic origin, although the work of Goldschmidt and Peters has shown that this is not necessarily true. In the case of the occurrence of scapolite in rocks, a metasomatic origin is also accepted—perhaps too often. The content of the "fugitive" Cl, CO_2, and SO_3 in the scapolite lattices gives rise to this prejudice. Sundius (1916) has pointed out that, in many cases, scapolite may develop in rocks by simple recrystallization of impure limestones, because the fugitive constituents are often originally present in such rocks. Numerous studies of regionally metamorphosed limestones have shown that such a nonmetasomatic origin of scapolite is rather common.

On the other hand, a metasomatic origin of scapolite is common in some regionally altered rocks. The chemical formula of scapolite corresponds to plagioclase, with some CO_2, Cl, SO_3, and H_2O in addition. We know that these constituents are mobile under regional metamorphism, and it is clear also that such fugitive constituents may be liberated by several possible processes: CO_2 is set free during reactions between carbonate rocks and silicates; SO_3 may form, for example, by oxidation of sulfide minerals or by dissociation of gypsum in sediments; Cl is available also, for instance, from watery solutions in porous sediments, from metamorphosed rock salt sediments, or from crystallizing magmas. A large-scale scapolitization during regional metamorphism should therefore offer no difficulties in reconciling it with our geologic knowledge. Another common form of CO_2 metasomatism is found connected with the recrystallization of basic igneous rocks in the green schist facies. At this low temperature many calc silicates are unstable: they will split up into quartz and calcite or dolomite, provided that CO_2 is available. The fact that this process commonly takes place in low-grade greenstones shows that CO_2 is able to migrate toward those places where the stabilized minerals require it.

REGIONAL WATER METASOMATISM

Regionally metamorphosed rocks always contain water as a constituent of some of the mineral lattices, as inclusions in the mineral grains, and/or as adsorbed molecules in the intergranular system. Because of the significance of water as a catalyst in rock alteration, the migration of water in rocks during regional metamorphism has been discussed elsewhere (p. 180). However, not only does water function as a catalyst, but it may also be a significant rock-building constituent. Hydrous minerals are abundant in low- and medium-grade rocks. Chlorite, talc, serpentine, epidote, and muscovite, as well as biotite and hornblendes, are all common minerals in low-grade rocks. Such minerals have HO in their lattices and cannot form unless H_2O is available. In many cases it is clear that the necessary water was introduced into the rocks during their formation. Examples are the soapstones formed by water metasomatism of dunites and other ultrabasites, and the greenstones developed by water metasomatism of gabbroic igneous rocks.

It is very interesting that water appears to be available wherever needed to form the hydrous minerals stable in the green schist and epidote amphibolite facies. Two important conclusions may be drawn from this experience: (1) H_2O is very mobile in common rocks during regional metamorphism; (2) the migration of H_2O is importantly controlled by its chemical potential gradient: The relatively low chemical potential of H_2O in the ultrabasite which slowly alters to a soapstone makes H_2O migrate into the growing soapstone from the surroundings. It is chiefly the stability conditions of the minerals which direct the motion of the several constituents during metasomatism.

TEMPERATURE IN REGIONAL METAMORPHISM

THE high temperature at which contact metamorphism occurs is readily understood in view of the heat liberated when the hot intruded magma crystallizes. The temperature rise which causes progressive regional recrystallization, however, is not so self-evident unless the temperature zoning in regional metamorphism is explainable as a consequence of depth of burial and the "normal" geothermal gradient. Although hard to prove, it is commonly believed that the depth of burial combined with a "normal" geothermal gradient is unable to account for the comparatively high temperature of regional recrystallization. Let us briefly check this assumption. The average "normal" geothermal gradient may be some 25° C/km in the outermost 3–4 km. of the earth's crust. Not knowing how the gradient changes with depth, we may reasonably guess that it is less rather than greater, than 25° C/km. Now what is the average depth of burial of the several regional metamorphic facies? The granulite facies, representing the highest P,T-conditions of regional metamorphism, is a crucial one to consider in this connection. If the pressure represented by the granulite facies mineral assemblages were known, the depth could have been determined rather accurately. This pressure is unknown, however, so we shall attempt to arrive at the depth of burial by some other means. Our answer on this question will depend most decidedly upon the theory of the architecture of folded mountain ranges. We shall accept the viewpoint that folded mountain ranges have a "root" reaching down into the substratum. Let us assume that the mean density of the folded mountain is some 2.7 g/cc and that of the substratum 3.1 g/cc; that the granitic shell is 10 km. thick; and that the newly born, but isostatically balanced, mountain range has an average height of 4 km. The depth of the bottom of the "root" will then be some 27 km. below the "normal" bottom of the gneiss-granitic shell (Fig. 130).

In the ancient gneiss-granitic peneplainized shields where granulite facies is situated, the present surface represents a considerably deep cut through the old folded mountains. Since the old peneplains are isostatically pretty well balanced, we may conclude that they do not have a "root" down in the substratum. The root of the old leveled mountains must therefore have vanished in the course of time, by erosion and elevation of the orogeny and/or by plastic lateral flow. In case the vanishing of the "root" is due solely to erosion and consequent elevation, it follows that the surface of the present peneplains is some 25± km. below the surface of the mountain ranges when they were at their highest (Fig. 130). A depth of 25 km. and a "normal" geothermal gradient of some 20° C/km would give some 500° C. for the granulite facies. This value is somethat too low (granulite facies=600°–700° C., p. 137), and it becomes still lower, for it is unlikely that the vanishing of the mountain roots during peneplainization is due solely to elevation. Some lateral plastic flow must have taken place deep in the sialic roots, such that the eroded top part of the mountains is less than 25 km. thick. The common assumption that the temperature in regional metamorphism is too high to correspond to a normal geothermal gradient seems, therefore, valid.

There are several possible explanations for a regional rise in crustal temperature under orogenic evolution: (1) Heat may be introduced by in-

FIG. 130.—Idealized cross-section of a folded mountain range with sialic root in sima (see text).

truded hot magmas, either early swarms of diabase (Wegmann, 1948) or syn- and late-orogenic quartzo-feldspathic melts. (2) Heat may be evolved by the intense mechanical shearing and deformation of the rocks in the orogenic belts. (3) Radioactive decay in the thickened sial in orogenies may be intensified enough to account for the higher temperature here than in the thinner "normal" parts of sial. (4) Exothermic chemical reactions may play a part. (5) Convection currents in the substratum may displace hot portions of the substratum toward the zones of geosynclines and folded mountains. (6) The net effect of the complex process of folded mountain formation may be exothermic, because this geologic event is a large-scale equilibriopetal process (Ramberg, 1945; van Bemmelen, 1949), aiming at an establishment of thermodynamic and mechanical equilibrium out of the nonequilibrial situation which precedes and causes the orogeny.

We shall consider briefly a few of these points. Point 1 offers a simple and apparently sound explanation of the rise of temperature in orogenies. However, a closer checkup makes point 1 an invalid explanation for the *regional* rise of temperature, although it offers an explanation for some displacement of temperature *within* the orogenic zones. We have here in mind the fact that rising magmas will bring high temperature and heat from deeper to shallower levels within the orogeny. On several earlier occasions in this book we have shown that field and thermodynamic data are in disharmony with the assumption that regional metamorphism is a huge "contact" reaction zone around hot magmatic melts.

Here we shall bring in another point, namely, the well-established, intimate association between *orogenies* and *volcanism*, in particular, acidic and calcalkalic volcanism. What is the causal relationship here? Is the whole orogenic tectonic-metamorphic event caused by the magmatic activity confined to such zones, is the occurrence of magmatic melts caused by the orogenies, or are the orogeny and the calcalkalic magmas created by some common agency? According to the several theories of folded mountain formation, there is no reason to believe that intrusion of magmas in the orogenic zones is the driving force of orogenies, although displacement of magmas may locally cause some foldings. The causal sequence is most likely the other way around: *the event of orogenic evolution creates magmas*. That is, during the evolution of orogenies, the temperature rises above "normal"; and, at the peak of the temperature rise, quartzo-feldspathic melts occur by refusion of parts of the crust. The magmas are of "secondary" origin. For the reasons mentioned above, there are no grounds for believing that regional metamorphism is chiefly caused by the heat liberated from intruded melts.

As to points 2, 3, and 4, all of them most reasonably contribute somewhat to the elevated regional temperature in orogenies.

The heat transfer connected with subcrustal convection currents is indeed questionable, in particular, because the whole convection-current theory of folded mountains is as yet only hypothetical (van Bemmelen, 1949–52).

Let us then consider point 6 in some detail. It is very likely that the whole sequence of geological processes which result in the folding and rise of folded mountain ranges is a large-scale event which proceeds spontaneously, tending to create a stable situation out of the unstable state of affairs of the geosynclinal stage. In the urge of nature to achieve stable equilibrium throughout the parts of the crust (and substratum) which are occupied by the geosyncline, displacement of matter on a small and large scale is crucial. Thus rock matter is displaced mechanically by faulting, folding, and plastic creep or flow of solid rocks and flowage of molten magmas. We know also that another kind of material transport is operating, namely, the diffusional migration creating metasomatism and growth of petroblastic rocks in the solid state. This latter kind of material displacement is intense and large-scale in deeper sections of folded mountains. During such a lively chemical exchange of rock-making matter, it is inevitable that significant heat effects occur. Since granitization and allied metasomatism largely mean consolidation of energy-rich or activated atomistic particles "at countless points throughout the rock," it is fair to assume that the net heat effect will be positive. That is, granitization and other kinds of metasomatic replacements should contribute much to the heat required, to account for not only the elevated temperature in regional metamorphism but also refusion of acidic magmas.

APPENDIX

CHEMICAL BONDS IN CRYSTALS: A CLUE TO PREDICTING RELATIVE STABILITY OF MINERALS AND CATION DISTRIBUTION AMONG THEM

In earlier chapters (pp. 5–75) we have studied how the scattered observations from natural and synthetic mineral assemblages can be interpreted in view of the general laws of chemical equilibrium, thereby providing us with greatly expanded information as to the behavior of mineral associations under changing P,T,X-conditions.[1] In this procedure the field and laboratory data on the relative stability of minerals and the element distribution among them were necessary prerequisites on which the theoretical expansion of the information was founded. It is now highly valuable to dig deeper into the theory of crystal energy and the theory of chemical bonds in crystals, because these theories can provide us with information about the data prerequisite to thermodynamic handling.

The concept of lattice or crystal energy has been mentioned on several previous occasions. It will be studied in some detail in the following pages.

The ions or atoms in minerals are kept together by chemical bonds of ionic and/or covalent nature. The strength of such bonds can conveniently be measured by the energy which they contain as compared with a defined standard state. When comparing the binding energy in a series of ionic crystals, for example, it is convenient to consider a very rarefied ionic gas as the standard state of zero energy. According to classical electrostatics, the attractive energy of a 100 per cent ionic bond between a cation and an anion separated by a distance d is, then,

$$\epsilon = -\frac{z\,e \cdot y\,e}{d}, \qquad (155)$$

where z is the number of charges on the one ion, y is the number of charges on the oppositely charged ion, and e is the unit charge per electron. The foregoing expression has to be multiplied by certain factors in order to give the attractive energy in calories, ergs, or any other energy unit. The larger the *numerical* value of ϵ, the *lower* is the energy of the combined ions, and the *stronger* is the electrostatic bond between them.

Two equally charged ions (i.e., two anions or two cations) which are forced to approach each other represent a positive energy, as compared with the ionic gas of zero energy. The general formula is the same as above, but with opposite sign, indicating that the two equally charged ions will repel each other and that the dilute gas of equally charged ions possesses less energy and is more stable than the condensed ions. By thus considering the attractive (negative) energy between unlike ions and the repulsive (positive) energy between like ions in ionic crystals, it is possible to give a general formula for the crystal (or lattice) energy of a number of types of simple ionic crystals:

$$E = -N \cdot A \cdot \frac{e^2 \cdot z^2}{d}\left(1 - \frac{1}{n}\right), \qquad (156)$$

1. X means bulk chemical composition of the mineral assemblages.

where E is the energy per mol of the crystal; N is Avogadro's number (6.03×10^{23}); e, z, and d are explained above; n is a certain number, often equal to 9, characterizing the small repulsion between unlike ions; and A is the so-called *Madelung constant*, which is solely determined by the geometry of the crystalline structure, i.e., the relative spatial positions of the several ions. The calculated Madelung constants for some simple types of ionic structures are given in Table 21 (see Pauling, 1949).

TABLE 21

Structure Type	Madelung Constant A	Structure Type	Madelung Constant A
Sodium chloride, M^+, X^-	1.748	Cuprite, $2M^+$, X^{2-}	4.116
Cesium chloride, M^+, X^-	1.763	Rutile, M^{2+}, $2X^-$	4.816
Sphalerite, M^+, X^-	1.638	Cadmium iodide, M^{2+}, $2X^-$	4.710
Wurtzite, M^+, X^-	1.641	β-quartz, M^{2+}, $2X^-$	4.439
Fluorite, M^{2+}, $2X^-$	5.039	Corundum, $2M^{3+}$, $3X^{2-}$	25.031
Antifluorite, $2M^+$, X^{2-}	5.039		

The crystal energy of NaCl with $d = 2.81$ A is then, for example, equal to:

$$E_{\mathrm{Na}} = -1.748 \cdot N \cdot \frac{e^2 \cdot 1^2}{2.81 \cdot 10^{-8}} \cdot \left(1 - \frac{1}{9}\right).$$

If we wished to calculate the crystal energy of MgO or CaS, which also have the sodium chloride structure, we should remember that z is now equal to 2. The crystal energy of quartz would theoretically be:

$$E_{\mathrm{SiO_2}} = -4.439 \cdot N \cdot \frac{e^2 \cdot 2^2}{1.62 \cdot 10^{-8}} \cdot \left(1 - \frac{1}{n}\right).$$

Also here $z = 2$ because the Madelung constant, 4.439, refers to a quartz-type lattice of divalent cations and univalent anions. Likewise, if the crystal energy of sphalerite, ZnS, should be evaluated, we would have to give z the value of 2 because the Madelung constant as given above refers to crystals of univalent ions. Some nitrides of trivalent metals, like AlN and GaN, adopt the wurtzite structure; in such cases z takes the value of 3, etc.

The theoretical crystal energy calculated from the above formula comes rather close to the actual calorimetrically measured energy in cases where the ions are not appreciably *polarized* or where *covalent* (or metallic) bonds are not significantly present in the crystals. *Polarization* in ionic compounds means chiefly deformation of the *anions* in the electrostatic field of the neighboring *cations*, because the anions are much more easily deformable in an electric field than are most cations. The stronger the electrostatic field of a cation, the higher is its polarizing power. The deformability or polarizability of the anions generally increases with increasing size of the ion and with increasing number of negative charges. Hence the polarizability increases in the sequences: F^-, Cl^-, Br^-, I^-, and O^{2-}, S^{2-}, Se^{2-}, Te^{2-}, and the doubly charged ions are more polarizable than are the singly charged halogens (Table 22). It is of great consequence that polarization always strengthens the chemical bonds between ions by an amount increasing with the increasing polarizing power of the cation and with the increasing polarizability of the anion, but also depending upon the spatial arrangement of the ions in the crystal or molecule. The ability of polarization to strengthen the binding between ions depends upon the geometry of the structure, in such a way, generally, that the less symmetrical the arrangement of cations around the anions and the lower the co-ordination number of the anions with respect to the cations, the more effective is the polarization as a bond-strengthening factor. The effect of polarization becomes most significant in molecules in which the anions are attached to a cation on one side only, as, for example, Cl in $\mathrm{Cl\overset{Cl}{\underset{Cl}{Si}}Cl}$, or I in LiI molecules. It becomes less effective in co-ordination lattices where the anions are symmetrically surrounded by cations, which therefore, to some extent, cancel each other's polarizing power.

Together with polarization, a covalent share of electron pairs between ions occurs to various extents in many ionic crystals. This also helps to increase the strength of the interionic bond.

We shall consider some of the consequences of these introductory statements by

TABLE 22

POLARIZABILITY OF SOME ANIONS*

Ion	Polarizability	Ion	Polarizability
Fe^{-}	$9.8\times10^{-25}cm^3$	O^{2-}	$27.5\times10^{-25}cm^3$
Cl^{-}	$35.3\times10^{-25}cm^3$	S^{2-}	$86\ \times10^{-25}cm^3$
Br^{-}	$49.7\times10^{-25}cm^3$	Se^{2-}	$112\ \times10^{-25}cm^3$
I^{-}	$75.5\times10^{-25}cm^3$	Te^{2-}	$157\ \times10^{-25}cm^3$

* From Landolf-Börnstein *Tabellen*.

discussing examples of chemical reactions among ionic crystals. Double reactions of the following general types are very frequent among minerals:

$$\mathrm{MgS} + \mathrm{BaO} \rightleftharpoons \mathrm{MgO} + \mathrm{BaS} + \Delta H\,. \tag{157}$$

The theory of thermodynamics cannot predict which side of this equation is stable at given P,T-conditions unless, for example, the heat of reaction and the entropies of the compounds are known.

The heat of reaction is, in particular, a valuable quantity to know, even if only as a rough approximation, because the sign of ΔH decides whether the right or the left side of the equation would be stable below a certain temperature limit. The magnitude of ΔH indicates whether the temperature limit is high or low. Specifically, if ΔH is negative, i.e., if the reaction proceeds toward the right exothermally, then the right-hand side of the reaction is stable below a certain temperature limit, the value of which almost invariably rises with the increasing numerical magnitude of ΔH. A positive ΔH shows that the left hand side is the stable one from low temperature up to a certain limit.

The value of ΔH is related to the crystal energy of the reactants (MgS + BaO) and the products (MgO + BaS), as follows: $\Delta H = -E_{\mathrm{MgS}} - E_{\mathrm{BaO}} + E_{\mathrm{MgO}} + E_{\mathrm{BaS}}$; hence the theory of crystal energetics is in a position to give approximate ΔH values on the basis of simple knowledge of the ionic charges and sizes, as well as the geometry of the crystal structures involved.

This is true only if the crystals are essentially ionic. All the crystals in the reaction above are ionic and have the sodium chloride structure. The ions involved are doubly charged, so that the differences in crystal energy are essentially determined by the interionic distances, $d = r_M + r_X$, being different in the several crystals.

$$\begin{aligned}\Delta H = -E_{\mathrm{MgS}} - E_{\mathrm{BaO}} + E_{\mathrm{MgO}} + E_{\mathrm{BaS}} &= -N\cdot 1{,}748\cdot e^2\cdot 2^2\cdot\left(1-\frac{1}{9}\right)\\ &\cdot\left(-\frac{1}{r_{\mathrm{Mg}}+r_{\mathrm{S}}}-\frac{1}{r_{\mathrm{Ba}}+r_{\mathrm{O}}}+\frac{1}{r_{\mathrm{Mg}}+r_{\mathrm{O}}}+\frac{1}{r_{\mathrm{Ba}}+r_{\mathrm{S}}}\right),\end{aligned} \tag{158}$$

where r_{Mg}, r_S, etc., are the ionic radii of the corresponding ions. The sign of ΔH is determined by the sign of the term

$$\left(\frac{1}{r_{Mg} + r_O} + \frac{1}{r_{Ba} + r_S} - \frac{1}{r_{Mg} + r_S} - \frac{1}{r_{Ba} + r_O}\right),$$

which can be shown to be positive if $r_{Mg} < r_{Ba}$ at the same time that $r_O < r_S$, or if $r_{Mg} > r_{Ba}$ at the same time that $r_S < r_O$.[2] In general, the term in the parentheses is positive, and hence ΔH is *negative*, if the crystals on the right-hand side of the reaction contain *small negative ions combined with small positive ions, and large negative ions combined with large positive ions*, and the crystals on the left-hand side contain small negative ions combined with large positive ions and small positive ions with large negative ions. But the term

$$\left(\frac{1}{r_{Mg} + r_O} \cdots - \frac{1}{r_{Ba} + r_O}\right)$$

is negative, and hence ΔH is positive, if $r_{Mg} < r_{Ba}$ at the same time that $r_O > r_S$ or if $r_{Mg} > r_{Ba}$ at the same time that $r_O < r_S$. In this case the reaction will go toward the left below a certain temperature limit, because now the left-hand side contains small ions combined in one crystal and large ions combined in the other crystal. *In other words, in a condensed system of positive and negative ions of different sizes but the same number of charges, the stable assemblage (below certain temperature limits) is one in which small ions are combined with small ions and large ions with large ions.* We shall not venture to estimate quantitative values of ΔH, but we shall list several examples of the correctness of this simple principle. In the equations below, the experimental ΔH values in kcal/mol are given (based upon the difference in heat of formation of the compound from the elements in their standard state as given underneath each compound), as well as the ionic radii which are indicated in A units above each element:[3]

$$\underset{-56.1}{\overset{0.31\ 1.84}{\mathrm{BeS}}} + \underset{-144}{\overset{0.65\ 1.40}{\mathrm{MgO}}} \rightarrow \underset{-147.3}{\overset{0.31\ 1.40}{\mathrm{BeO}}} + \underset{-84.2}{\overset{0.65\ 1.84}{\mathrm{MgS}}} \quad \underset{-31.4}{\Delta H}, \tag{159}$$

$$\underset{-56.1}{\overset{0.31\ 1.84}{\mathrm{BeS}}} + \underset{-152}{\overset{0.99\ 1.40}{\mathrm{CaO}}} \rightarrow \underset{-147.3}{\overset{0.31\ 1.40}{\mathrm{BeO}}} + \underset{-114.3}{\overset{0.99\ 1.84}{\mathrm{CaS}}} \quad \underset{-53.5}{\Delta H}, \tag{160}$$

$$\underset{-56.1}{\overset{0.31\ 1.84}{\mathrm{BeS}}} + \underset{-141}{\overset{1.13\ 1.40}{\mathrm{SrO}}} \rightarrow \underset{-147.3}{\overset{0.31\ 1.40}{\mathrm{BeO}}} + \underset{-113}{\overset{1.13\ 1.84}{\mathrm{SrS}}} \quad \underset{-60.2}{\Delta H}, \tag{161}$$

$$\underset{-56.1}{\overset{0.31\ 1.84}{\mathrm{BeS}}} + \underset{-126.1}{\overset{1.35\ 1.40}{\mathrm{BaO}}} \rightarrow \underset{-147.3}{\overset{0.31\ 1.40}{\mathrm{BeO}}} + \underset{-111}{\overset{1.35\ 1.84}{\mathrm{BaS}}} \quad \underset{-75.1}{\Delta H}, \tag{162}$$

$$\underset{-84.2}{\overset{0.65\ 1.84}{\mathrm{MgS}}} + \underset{-152}{\overset{0.99\ 1.40}{\mathrm{CaO}}} \rightarrow \underset{-144}{\overset{0.65\ 1.40}{\mathrm{MgO}}} + \underset{-114.3}{\overset{0.99\ 1.84}{\mathrm{CaS}}} \quad \underset{-22.1}{\Delta H}, \tag{163}$$

$$\underset{-84.2}{\overset{0.65\ 1.84}{\mathrm{MgS}}} + \underset{-141}{\overset{1.13\ 1.40}{\mathrm{SrO}}} \rightarrow \underset{-147}{\overset{0.65\ 1.40}{\mathrm{MgO}}} + \underset{-113}{\overset{1.13\ 1.84}{\mathrm{SrS}}} \quad \underset{-31.8}{\Delta H}, \tag{164}$$

2. The writer is obliged to Mr. John C. Jamieson for the mathematical proof of this.

3. The energy values throughout this chapter are chiefly from the compilation by Bischowsky and Rossini (1936) and from Evans and Kubaschewski (1951).

$$\underset{-84.2}{\overset{0.65\ 1.84}{\mathrm{MgS}}} + \underset{-126.1}{\overset{1.35\ 1.40}{\mathrm{BaO}}} \rightarrow \underset{-144}{\overset{0.65\ 1.40}{\mathrm{MgO}}} + \underset{-111}{\overset{1.35\ 1.84}{\mathrm{BaS}}} \quad \underset{-44.7}{\Delta H}, \tag{165}$$

$$\underset{-114.3}{\overset{0.99\ 1.84}{\mathrm{CaS}}} + \underset{-141}{\overset{1.13\ 1.40}{\mathrm{SrO}}} \rightarrow \underset{-152}{\overset{0.99\ 1.40}{\mathrm{CaO}}} + \underset{-113}{\overset{1.13\ 1.84}{\mathrm{SrS}}} \quad \underset{-9.7}{\Delta H}, \tag{166}$$

$$\underset{-114.3}{\overset{0.99\ 1.87}{\mathrm{CaS}}} + \underset{-126.1}{\overset{1.35\ 1.40}{\mathrm{BaO}}} \rightarrow \underset{-152}{\overset{0.99\ 1.40}{\mathrm{CaO}}} + \underset{-111}{\overset{1.35\ 1.84}{\mathrm{BaS}}} \quad \underset{-21.6}{\Delta H}, \tag{167}$$

$$\underset{-113}{\overset{1.13\ 1.84}{\mathrm{SrS}}} + \underset{-126.1}{\overset{1.35\ 1.40}{\mathrm{BaO}}} \rightarrow \underset{-141}{\overset{1.13\ 1.40}{\mathrm{SrO}}} + \underset{-111}{\overset{1.35\ 1.84}{\mathrm{BaS}}} \quad \underset{-12.9}{\Delta H}. \tag{168}$$

In all these reactions ΔH is negative, showing that the right-hand combinations represent the lowest energy and are stable from low temperature up to certain limits which perhaps are determined by the melting of one or more of the members. In all cases the stable assemblages contain the combination of the smallest cation with the smallest anion and the largest cation with the largest anion. We also see that the greater the difference in radii of the two cations involved in each equation, the larger is the numerical value of ΔH, in close agreement with what one would expect from the theory of the crystal energy as expressed by equations (156) and (158).

These remarkably simple relations can serve as important models when mineral reactions in rocks are considered. We shall therefore find it advisable to show some other examples of the stability rules of combinations of ions of different size. Univalent positive and negative ions behave equally nicely and regularly as do the doubly charged ions:

$$\underset{-97.5}{\overset{0.60\ 1.81}{\mathrm{LiCl}}} + \underset{-133}{\overset{0.95\ 1.36}{\mathrm{NaF}}} \rightarrow \underset{-145.3}{\overset{0.60\ 1.36}{\mathrm{LiF}}} + \underset{-98.2}{\overset{0.95\ 1.81}{\mathrm{NaCl}}} \quad \underset{-13.0}{\Delta H}, \tag{169}$$

$$\underset{-97.5}{\overset{0.60\ 1.81}{\mathrm{LiCl}}} + \underset{-134.5}{\overset{1.33\ 1.36}{\mathrm{KF}}} \rightarrow \underset{-145.3}{\overset{0.60\ 1.36}{\mathrm{LiF}}} + \underset{-104.5}{\overset{1.33\ 1.81}{\mathrm{KCl}}} \quad \underset{-17.8}{\Delta H}, \tag{170}$$

$$\underset{-97.5}{\overset{0.60\ 1.81}{\mathrm{LiCl}}} + \underset{-133}{\overset{1.48\ 1.36}{\mathrm{RbF}}} \rightarrow \underset{-145.3}{\overset{0.60\ 1.36}{\mathrm{LiF}}} + \underset{-105.2}{\overset{1.48\ 1.81}{\mathrm{RbCl}}} \quad \underset{-20}{\Delta H}, \tag{171}$$

$$\underset{-97.5}{\overset{0.60\ 1.81}{\mathrm{LiCl}}} + \underset{-131.7}{\overset{1.69\ 1.36}{\mathrm{CsF}}} \rightarrow \underset{-145.3}{\overset{0.60\ 1.36}{\mathrm{LiF}}} + \underset{-106.2}{\overset{1.69\ 1.81}{\mathrm{CsCl}}} \quad \underset{-22.3}{\Delta H}, \tag{172}$$

$$\underset{-98.2}{\overset{0.95\ 1.81}{\mathrm{NaCl}}} + \underset{-134.5}{\overset{1.33\ 1.36}{\mathrm{KF}}} \rightarrow \underset{-133}{\overset{0.95\ 1.36}{\mathrm{NaF}}} + \underset{-104.5}{\overset{1.33\ 1.81}{\mathrm{KCl}}} \quad \underset{-4.8}{\Delta H}, \tag{173}$$

$$\underset{-98.2}{\overset{0.95\ 1.81}{\mathrm{NaCl}}} + \underset{-133}{\overset{1.48\ 1.36}{\mathrm{RbF}}} \rightarrow \underset{-133}{\overset{0.95\ 1.36}{\mathrm{NaF}}} + \underset{-105.2}{\overset{1.48\ 1.81}{\mathrm{RbCl}}} \quad \underset{-7.0}{\Delta H}, \tag{174}$$

$$\underset{-98.2}{\overset{0.95\ 1.81}{\mathrm{NaCl}}} + \underset{-131.7}{\overset{1.69\ 1.36}{\mathrm{CsF}}} \rightarrow \underset{-133}{\overset{0.95\ 1.36}{\mathrm{NaF}}} + \underset{-106.2}{\overset{1.69\ 1.81}{\mathrm{CsCl}}} \quad \underset{-9.3}{\Delta H}, \tag{175}$$

$$\underset{-104.5}{\overset{1.33\ 1.81}{\mathrm{KCl}}} + \underset{-133}{\overset{1.48\ 1.36}{\mathrm{RbF}}} \rightarrow \underset{-134.5}{\overset{1.33\ 1.36}{\mathrm{KF}}} + \underset{-105.2}{\overset{1.48\ 1.81}{\mathrm{RbCl}}} \quad \underset{-2.2}{\Delta H}, \tag{176}$$

$$\underset{-104.5}{\overset{1.33\ 1.81}{\mathrm{KCl}}} + \underset{-131.7}{\overset{1.69\ 1.36}{\mathrm{CsF}}} \rightarrow \underset{-134.5}{\overset{1.33\ 1.36}{\mathrm{KF}}} + \underset{-106.2}{\overset{1.69\ 1.81}{\mathrm{CsCl}}} \quad \underset{-4.5}{\Delta H}, \tag{177}$$

$$\underset{-105.2}{\overset{1.48\ 1.81}{\mathrm{RbCl}}} + \underset{-131.7}{\overset{1.96\ 1.36}{\mathrm{CsF}}} \rightarrow \underset{-133}{\overset{1.48\ 1.36}{\mathrm{Rb}}} + \underset{-106.2}{\overset{1.69\ 1.81}{\mathrm{CsCl}}} \quad \underset{-2.3}{\Delta H}. \tag{178}$$

The minus sign of all the ΔH's in these sets of chemical reactions shows that the right-hand-side assemblages represent the lowest energy. At the same time, the right-hand-side assemblages contain the combinations of small cations with small anions and of large cations with large anions. Furthermore, the greater the difference in ionic radii of the two metals involved in each assemblage, the greater the numerical value of ΔH. Note, for example, reaction (172) in contrast to reaction (176).

We shall also exhibit some examples of ionic compounds in which the valence of the cations is different from the valence of the anions:

$$\underset{-88.2}{\overset{\mathrm{S\ L}}{\mathrm{Na_2S}}} + \underset{-86}{\overset{\mathrm{L\ S}}{\mathrm{K_2O}}} \rightarrow \underset{-92}{\overset{\mathrm{L\ L}}{\mathrm{K_2S}}} + \underset{-103}{\overset{\mathrm{S\ S}}{\mathrm{Na_2O}}} \quad \underset{-20.8}{\Delta H}, \tag{179}$$

$$\underset{-88.2}{\overset{\mathrm{S\ L}}{\mathrm{Na_2S}}} + \underset{-82}{\overset{\mathrm{L\ S}}{\mathrm{Cs_2O}}} \rightarrow \underset{-87}{\overset{\mathrm{L\ L}}{\mathrm{Cs_2S}}} + \underset{-103}{\overset{\mathrm{S\ S}}{\mathrm{Na_2O}}} \quad \underset{-19.8}{\Delta H}, \tag{180}$$

$$\underset{-153.2}{\overset{\mathrm{S\ L}}{\mathrm{MgCl_2}}} + \underset{-290}{\overset{\mathrm{L\ S}}{\mathrm{CaF_2}}} \rightarrow \underset{-190}{\overset{\mathrm{L\ L}}{\mathrm{CaCl_2}}} + \underset{-261.5}{\overset{\mathrm{S\ S}}{\mathrm{MgF_2}}} \quad \underset{-8.3}{\Delta H}, \tag{181}$$

$$\underset{-153.2}{\overset{\mathrm{S\ L}}{\mathrm{MgCl_2}}} + \underset{-287.9}{\overset{\mathrm{L\ S}}{\mathrm{BaF_2}}} \rightarrow \underset{-205.3}{\overset{\mathrm{L\ L}}{\mathrm{BaCl_2}}} + \underset{-261.5}{\overset{\mathrm{S\ S}}{\mathrm{MgF_2}}} \quad \underset{-25.7}{\Delta H}, \tag{182}$$

$$\underset{-39}{\overset{\mathrm{S\ L}}{\mathrm{SiS_2}}} + \underset{-219}{\overset{\mathrm{L\ S}}{\mathrm{TiO_2}}} \rightarrow \underset{-206}{\overset{\mathrm{S\ S}}{\mathrm{SiO_2}}} + \underset{-80\pm 20}{\overset{\mathrm{L\ L}}{\mathrm{TiS_2}}} \quad \underset{-28\pm 20}{\Delta H}, \tag{183}$$

$$\underset{-39}{\overset{\mathrm{S\ L}}{\mathrm{SiS_2}}} + \underset{-130}{\overset{\mathrm{L\ S}}{\mathrm{MoO_2}}} \rightarrow \underset{-206}{\overset{\mathrm{S\ S}}{\mathrm{SiO_2}}} + \underset{-56}{\overset{\mathrm{L\ L}}{\mathrm{MoS_2}}} \quad \underset{-93}{\Delta H}, \tag{184}$$

$$\underset{-39}{\overset{\mathrm{S\ L}}{\mathrm{SiS_2}}} + \underset{-134.5}{\overset{\mathrm{L\ S}}{\mathrm{WO_2}}} \rightarrow \underset{-206}{\overset{\mathrm{S\ S}}{\mathrm{SiO_2}}} + \underset{-46}{\overset{\mathrm{L\ L}}{\mathrm{WS_2}}} \quad \underset{-78.5}{\Delta H}. \tag{185}$$

The letters "L" and "S" above the ions indicate the large and the small ions of the two like ions involved in the individual cases above. Also here are the stable sides of the equations, those in which large cation goes to large anion and small cation to small anion.

In each of the examples above we have selected two cations of the same charge but different size and two anions of same charge but different size. In these cases the ionic sizes were decidedly important in determining the relative stability. We shall now compare salt pairs with positive ions of closely equal size but different charge (e.g., Li^+ [0.60 A], Mg^{2+} [0.65 A]), and negative ions of similar size but different charge (e.g., F^- [1.36 A], O^{2-} [1.40 A]):

$$\underset{-263.8}{\mathrm{MgF_2}} + \underset{-142.3}{\mathrm{Li_2O}} \rightarrow \underset{-146}{\mathrm{MgO}} + \underset{2\cdot -145.5}{2\ \mathrm{LiF}} \quad \underset{-30.9}{\Delta H}, \tag{186}$$

$$\underset{-290.2}{BaF_2} + \underset{-99.5}{Na_2O} \rightarrow \underset{-152}{CaO} + \underset{2\cdot-135.9}{2\,NaF} \quad \underset{-34.1}{\Delta H}, \tag{187}$$

$$\underset{-289}{SrF_2} + \underset{-86.2}{K_2O} \rightarrow \underset{-140.8}{SrO} + \underset{2\cdot-134.5}{2\,KF} \quad \underset{-34.6}{\Delta H}, \tag{188}$$

$$\underset{-287.9}{BaF_2} + \underset{-82.9}{Rb_2O} \rightarrow \underset{-133}{BaO} + \underset{2\cdot-133.2}{2\,RbF} \quad \underset{-28.6}{\Delta H}, \tag{189}$$

$$\underset{-190.6}{CaCl_2} + \underset{-89.8}{Na_2S} \rightarrow \underset{-113.7}{CaS} + \underset{2\cdot-98.3}{2\,NaCl} \quad \underset{-29.6}{\Delta H}, \tag{190}$$

$$\underset{-197.8}{SrCl_2} + \underset{-100.0}{K_2S} \rightarrow \underset{-113.1}{SrS} + \underset{2\cdot-104.4}{2\,KCl} \quad \underset{-24.1}{\Delta H}, \tag{191}$$

$$\underset{-205.3}{BaCl_2} + \underset{-88}{Rb_2S} \rightarrow \underset{-111.2}{BaS} + \underset{2\cdot-105.1}{2\,RbCl} \quad \underset{-28.1}{\Delta H}, \tag{192}$$

$$\underset{2\cdot-47.0}{2\,Li_3N} + \underset{3\cdot-84.2}{3\,MgS} \rightarrow \underset{-110.3}{Mg_3N_2} + \underset{3\cdot-100\pm10.0}{3\,Li_2S} \quad \underset{-63.4\pm30}{\Delta H}, \tag{193}$$

$$\underset{2\cdot-47.0}{2\,Li_3N} + \underset{3\cdot-84.2}{3\,MgCl_2} \rightarrow \underset{-110.3}{Mg_3N_2} + \underset{6\cdot-97.5}{6\,LiCl} \quad \underset{-141.2}{\Delta H}, \tag{194}$$

$$\underset{2\cdot-234.8}{2\,YCl_3} + \underset{-105}{Ca_3N_2} \rightarrow \underset{2\cdot-71.5}{2\,YN} + \underset{3\cdot-190.4}{3\,CaCl_2} \quad \underset{-139.4}{\Delta H}. \tag{195}$$

In these examples the radii of the two anions and the two cations, respectively, are very similar in each of the assemblages. It is a general rule that *the low-energy assemblages (on the right-hand sides) contain crystals in which the most highly charged cations are combined with the most highly charged anions, at the same time that the lower-charged cations are combined with the lower-charged anions.*

So far we have demonstrated two important stability rules of assemblages of simple ionic crystals, the one showing how the relative radii determine the stable assemblages, the other showing how different charges influence the stable assemblages (Arkel, 1949). We shall refer to the first rule as the "size-effect" and to the second as the "charge-effect."

These two effects usually operate simultaneously, because the ions present in natural systems often have different sizes as well as different charges. In some cases, then, the effect of the relative size pulls in the same direction as the effect of relative charge; in other cases they pull in opposite directions. We shall consider a few examples:

$$\underset{-153.2}{\overset{S\ \ L}{MgCl_2}} + \underset{-86.2}{\overset{L\ S}{K_2O}} \rightarrow \underset{-146}{\overset{S\ \ S}{MgO}} + \underset{2\cdot-104.4}{\overset{L\ L}{2\,KCl}} \quad \underset{-115.4}{\Delta H}, \tag{196}$$

$$\underset{-197.8}{\overset{L\ L}{SrCl_2}} + \underset{-142.3}{\overset{S\ S}{Li_2O}} \leftarrow \underset{-140.8}{\overset{L\ S}{SrO}} + \underset{2\cdot-97.5}{\overset{S\ L}{2\,LiCl}} \quad \underset{+4.3}{\Delta H}. \tag{197}$$

In the first example the combination of small cation with small anion and large cation with large anion also means combination of highly charged cation with highly charged anion and less charged cation with less charged anion. That is, both the size- and the charge-effect drive in the same direction, and, as a result, $-\Delta H$ has a very large value. In the second example, however, the largest cation has the highest charge (Sr^{2+}), and the smallest anion (O^{2-}) has the highest charge. Hence the effect of charge and size counteract each other, and the result is a small ΔH with a sign indicating that the size-effect is slightly larger than the charge-effect in this particular example. By choosing a smaller size difference between the two cations, however, we can make the charge-effect determine the direction of the reaction:

$$\underset{-197.8}{\overset{\text{L L}}{SrCl_2}} + \underset{-99.5}{\overset{\text{S S}}{Na_2O}} \rightarrow \underset{-140.8}{\overset{\text{L S}}{SrO}} + \underset{2\cdot -98.3}{\overset{\text{S L}}{2\,NaCl}} \quad \underset{-40.1}{\Delta H} . \tag{198}$$

We may consider the general case of the combination of two positive (or negative) ions of the same charge but different size with two negative (or positive) ions of the same size but different charge. Examples:

$$\underset{-263.8}{\overset{0.65\ 1.36}{MgF_2}} + \underset{-152}{\overset{0.99\ 1.40}{CaO}} \rightarrow \underset{-146}{\overset{0.65\ 1.40}{MgO}} + \underset{-290.2}{\overset{0.99\ 1.36}{CaF_2}} \quad \underset{-20.4}{\Delta H} , \tag{199}$$

$$\underset{-190}{\overset{0.99\ 1.81}{CaCl_2}} + \underset{-111}{\overset{1.35\ 1.84}{BaS}} \rightarrow \underset{-114.3}{\overset{0.99\ 1.84}{CaS}} + \underset{-205.3}{\overset{1.35\ 1.81}{BaCl_2}} \quad \underset{-18.6}{\Delta H} , \tag{200}$$

$$\underset{2\cdot -133}{\overset{0.95\ 1.36}{2\,NaF}} + \underset{-86.2}{\overset{1.33\ 1.40}{K_2O}} \rightarrow \underset{-99.5}{\overset{0.95\ 1.40}{Na_2O}} + \underset{2\cdot -134.5}{\overset{1.33\ 1.36}{2\,KF}} \quad \underset{-16.3}{\Delta H} , \tag{201}$$

$$\underset{-400}{\overset{0.50\ 1.40}{Al_2O_3}} + \underset{2\cdot -257}{\overset{0.2\ 1.36}{2\,BF_3}} \rightarrow \underset{-302}{\overset{0.2\ 1.40}{B_2O_3}} + \underset{2\cdot -329}{\overset{0.50\ 1.36}{2\,AlF_3}} \quad \underset{-46}{\Delta H} . \tag{202}$$

These examples show that the energy is at a minimum if the smallest of the two equally charged cations combines with the most highly charged of the two equally sized anions.

$$\underset{-99.5}{\overset{0.95\ 1.40}{Na_2O}} + \underset{-114.3}{\overset{0.99\ 1.84}{CaS}} \rightarrow \underset{-89.8}{\overset{0.95\ 1.84}{Na_2S}} + \underset{-152}{\overset{0.99\ 1.40}{CaO}} \quad \underset{-28}{\Delta H} , \tag{203}$$

$$\underset{2\cdot -145.3}{\overset{0.60\ 1.36}{2\,LiF}} + \underset{-153.2}{\overset{0.65\ 1.81}{MgCl_2}} \rightarrow \underset{2\cdot -97.5}{\overset{0.60\ 1.81}{2\,LiCl_2}} + \underset{-261.5}{\overset{0.65\ 1.36}{MgF_2}} \quad \underset{-12.7}{\Delta H} . \tag{204}$$

These two examples show that the energy is at a minimum if the smallest of the two equally charged anions combines with the most highly charged of the two equally sized cations. If the cation with the lowest charge is also the largest, it will certainly always tend to combine with the largest anion. This is of particular interest for silicate chemistry, because, in competition for oxygen and sulfur among silicon and earth-alkali metals and alkali metals, silicon will combine with oxygen, and the larger and low-charged ions with sulfur:

$$\underset{2\cdot -146}{\overset{\text{L S}}{2\,MgO}} + \underset{-35}{\overset{\text{S L}}{SiS_2}} \rightarrow \underset{-206}{\overset{\text{S S}}{SiO_2}} + \underset{2\cdot -84}{\overset{\text{L L}}{2\,MgS}} \quad \underset{-35}{\Delta H} . \tag{205}$$

We have so far explored some of the rather strikingly simple rules which determine the sign, and in a general way also the relative magnitude, of the heat of reaction among essentially ionic crystalline compounds. But the natural minerals are not always essentially ionic in character; polarization and partly covalent types of bonds play a significant role in, for example, sulfides and silicates. The strengthening of bonds in crystals by means of polarization has a tendency to work directly against some of the stability rules which are valid for nonpolarized ionic crystals. This is readily recognized if one remembers that a small positive ion generally has greater polarizing power than a large ion of the same charge and that a large negative ion usually is more polarizable than a small one with the same number of charges (Table 22, p. 277). The additional bond strength due to polarization therefore tends to make small cations combine with large anions, which is directly opposite to the situation of nonpolarized ions, as we have seen. But the stabilizing effect of polarization is not significantly large unless we consider cations which have large polarizing power and ions which are easily polarizable. The univalent alkali metals all have only a small polarizing power, and the relative stabilities of halogenides, oxides, or sulfides of these metals are not so drastically influenced by the polarization that the stability rules, as discussed above, are made invalid. The earth-alkali metals have a stronger polarizing power than do the alkali metals, but even here the relative stabilities of the oxides, halogenides, and sulfides are chiefly governed by the simple ionic size- and charge-effects, as discussed previously. However, if we consider compounds of the earth-alkali metals and the most polarizable anions like Se^{2-} and Te^{2-}, the effect of polarization is indeed strong enough to make the smallest metal ion combine with the largest and most polarizable anion:

$$\underset{-84.3}{\overset{\mathrm{S\ S}}{\mathrm{MgS}}} + \underset{-59}{\overset{\mathrm{L\ L}}{\mathrm{BaTe}}} \rightarrow \underset{-50}{\overset{\mathrm{S\ L}}{\mathrm{MgTe}}} + \underset{-111}{\overset{\mathrm{L\ S}}{\mathrm{BaS}}} \quad \underset{-17.3}{\Delta H}\,, \tag{206}$$

$$\underset{-84.3}{\overset{\mathrm{S\ S}}{\mathrm{MgS}}} + \underset{63}{\overset{\mathrm{L\ L}}{\mathrm{SrTe}}} \rightarrow \underset{-50}{\overset{\mathrm{S\ L}}{\mathrm{MgTe}}} + \underset{-113}{\overset{\mathrm{L\ S}}{\mathrm{SrS}}} \quad \underset{-16.7}{\Delta H}\,, \tag{207}$$

$$\underset{-84.3}{\overset{\mathrm{S\ S}}{\mathrm{MgS}}} + \underset{-63}{\overset{\mathrm{L\ L}}{\mathrm{CaTe}}} \rightarrow \underset{-50}{\overset{\mathrm{S\ L}}{\mathrm{MgTe}}} + \underset{-114.3}{\overset{\mathrm{L\ S}}{\mathrm{CaS}}} \quad \underset{-17.1}{\Delta H}\,. \tag{208}$$

The very high polarizability of Te^{2-} makes the stabilizing effect of polarization so strong that the smallest cation (Mg^{2+}) with the largest polarizing power will combine with the large Te^{2-} ion. In other words, the stability of MgTe is considerably greater than it would be if the ions were undeformable spheres.

Consider next the trivalent ions B^{3+} and Al^{3+}. These ions have larger polarizing power than the earth-alkali metals; as a consequence, the polarizability of sulfur is large enough to make the small boron combine with the large polarizable sulfur and the large aluminum with the smaller and less polarizable oxygen:

$$\underset{-121.6}{\overset{0.50\ 1.84}{\mathrm{Al_2S_3}}} + \underset{-302}{\overset{0.20\ 1.40}{\mathrm{B_2O_3}}} \rightarrow \underset{-400}{\overset{0.50\ 1.40}{\mathrm{Al_2O_3}}} + \underset{-112.3}{\overset{0.20\ 1.84}{\mathrm{B_2S_3}}} \quad \underset{-88.7}{\Delta H}\,. \tag{209}$$

This effect of polarization and covalency is also prominent in the oxides and sulfides of silicon and carbon:

$$\underset{-39}{\overset{0.40\ 1.84}{\mathrm{SiS_2}}} + \underset{-94.0}{\overset{0.19\ 1.40}{\mathrm{CO_2}}} \rightarrow \underset{-206}{\overset{0.40\ 1.40}{\mathrm{SiO_2}}} + \underset{+20.9}{\overset{0.19\ 1.84}{\mathrm{CS_2}}} \quad \underset{-52.4}{\Delta H}\,. \tag{210}$$

Thus, when strongly polarizing cations or anions with large polarizability are involved, the simple rules for relative stability of ionic crystals do not hold. However, since the polarizability of the common anions is known and the relative polarizing power of the cations is well understood, it is also possible to estimate the degree of polarization which occurs, and hence in which direction a reaction will proceed.

The cations considered up to now have all been of the so-called "noble-gas" type, with 8 electrons in their outer shell. Such ions have less polarizing power than ions of the same size and the same charge with nonnoble-gas structure, such as the heavy metals: Cu^{+}, Cu^{2+}, Ag^{+}, Ag^{2+}, Au^{3+}, Cd^{2+}, Zn^{2+}, Hg^{2+}, Pb^{2+}, Fe^{2+}, Ni^{2+}, Co^{2+}, Mn^{2+}, and others of the elements so important in ore minerals and some silicates. These latter ions therefore possess a very pronounced tendency to combine with the most polarizable anions in competition with noble-gas cations of similar size:

$$\underset{-69.3}{\mathrm{NaI}} + \underset{-32.6}{\mathrm{CuCl}} \rightarrow \underset{-16.3}{\mathrm{CuI}} + \underset{-98.2}{\mathrm{NaCl}} \quad \underset{-12.6}{\Delta H}\ . \tag{211}$$

Although Na^{+} and Cu^{+} have closely similar radii (0.95 and 0.96 A, respectively), the highly polarizing copper prefers to combine with the highly polarizable iodine ion.

$$\underset{-78.8}{\mathrm{KI}} + \underset{-32.6}{\mathrm{CuCl}} \rightarrow \underset{-16.3}{\mathrm{CuI}} + \underset{-104.2}{\mathrm{KCl}} \quad \underset{-9.1}{\Delta H}\ . \tag{212}$$

In this case the increased stability of CuI due to strong polarization is even able to hinder the large potassium from combining with the large iodine, which would be the case if the compounds were unpolarized and strictly ionic in nature.

$$\overset{1.36}{\underset{-78.8}{\mathrm{KI}}} + \overset{1.26}{\underset{-30.1}{\mathrm{AgCl}}} \rightarrow \underset{-15.3}{\mathrm{AgI}} + \underset{-104.2}{\mathrm{KCl}} \quad \underset{-10.6}{\Delta H}\ , \tag{213}$$

$$\overset{1.48}{\underset{-80.9}{\mathrm{RbI}}} + \overset{1.26}{\underset{-30.1}{\mathrm{AgCl}}} \rightarrow \underset{-15.3}{\mathrm{AgI}} + \underset{-105.2}{\mathrm{RbCl}} \quad \underset{-9.5}{\Delta H}\ , \tag{214}$$

$$\overset{1.69}{\underset{-83.7}{\mathrm{CsI}}} + \overset{1.26}{\underset{-30.1}{\mathrm{AgCl}}} \rightarrow \underset{-15.3}{\mathrm{AgI}} + \underset{-106.2}{\mathrm{CsCl}} \quad \underset{-7.7}{\Delta H}\ . \tag{215}$$

In the last series of reactions the energy driving the reactions toward the right side decrease with increasing difference in size of the alkali metals and the silver ion, because the effect of combining large with large and small with small ions, which pulls in the opposite direction, becomes gradually stronger.

Hg^{2+}, Zn^{2+}, Fe^{2+}, and Pb^{2+} all have stronger polarizing power than earth-alkali metals of comparable size. This is reflected in the relative stability of oxides and sulfides and very decidedly in sulfides and tellurides. Some examples of oxides and sulfides are given below:

$$\overset{0.83}{\underset{-83.4}{\mathrm{ZnO}}} + \overset{0.99}{\underset{-114.3}{\mathrm{CaS}}} \rightarrow \underset{-48}{\mathrm{ZnS}} + \underset{-152}{\mathrm{CaO}} \quad \underset{-2.3}{\Delta H}\ , \tag{216}$$

$$\overset{0.83}{\underset{-83.4}{\mathrm{ZnO}}} + \overset{0.65}{\underset{-84.3}{\mathrm{MgS}}} \rightarrow \underset{-48}{\mathrm{ZnS}} + \underset{-144}{\mathrm{MgO}} \quad \underset{-24.6}{\Delta H}\ , \tag{217}$$

$$\overset{1.12}{\underset{-01.6}{\mathrm{HgO}}} + \overset{1.13}{\underset{-113.0}{\mathrm{SrS}}} \rightarrow \underset{-14.0}{\mathrm{HgS}} + \underset{-141}{\mathrm{SrO}} \quad \underset{-20.4}{\Delta H}\ , \tag{218}$$

$$\overset{1.12}{\underset{-21.6}{\mathrm{HgO}}} \rightarrow \overset{1.35}{\underset{-111}{\mathrm{BaS}}} \rightarrow \underset{-14.0}{\mathrm{HgS}} + \underset{-126.1}{\mathrm{BaS}} \quad \underset{-8.5}{\Delta H}\ , \tag{219}$$

$$\underset{-57.5}{\mathrm{CoO}} + \underset{-114.3}{\mathrm{CaS}} \rightarrow \underset{-22.3}{\mathrm{CoS}} + \underset{-152}{\mathrm{CaO}} \quad \underset{-2.5}{\Delta H}\,, \tag{220}$$

$$\underset{-58.4}{\mathrm{NiO}} + \underset{-114.3}{\mathrm{CaS}} \leftarrow \underset{-20.7}{\mathrm{NiS}} + \underset{-152}{\mathrm{CaO}} \quad \underset{+0.3}{\Delta H}\,, \tag{221}$$

$$\underset{-62.2}{\mathrm{CdO}} + \underset{-114.3}{\mathrm{CaS}} \rightarrow \underset{-34.5}{\mathrm{CdS}} + \underset{-152}{\mathrm{CaO}} \quad \underset{-10.0}{\Delta H}\,. \tag{222}$$

We shall then consider some examples of how the stability rules, as explained above, can be applied to the complex natural minerals. To this end, more information about the general behavior of oxysalts is needed. The oxysalts, among them the silicates, consist of positive metal ions which bind together the negatively charged oxyanions like BO_3^{3-}, CO_3^{-}, NO_3^{1-}, SiO_4^{4-}, PO_4^{3-}, SO_4^{2-}, ClO_4^{1-}, etc. These negative oxyions occur as more or less independent units in the structures, because the central positive ions in the oxyions (B, C, P, S, etc.) are very small and highly charged, as compared with the metal cations in the structure, so that the bonds between oxygen and the central ion are strong and highly polarized and possess a high degree of covalency.

It can be shown that the average negative charge on each oxygen in the oxyion is the most important single factor in determining the distribution of cations among different oxysalts. The average negative charge on each oxygen changes from $1e$ to $\frac{1}{3}e$ through the sequence BO_3^{3-}, CO_3^{2-}, NO_3^{1-}; and from $1e$ to $\frac{1}{4}e$ through the sequence SiO_4^{4-}, PO_4^{3-}, SO_4^{2-}, ClO_4^{1-}.

It has previously been demonstrated that if two (or more) differently charged anions of the same size (i.e., F^-, O^{2-}) compete for cations of different charge or different size, then the anion with the lowest charge will combine with the cation with the largest size or the lowest charge, at the same time that the anion with the largest charge combines with the cation with the smallest size or largest charge. This principle also remains valid, with surprisingly few exceptions, for the complex oxysalts and for the relationship between them and oxides and sulfides (Ramberg, 1952). Some examples are given below:

$$\underset{-152}{\overset{\mathrm{L}}{\mathrm{CaO}}} + \underset{-262}{\overset{\mathrm{S}}{\mathrm{MgCO_3}}} \rightarrow \underset{-288.4}{\overset{\mathrm{L}}{\mathrm{CaCO_3}}} + \underset{-144}{\overset{\mathrm{S}}{\mathrm{MgO}}} \quad \underset{-28.4}{\Delta H}\,, \tag{223}$$

$$\underset{-288.4}{\overset{\mathrm{L}}{\mathrm{CaCO_3}}} + \underset{-305.0}{\overset{\mathrm{S}}{\mathrm{MgSO_4}}} \rightarrow \underset{-340.5}{\overset{\mathrm{L}}{\mathrm{CaSO_4}}} + \underset{-262}{\overset{\mathrm{S}}{\mathrm{MgCO_3}}} \quad \underset{-9.1}{\Delta H}\,, \tag{224}$$

$$\underset{-510}{\overset{\mathrm{L}}{\mathrm{Ca_2SiO_4}}} + \underset{2.-262}{\overset{\mathrm{S}}{\mathrm{2MgCO_3}}} \rightarrow \underset{-509}{\overset{\mathrm{S}}{\mathrm{Mg_2SiO_4}}} + \underset{2.-288.4}{\overset{\mathrm{L}}{\mathrm{2CaCO_3}}} \quad \underset{-51.8}{\Delta H}\,, \tag{225}$$

$$\underset{-86}{\overset{\mathrm{L}}{\mathrm{K_2O}}} + \underset{-289.5}{\overset{\mathrm{S}}{\mathrm{Li_2CO_3}}} \rightarrow \underset{-275.1}{\overset{\mathrm{L}}{\mathrm{K_2CO_3}}} + \underset{-142.3}{\overset{\mathrm{S}}{\mathrm{Li_2O}}} \quad \underset{-41.9}{\Delta H}\,, \tag{226}$$

$$\underset{-275.1}{\overset{\mathrm{L}}{\mathrm{K_2CO_3}}} + \underset{-341.5}{\overset{\mathrm{S}}{\mathrm{Li_2SO_4}}} \rightarrow \underset{-342.6}{\overset{\mathrm{L}}{\mathrm{K_2SO_4}}} + \underset{-289.5}{\overset{\mathrm{S}}{\mathrm{Li_2CO_3}}} \quad \underset{-15.5}{\Delta H}\,, \tag{227}$$

$$\underset{-365.5}{\overset{\mathrm{L}}{\mathrm{K_4SiO_4}}} + \underset{2\cdot-289.5}{\overset{\mathrm{S}}{\mathrm{2\,Li_2CO_3}}} \rightarrow \underset{2\cdot-275.1}{\overset{\mathrm{L}}{\mathrm{2\,K_2CO_3}}} + \underset{-439.6}{\overset{\mathrm{S}}{\mathrm{Li_4SiO_4}}} \quad \underset{-45.3}{\Delta H}\,. \tag{228}$$

The energy data are from Kubaschewski and Evans (1951).

The common minerals in metamorphic rocks may, for the present purpose, be grouped into (1) *silicates* of different degree of polymerization; (2) *complex oxides* (in which none of the cations are so small or so highly charged that they polarize the surrounding oxygen into tight oxyanion units); (3) nonsilicate *oxysalts*, like borates, carbonates, phosphates, and sulfates; and (4) *sulfides*. We are now in a position to handle theoretically the distribution of cations among these compounds. To start with a random example: Why is calcite so much more common a carbonate than magnesite? This follows directly from the principle discussed above, the small Mg^{2+} ion tends to combine with oxygen in the silicate anions because of the relatively large negative effective charge of this oxygen, whereas the large Ca^{2+} ions tends to combine with the oxygen in the carbonate ion which has a lower effective charge:

$$\underset{\text{magnesite}}{2\,MgCO_3} + \underset{\text{larnite}}{Ca_2SiO_4} \rightarrow \underset{\text{forsterite}}{Mg_2SiO_4} + \underset{\text{calcite}}{2CaCO_3}\,. \tag{229}$$

The natural distribution of cations between oxides, on the one side, and silicates (as a group), on the other side, is chiefly governed by the above principle. Al^{3+} and Fe^{3+} are the most important trivalent metals in the rock-forming minerals; these two elements occur commonly as oxides (Al_2O_3 and Fe_2O_3) or in complex oxides (Fe_3O_4, $MgAl_2O_4$, etc.) in association with silicates or carbonates. The relatively highly charged and small ions, Al^{3+} and Fe^{3+}, in competition with divalent and univalent ions, go to the doubly charged oxygen in oxides rather than to the singly or less charged oxygen in the silicates and other oxysalts.

It is structurally possible to conceive silicates of the following formula:

$$Al_4(SiO_4)_3 \text{ and } Fe_4(SiO_4)_3\,. \tag{230}$$

Such silicates are highly unstable in association with, for example, MgO, CaO, and FeO:

$$6\,MgO + Fe_4(SiO_4)_3 \rightarrow 3\,\underset{\text{olivine}}{Mg_2SiO_4} + 2\,\underset{\text{hematite}}{Fe_2O_3}\,, \tag{231}$$

and

$$6\,MgO + Al_4(SiO_4)_3 \rightarrow 3\,\underset{\text{olivine}}{Mg_2SiO_4} + \underset{\text{corundum}}{2Al_2O_3}\,. \tag{232}$$

The right-hand sides of these reactions are the stable ones in rocks. Because of the strong bonds between Fe^{3+} or Al^{3+} and O^{2-} in oxides, the compound $Fe_4(SiO_4)_3$ and $Al_4(SiO_4)_2$ are actually highly unstable, even if not in contact with MgO or CaO:

$$Fe_4(SiO_4)_3 \rightarrow 2\,Fe_2O_3 + 3\,SiO_2\,. \tag{233}$$

Fe^{3+} tends so strongly to combine with doubly charged oxygen ions that oxygen becomes released from silicon, and Fe_2O_3 forms upon liberation of silica.

$$Al_4(SiO_4)_3 \rightarrow 2\,Al_2O_3 + 3\,SiO_2\,. \tag{234}$$

In this case the right-hand side would be the stable one, if it were not for the fact that aluminum is able to make some very stable *basic* silicates—kyanite, andalusite, and sillimanite of the formula Al_2OSiO_4. These silicates are halfway between orthosilicate and aluminum oxide.

We have thus seen how Fe^{3+} and Al^{3+} tend to make oxides in preference to silicates, although the less polarizing Al^{3+} ion is able to make basic silicates with excess of oxygen (kyanite, etc.).

Ti^{4+} is a common ion in minerals. Being tetravalent and rather small, it has a strong polarizing power, thus showing little tendency to form pure titanium silicates:

$$TiSiO_4 + 2\ MgO \rightarrow \underset{\text{olivine}}{Mg_2SiO_4} + TiO_2 . \tag{235}$$

This reaction is displaced completely toward the right side in rocks. It is interesting to note that sphene ($CaTiOSiO_4$), which is a basic orthosilicate of Ti and Ca, is very stable indeed. Apparently, the excess of oxygen ions not bound to the silicon stabilizes sphene in a similar way as andalusite, kyanite, and sillimanite are stabilized. Much like ferri-orthosilicate, $TiSiO_4$ is in itself unstable because of the strong tendency of Ti to combine with oxygen under liberation of silica:

$$TiSiO_4 \rightarrow SiO_2 + \underset{\text{rutile}}{TiO_2} . \tag{236}$$

Thus the fact that we find TiO_2, Fe_2O_3, and Al_2O_3, but not MgO, CaO, or FeO, as common minerals in rocks is in agreement with theoretical expectations.

Al_2O_3, Fe_2O_3, and TiO_2 all have some tendency to form complex oxides with divalent metals (spinels, magnetite, ilmenite); and it is of interest to consider how the divalent metals would distribute themselves among the complex oxides, on the one side, and the silicates (or other oxysalts) as a group, on the other side.

The discussion throughout this chapter would make us conclude that, generally, the small and relatively highly charged ions will go to the complex oxides, where the oxygen has a higher average charge than in the silicates, and the larger and less charged ions would go to the silicates, carbonates, and other oxysalts. In this connection we should also be aware of the fact that the polarizability of oxygen is much higher in the simple or complex metal oxides than in the oxysalts. Actually, the polarizability of oxygen decreases stepwise through the sequence: ionic oxides or complex oxides → orthosilicates → single-chain metasilicates → double-chain silicates → phyllosilicates → tectosilicates. That will make highly polarizing divalent ions prefer to go to the oxides rather than to silicates or other oxysalts.

The distribution of divalent iron and magnesium in minerals is informative. Iron is a little bigger than magnesium, but yet it has a larger polarizing power, because of its nonnoble-gas ionic structure. For that reason magnesium tends to go to the silicates, divalent iron to the complex oxides:

$$\left.\begin{aligned} 2\ MgFe_2O_4 + Fe_2SiO_4 &\rightarrow Mg_2SiO_4 + 2\ FeFe_2O_4 , \\ 2\ MgTiO_3 + Fe_2SiO_4 &\rightarrow Mg_2SiO_4 + 2\ FeTiO_3 , \\ 2\ MgAl_2O_4 + Fe_2SiO_4 &\rightarrow Mg_2SiO_4 + 2\ FeAl_2O_4 . \end{aligned}\right\} \tag{237}$$

Some of these equations have been discussed before in thermodynamic terms (pp. 35 ff.), here we see how the character of the chemical bonds causes the right-hand sides to be the stable, low-energy sides. In an analogous manner the heavy metals which form polarizing ions with nonnoble-gas structure are generally likely to build oxides or complex oxides rather han silicates if competing with less polarizing cations. In competition with Mg or Ca, or even Fe^{2+}, for example, Zn^{2+} will leave the silicate phase in favor of any of the other three divalent cations:

$$2MgAl_2O_4 + Zn_2SiO_4 \rightarrow Mg_2SiO_4 + 2ZnAl_2O_4 . \tag{238}$$

The relationships between sulfides and silicates are illuminating. In competition with cations of less power of polarization, the heavy metals which have high polarizing power strive strongly to combine with the polarizable sulfur, not with the much less polarizable oxygen in silicates or other oxysalts:

$$\left.\begin{array}{llll} 2\,MgS & +\,Fe_2SiO_4 & \rightarrow 2\,FeS & +\,Mg_2SiO_4\,, \\ 2\,MgS & +\,Zn_2SiO_4 & \rightarrow 2\,ZnS & +\,Mg_2SiO_4\,, \\ 2\,MgS & +\,Cu_2SiO_4 & \rightarrow 2\,CuS & +\,Mg_2SiO_4\,, \\ CaS & +\,PbCO_3 & \rightarrow PbS & +\,CaCO_3\,, \\ CaS & +\,ZnCO_3 & \rightarrow ZnS & +\,CaCO_3\,. \end{array}\right\} \qquad (239)$$

These equations are displaced very strongly toward the right side in the natural occurrences of ore and silicate rocks.

The silicates have so far been treated as a single group in contradistinction to the oxides or sulfides. However, the distribution of cations among the different silicates is of particular consequence for the study of metamorphic rocks. These minerals have complicated structures in which the "anions" usually are polymerized Si-O networks extending over the whole crystal in *one dimension* for the single-chain and double-chain silicates; in *two dimensions* for the sheet-type phyllosilicates; and in *three dimensions* for the tectosilicates. For these reasons, the chief role which cationic size plays in determining the relative stability of crystals is somewhat different in the complex polymerized silicates and in the simple oxysalts and orthosilicates. In the latter compounds the size of the cations is an important factor in determining bond strength, because the ionic potential (charge/radius) depends upon the ionic radii. The individual ions in the sulfides, oxides, and simple oxysalts, including orthosilicates, can arrange themselves rather unrestrictedly in space so that the crystals adopt the structure which is agreeable to whatever size the ions happen to have. The dominating factor which determines which ions shall combine are the fields of valence forces around the ions.

The various polymerized Si-O "anions," however, are rigid and complicated in building; they cannot, therefore, arrange themselves unrestrictedly in space to accommodate cations of any size. For that reason a complex silicate anion cannot always combine with the cations, the valence forces of which actually would give the strongest bonds with the oxygen ions in the given Si-O "anion."

With this restriction in mind, the stability rules valid for simple compounds can be rather successfully applied to the complex polymerized silicates. A crucial point is that the electrostatic field around the oxygen ions in silicates depends significantly upon the degree of polymerization, in such a manner that the effective negative charges on the oxygen ions in the Si-O network decreases stepwise through the sequence: orthosilicates (SiO_4^{4-}) → pyrosilicates ($Si_2O_7^{6-}$) → single chain metasilicates ($Si_nO_{3n}^{2n-}$) → double-chain silicates ($Si_4O_{11}^{6-}$) → phyllosilicates ($Si_2O_5^{2-}$) → tectosilicates [(Si,Al) O_2^{n-}]. This has been shown by the author in recent papers (Ramberg, 1952). These facts, in view of the previous discussion, explain why tectosilicates and phyllosilicates preferably take large cations with low charge outside the silicon positions (alkali metals), the orthosilicates take smaller cations with higher charge (Mg, Fe), and the metasilicates distinguish less decidedly among the cations.

The distribution of Fe^{2+} and Mg^{2+} among the different silicates is governed by this rule. Mg^{2+} and Fe^{2+} have much the same radii, but the field strength around Fe^{2+} is larger than that around Mg^{2+}; hence Fe^{2+}, so to speak, behaves chemically as if it were smaller than Mg^{2+} (see above). The relative distribution of Fe^{2+} and Mg^{2+} in olivine (orthosilicate), hypersthene (metasilicate), anthophyllite (double-chain metasilicate), and talc (phyllosilicate) is as follows: Olivine has a higher Fe/Mg ratio than does coexisting pyroxene; pyroxene has higher Fe/Mg than coexisting anthophyllite; and anthophyllite has higher Fe/Mg ratio than coexisting talc. That is, Fe tends to go to the oxygen which has the highest negative charge, namely, that in olivine, whereas Mg tends to combine with the oxygen with the lowest effective charge, namely, that in talc. It is also an important effect that the polarizability of oxygen increases from the talc structure to the olivine structure and that the strongest polarizing ion, namely, Fe^{2+}, tends to combine with the most polarizable oxygen, namely, that in the orthosilicate.

At this point we shall recall the splitting-up of ferrosillite into fayalite and quartz (p. 33):

$$2\ FeSiO_3 \rightarrow Fe_2SiO_4 + SiO_2\ , \tag{240}$$

which takes place at a temperature well below melting.

The comparable reaction with enstatite requires much more energy and hence does not take place before melting starts:

$$2\ MgSiO_3 \rightarrow Mg_2SiO_4 + SiO_2 + 2.26\ \text{kcal}\ , \tag{241}$$

(see Sahama and Torgeson, 1949). The reason for the energy difference between the Mg-silicates and the Fe-silicate follows readily from the theory. Fe tends strongly to change its environment from the less polarizable and lower-charged oxygen in the metasilicate structure to the more polarizable and higher-charged oxygen in the orthosilicate. Mg tends much less strongly to do the same.

The aluminum silicate, Al_2OSiO_4, behaves in a very similar way. Heating kyanite to 1,000° C. makes it release some silica upon forming a more basic aluminum silicate, mullite (Hedvall, 1942*a*):

$$\underset{\text{kyanite}}{3\ Al_2OSiO_4} \rightarrow \underset{\text{mullite}}{2\ Al_3O_{2.5}SiO_4} + \underset{\text{silica}}{SiO_2}\ . \tag{242}$$

In mullite there are more oxygen ions outside the Si-O tetrahedra than in kyanite. Al will hence try to change its environment from that in kyanite to that in mullite, in conformity with its high tendency to combine with independent oxygen ions rather than with oxygen in SiO_4 groups.

Orthosilicates of calcium are very rare in rocks. In competition with Mg^{2+} and Fe^{2+}, the large Ca^{2+} prefers to combine with the oxygen ions with low effective charge in metasilicates:

$$\underset{\text{larnite}}{Ca_2SiO_4} + \underset{\text{enstatite}}{4\ MgSiO_3} \rightarrow \underset{\text{diopside}}{2\ CaMg(SiO_3)_2} + \underset{\text{forsterite}}{Mg_2SiO_4}\ . \tag{243}$$

The right side is the low-energy side, which is stable in rocks. The reaction may go a step further, so that Ca^{2+} enters the double-chain metasilicate (hornblende) in which the oxygen has a still weaker electric field:

$$\underset{\text{diopside}}{2CaMg(SiO_3)_2} + \underset{\text{anthophyllite}}{Mg_7Si_8O_{22}(OH)_2} \rightarrow \underset{\text{hornblende}}{Ca_2Mg_5Si_8O_{22}(OH)_2} + \underset{\text{enstatite}}{4MgSiO_3}\ . \tag{244}$$

The right-hand side is rather common in rocks, the left side is hardly known.

The last example illustrates the inability of complex polymerized silicate anions to accommodate cations of certain sizes. The double $Si_{4n}O_{11n}$ chains apparently cannot arrange themselves so that seven large Ca^{2+} ions can be placed between the oxygen ions instead of the seven small Mg^{2+} ions. For that reason there exist no natural silicates of the formula $Ca_7Si_8O_{22}(OH)_2$.

This restriction due to size of ions and space available in silicate structures is very important for the phyllosilicates like talc, serpentine, chlorites, etc. The $(Si_2O_5)^{-2}$ sheets in these silicates are such that small ions like Mg^{2+}, Fe^{2+}, Li^+, etc., can fit nicely in between the unshared oxygen ions of two parallel $Si_2O_5^{-2}$ sheets (each SiO_4 tetrahedron shares three of its four oxygen ions with adjacent silicon atoms). It is geometrically impossible, or at least very difficult, to arrange the sheets so that the large Ca^{2+} ions (or Ba^{2+}, Na^+, K^+, etc.) can fit in between the *unshared* oxygen ions of the sheets. The unshared oxygen, however, is actually of a type (relatively weak electrostatic field around it, see above) which will prefer to take Ca^{2+} ions relative to Mg^{2+} or Fe^{2+} more than the oxygen in metasilicates. For purely spatial reasons, therefore, there is no common sheet-type silicate with calcium surrounded by *unshared* oxygen ions.

The oxygen shared between two silicon atoms or aluminum and silicon (Si-O-Si; Al-O-Si) has less effective negative charge or less field stength than any of the unshared oxygen in silicates. The shared oxygen is most abundant in tectosilicates and phyllosilicates. The large divalent earth-alkali metals (Ca^{2+}, Sr^{2+}, Ba^{2+}) and the univalent alkali metals are therefore the cations which, in harmony with the stability rules, must take the place between the shared oxygen in micas, feldspars, and other tectosilicates like leucite, nepheline, zeolites, etc.

It should be noticed in conclusion that the effective charge and field strength of oxygen in the complex Si-O anions depend significantly upon the substitution of Al for Si. In the silicon position Al increases the effective negative charge of oxygen because of the low positive charge and large size of Al as compared with Si. This condition will influence the distribution of cations among the several silicates.

The reason for not having tectosilicates of Fe^{2+}, Mg^{2+}, Cu^{2+}, Cu^+, Ag^+, and several other heavy metals is chiefly due to the relatively strong polarizing power of these ions. These ions will strongly tend to change their surroundings from the shared, unpolarizable, and very weakly charged oxygen in the tectosilicates to the unshared and higher-charged oxygen in meta- or orthosilicates or even to the independent doubly charged oxygen in simple and complex oxides (spinel, etc.).

A copper tectosilicate is, for example, conceivable structurally, but it is highly unstable:

$$2CuAl_2Si_2O_8 \rightarrow Cu_2SiO_4 + 2Al_2OSiO_4 + SiO_2 . \qquad (245)$$

The matter discussed in this Appendix has, it is hoped, shown how valuable the knowledge of chemical bonds and properties of ions is as an aid in the interpretation of mineral assemblages.

REFERENCES

ADAM, N. K. (1941) The physics and chemistry of surfaces. 3d ed. London: Oxford University Press, H. Milford.

ADAMS, F. D. (1929) The geology of Ceylon. Canad. J. Research, **1,** 425.

AMINOFF, G. (1927*a*) Über Verdämpfungs- und Lösungserscheinungen bei Zink. Ztschr. Kristallog., **65,** 23.

———. (1927*b*) Über Verdämpfungen von rhombischem Schwefel. *Ibid.*, p. 632.

ANDERSEN, O. (1931) Discussion on certain phases of the genesis of pegmatites. Norsk. geol. Tids., **13,** 1.

ANDERSON, E. M. (1948) On lineation and petrofabric structure, and the shearing movement by which they have been produced. Quart. J. Geol. Soc. London, **104,** Part I, 99.

ARKEL, A. E. VON. (1949) Molecules and crystals. London: Butterworths Scientific Publications; New York: Interscience Publishers, Inc.

BACKLUND, H. G. (1936) Zur genetischen Deutung der Eklogite. Geol. Rundschau, **27,** 47.

BAILEY, E. B., and GRABHAM, G. W. (1909) Albitization of basic plagioclase feldspars. Geol. Mag., **46,** 250.

BALK, R. (1949) Structure of Grand Saline salt dome, Van Zandt County, Texas. Bull. Am. Assoc. Pet. Geol., **33,** No. 11, 1791.

BALSLEY, J. R. (1941) Deformation of marble under tension at high pressure. Trans. Am. Geophys. Union, **22,** Part II, 519.

BANDY, M. C. (1940) A theory of mineral sequence in hypogene ore deposits. Econ. Geol., **35,** 359, 546.

BARRER, R. M. (1941) Diffusion in and through solids. Cambridge: At the University Press.

BARRETT, C. S. (1943) Structure of metals, crystallographic methods, principles, and data. 1st ed. New York and London: McGraw Hill Book Co., Inc.

BARROW, G. (1893) On an intrusion of muscovite-biotite gneiss in the Southeast Highlands of Scotland. Quart. J. Geol. Soc. London, **49,** 330.

———. (1912) On the geology of lower Dee-side and the Southern Highland border. Proc. Geol. Assoc., **23,** 268.

BEARTH, P. (1948) Über Albitisierung im Altkristallin des Monte Rosa. In: Festschrift Paul Niggli. Zurich: A. G. Gebr. Leemann & Co. Also in Schweiz. Min. Petr. Mitt., **28,** Heft 1, 140.

BEMMELEN, R. W. VAN. (1949) The geology of Indonesia. V. 1A, General geology; 1B, General geology; 2, Economic geology. (Special ed. of the Bureau of Mines in Indonesia.) The Hague: Government Printing Office, M. Nijhoff.

———. (1952) The endogenic energy of the earth. Am. J. Sci., **250,** 104.

BERKEY, C. P. (1924) The new petrology. Bull. New York State Museum, **251,** 105.

BERTELSEN, A. (1950) A pre-Cambrian dome structure at Tovqussaq, West Greenland. Medd. Dansk Geol. Foren., **11,** 558.

BICHOWSKY, F. R., and ROSSINI, F. D. (1936) The thermochemistry of the chemical substances. New York: Reinhold Publishing Corp.

Bowen, N. L., and Schairer, J. F. (1935*a*) Grünerite from Rockport, Mass., and a series of synthetic fluoramphiboles. Am. Mineral., **20**, 543.

———. (1935*b*) The system, MgO, FeO, SiO_2. Am. J. Sci., ser. 5, **29**, 151.

Bowen, N. L., and Tuttle, O. F. (1949) The system $MgO-SiO_2-H_2O$. Bull. Geol. Soc. Am., **60**, 439.

———. (1950) The system $NaAlSi_3O_8-KAlSi_3O_8-H_2O$. J. Geol., **58**, 489.

Breddin, H. (1930) Die Milchquartzgänge des rheinischen Schiefergebirges, eine Nebenerscheinung der Druckschieferung. Geol. Rundschau, **21**, 367.

Brewer, L. (1951) The equilibrium distribution of elements in the earth's gravitational field. J. Geol., **59**, 490.

Bridgman, P. W. (1936) Shearing phenomena at high pressure of possible importance for geology. J. Geol., **44**, 653.

———. (1937) Shearing phenomena at high pressure, particularly in inorganic compounds. Proc. Am. Acad. Arts and Sci., **77**, 387.

———. (1946) Recent work in the field of high pressures. Rev. Mod. Phys., **18**, 1.

Brindley, G. W., and Ridley, P. (1938) An X-ray study of lattice distortion in rhodium. Proc. Phys. Soc., **50**, 501.

Brönsted, J. N. (1938) Physical chemistry. New York: Chemical Publishing Co. of New York, Inc.

Brown, J. S. (1948) Ore genesis: a metallurgical interpretation, an alternative to the hydrothermal theory. 1st ed. Hopewell, N.J.: Hopewell Press.

Buerger, M. J. (1934) The lineage structure of crystals. Ztschr. Kristallog., **89**, 195.

———. (1948) The role of temperature in metamorphism. Am. Mineral., **33**, 101.

Bugge, J. A. W. (1943) Geological and petrological investigations in the Kongsberg-Bamble formation. Norg. Geol. Undersökelse, No. 160.

———. (1945) The geological importance of diffusion in the solid state. Avhandl. Norske Vidensk.-Akad. Oslo. I. Mat.-naturv. Kl., No. 13.

Bullen, K. E. (1947) An introduction to the theory of seismology. Cambridge: At the University Press.

Bunn, C. W. (1946) Chemical crystallography. Oxford: Clarendon Press.

Cameron, E. N.; Jahns, R. H.; McNair, A. H.; and Page, L. R. (1949) Internal structure of granitic pegmatites. Econ. Geol. Mono., No. 2. Urbana, Ill.: Econ. Geol. Pub. Co.

Chalmers, R. (1936) Micro plasticity in crystals of tin. Proc. Roy. Soc. London, A, **156**, 427.

Chao, S. H., and Taylor, W. H. (1940) Isomorphous replacement and superlattice structures in the plagioclase felspars. Proc. Roy. Soc. London, A, **176**, 76.

Chentsova, L. G., and Vedeneeva, N. E. (1950) Nature of inclusions in the structure of smoky quartz. Chem. Abstr., **44**, 2894.

Chudoba, K. F., and Frechen, J. (1950) Über die plastische Verformung von Olivin. Neues Jahrb. Mineral. Geol., Beilage-Bd. A, **81**, 183.

Clabaugh, S. E. (1950) Discussion: pegmatites of Montana. Econ. Geol., **45**, 254.

Clarke, F. W. (1924) The data of geochemistry. U.S. Geol. Survey Bull. 770.

Cloos, E. (1944) Correlation of lineation with rock-movement. Trans. Am. Geophys. Union, **25**, 660.

———. (1946) Lineation. Mem. Geol. Soc. Am. **18**.

———. (1947) Boudinage. Trans. Am. Geophys. Union, **28**, 626.

Correns, C. W. (1939) In: Barth, T. F. W.; Correns, C. W.; and Eskola, P., Die Entstehung der Gesteine. Berlin: Julius Springer.

Credner, H. (1875) Die granitischen Gänge des sächsischen Granulitgebirges. Ztschr. deutsch. geol. Gesellsch., **27**, 104.

Danielsson, A. (1950) Das Calsit-Wollastonitgleichgewicht. Geochim. et cosmochim. acta, **1**, 55.

Dehlinger, H. (1939) Chemische Physik der Metalle und Legierungen. Leipzig: Akademischer Verlagsgesellschaft.

Delesse. (1861) Études sur le métamorphisme des roches. Bull. Soc. géol. France, ser. 2, **18**, 541.

Dewey, H., and Flett, J. S. (1911) On some British pillow lavas and the rocks associated with them. Geol. Mag., **8**, 202, 241.

Drescher-Kaden, F. K. (1948) Die Feldspat-Quarz-Reaktionsgefüge der Granite und Gneise und ihre genetische Bedeutung. ("Mineralogie und Petrographie in Einzeldarstellungen, Vol. **1**.) Berlin, Göttingen, and Heidelberg: Springer-Verlag.

Durocher, J. (1846) Études sur le métamorphisme des roches. Bull. Soc. géol. France, ser. 2, **3**, 546.

Eckermann, H. von. (1922) The rocks and contact minerals of the Mansjö Mountain. Geol. Fören. i Stockholm Förh., **44**, 203.

———. (1923) The rocks and contact minerals of Tennberg. *Ibid.*, **45**, 465.

Elliot, J. (1947) The Marievale granophyre—a metamorphic rock. Trans. Proc. Geol. Soc. South Africa, **50**, 121.

Engel, A. E. J. (1948) The direction of flow of mineralizing solutions. Econ. Geol., **43**, 655.

Eskola, P. (1914) On the petrology of the Orijärvi region in southwestern Finland. Bull. comm. géol. Finlande, No. 40.

———. (1915) On the relation between chemical and mineralogical composition in the metamorphic rocks of the Orijärvi region. *Ibid.*, No. 44.

———. (1921*a*) The mineral facies of rocks. Norsk geol. Tids., **6**, 143.

———. (1921*b*) On the eclogites of Norway. Vidensk.-Skrifter. I. Mat.-naturv. Kl., Kristiania, No. 8.

———. (1932) On the principles of metamorphic differentiation. Bull. comm. géol. Finlande, **16**, No. 97, 68.

———. (1939) In: Barth, T. F. W.; Correns, C. W.; and Eskola, P., Die Entstehung der Gesteine. Berlin: Julius Springer.

Eskola, P.; Vuoristo, U.; and Rankama, K. (1937) An experimental illustration of the spilite reaction. Bull. comm. géol. Finlande, **20**, No. 119.

Evans, R. J. (1946) An introduction to crystal chemistry. Cambridge: At the University Press.

Faber, H. (1941) On the salt solutions in microscopic cavities in granites. Danmarks Geol. Undersökelse, Ser. II, No. 67.

Fairbairn, H. W. (1943) Packing of ionic minerals. Bull. Geol. Soc. Am., **54**, 1305.

———. (1949) Synthetic quartzite. *Ibid.*, **60**, 1886.

Fairbairn, H. W., and Chayes, F. (1949) Structural petrology of deformed rocks. Cambridge 42, Mass.: Addison-Wesley Press, Inc.

Fenner, C. N. (1940) The nature of ore-forming fluid: a discussion. Econ. Geol., **35**, 883.

Fischer, F. G. (1950) A method for determining the direction of flow of hydrothermal solutions. Econ. Geol., **45**, 62.

Fischer, R. P. (1950) Uranium-bearing sandstone deposits of the Colorado plateau. Econ. Geol., **45**, 1.

FLETT, J. S. (1909) The geology of the country around Bodmin and St. Austell. Geol. Survey England and Wales Mem.

FOSLIE, S. (1945) Hastingsites and amphiboles from the epidote amphibolite facies. Norsk geol. Tids., **25**, 74.

FREDERICKSON, A. F. (1951) Mechanism of weathering. Bull. Geol. Soc. Am., **62**, 221.

FRENKEL, J. (1946) Kinetic theory of liquids. Oxford: Clarendon Press.

FREUNDLICH, H. (1922) Colloid and capillary chemistry. Translated from 3d German ed. by H. STAFFORD HATFIELD. New York: E. P. Dutton Co.

GARRELS, R. M. (1944) Solubility of metal sulphides in dilute vein forming solutions. Econ. Geol., **39**, 472.

GARRELS, R. M.; DREYER, R. M.; and HOWLAND, A. H. (1949) Diffusion of ions through intergranular spaces in water-saturated rocks. Bull. Geol. Soc. Am., **60**, 1809.

GIBBS, J. W. (1928) Collected works, Vols. **1** and **2**. New Haven: Yale University Press.

GIBSON, R. E. (1938) On the effect of pressure on the solubility of solids in liquids. Am. J. Sci., ser. 5, **35**, A, 49.

GILBERT, C. M. (1949) Cementation of some California tertiary reservoir sands. J. Geol., **57**, 1.

GILBERT, G. (1924) The relation of hardness to the sequence of the ore minerals. Econ. Geol., **19**, 668.

GILLULY, J. (1935) Keratophyres of eastern Oregon, and the spilitic problem. Am. J. Sci., ser. 5, **29**, 225, 336.

GOLDSCHMIDT, V. M. (1911) Die Kontaktmetamorphose im Kristianiagebiete. Vidensk.-Skrifter. I. Mat.-naturv. Kl., Kristiania, No. 11.

———. (1912) Die Gesetze der Gesteinsmetamorphose, mit Beispielen aus der Geologie des südlichen Norwegens. *Ibid.*, No. 20.

———. (1921) Die Injektionsmetamorphose im Stavanger-Gebiete. *Ibid.*, No. 10.

GOLDSCHMIDT, V. M., and PETERS, C. (1932) Zur Geochemie des Bors. Nachr. Gesellsch. Wiss. Göttingen, Math.-phys. Kl., **4**, 377.

GOLDSMITH, J. (1952) Diffusion in zoned plagioclases. J. Geol., **60**, 288.

GOODSPEED, G. E. (1940) Dilation and replacement dikes. J. Geol., **48**, 175.

GORANSON, R. W. (1931) The solubility of water in granite magmas. Am. J. Sci., ser. 5, **22**, 481.

———. (1938) Silicate-water systems: phase equilibria in the $NaAlSi_3O_8$-H_2O and $KAlSi_3O_8$-H_2O systems at high temperatures and pressures. *Ibid.*, **35**, A, 71.

GRATON, L. C. (1940) Nature of the ore-forming fluid. Econ. Geol., **35**, 197.

GRIGGS, D. T. (1936) Deformation of rocks under high confining pressures. J. Geol., **44**, 541.

———. (1939) Creep of rocks. *Ibid.*, **47**, 225.

———. (1940) Experimental flow of rocks, etc. Bull. Geol. Soc., Am., **51**, 1001.

———. (1941) An experimental approach to dynamic metamorphism. Trans. Am. Geophys. Union, **22**, Part II, 526.

GROVES, A. W. (1935) The charnockite series of Uganda, East Africa. Quart. J. Geol. Soc. London, **91**, 150.

GRUNER, J. W. (1944*a*) The structure of stilpnomelane reexamined. Am. Mineral., **29**, 291.

———. (1944*b*) The composition and structure of minnesotaite, a common iron silicate in iron formations. *Ibid.*, p. 363.

HALL, A. H. (1941) The relation between chemical composition and refractive index in the biotites. Am. Mineral., **26**, 34.

Handbook of chemistry and physics. (1947) 30th ed. Cleveland, Ohio: Chemical Rubber Publishing Co.

HARKER, A. (1932) Metamorphism. London: Methuen.

HASKELL, N. A. (1935) The motion of a viscous fluid under a surface load. Physics, **6**, 265.

———. (1936) The motion of a viscous fluid under a surface load. Part II. *Ibid.*, **7**, 56.

———. (1937) The viscosity of the asthenosphere. Am. J. Sci., ser. 5, **33**, 22.

HEDVALL, J. A. (1938) Die Reaktionsfähigkeit fester Stoffe. Leipzig: J. A. Barth.

———. (1942*a*) Pulverreaktionen zwischen Kalk und entwässertem Ton. Chalmers Tek. Högskol. Handl. 2 (avd. kemi och kemisk teknologi).

———. (1942*b*) Various types of disturbances in crystal lattices and their influence on chemical reactions and surface activity. *Ibid.*, 4 (avd. kemi och kemisk teknologi).

HENNIKER, J. C. (1949) The depth of the surface zone of liquid. Rev. Mod. Phys., **21**, 322.

HEVESY, G. VON. (1928) Electrolytische Leitung in festen Körpern. Handb. d. Phys., **13**, 286.

———. (1933) Materietransport in festen Körpern. Ztschr. Elektrochem., **39**, 490.

HEY, M. H. (1935) Studies of the zeolites. VIII. A theory of the vapor pressure of the zeolites, and of the diffusion of water or gases in a zeolite crystal. Mineral. Mag., **24**, 99.

HOLLAND, T. H. (1900) The charnockite series: a group of Archean hypersthenic rocks in peninsular India. Mem. Geol. Survey India, **28**, Part 2, 119.

HOLMES, A, (1947) The age of the earth. Endeavour, **6**, 99.

HOUWINK, R. (1940) Elasticity, plasticity and structure of matter. Cambridge: At the University Press.

HUBBERT, KING M. (1940) The theory of ground-water motion. J. Geol., **48**, 785.

INGERSON, E., and MOREY, G. W. (1940) Nature of the ore-forming fluid: a discussion. Econ. Geol., **35**, 772.

INTERNATIONAL GEOLOGICAL CONGRESS, 18th, London, 1948, Part VII: The geology, paragenesis, and reserves of the ores of lead and zinc.

JAGITSCH, R. (1949) Geological diffusion in solid phases. Nature, **164**, 497.

JAMESON, A. H. (1937) An introduction to fluid mechanics. London: Longmans, Green & Co.

JOST, W. (1937) Diffusion und chemische Reaktion in festen Stoffen. Leipzig: Steinkopf.

KEILHAU, B. M. (1828) Ein vorläufiges Wort über Contactbildungen. Ann. Phys. u. Chem., **14**, 131.

———. (1838) Granitens og de övrige såkaldte massive Bergarters samt krystallinske Skiferes Theorie. Nytt Mag. Naturvidensk., No. 1, p. 1.

KELLEY, K. K. (1941) The specific heats at low temperature of ferrous silicate, manganous silicate, and zirconium silicate. J. Am. Chem. Soc., **63**, 2750.

———. (1948) Contributions to the data on theoretical metallurgy. XI. Entropies of inorganic substances. U.S. Dept. of the Interior, Bureau of Mines Bull. 477.

———. (1949) Contributions to the data on theoretical metallurgy. X. High temperature heat content, etc. U.S. Dept. of the Interior, Bureau of Mines Bull. 476.

KENNEDY, G. C. (1950) A portion of the system silica-water. Econ. Geol., **45**, 629.

KOCHENDÖRFER, A. (1941) Plastische Eigenschaften von Kristallen und metallischen Werkstoffen. Berlin: J. Springer.

KÖHLER, A. (1949) Recent results of investigations on the feldspars. J. Geol., **57**, 592.

KÖNIGSBERGER, J. (1901) Die Minerallagerstätten in Biotitprotogin des Aarmassivs. Neues Jahrb. Mineral. Geol., Beilage Bd. A, **14**, 43.

KÖNIGSBERGER, J., and MÜLLER, W. J. (1906) Ueber die Flüssigkeitseinschlüsse im Quarz alpiner Mineralklüfte. Centr. Mineral. Geol., Vol. **72**.

KOENIGSBERGER, J. G.; NIGGLI, P.; and PARKER, R. L. (1940) Die Mineralien der Schweizeralpen. (Der Geotechnischen Kommission der Schweizerischen naturforschenden Gesellschaft.) Basel: B. Wepf & Co.

KORDES, E. VON. (1935) Die Beziehungen zwischen den Dissoziationsdampfdrucken von Sulfiden, und ihrer Ausscheidungsfolge auf magmatogenen Erzlagerstätten. Mineral. u. petrog. Mitt., **46**, 256.

KUBASCHEWSKI, O. and EVANS, E. LL. (1951) Metallurgical Thermochemistry. London: Butterworth-Springer, Ltd.

KÜHNE, R. (1932) Die Ertzführung des Auerbacher Marmor. Chemie der Erde, **7**, 503.

LACROIX, A. (1893) Les enclaves des roches volcaniques. Mâcon: Protat.

LAITAKARI, A. (1921) Über die Petrographie und Mineralogie der Kalksteinlagerstätten von Parainen. Bull. comm. géol. Finlande, No. 54.

LANGMUIR, I. (1950) Phenomena, atoms and molecules. New York: Philosophical Library.

LAPADU-HARGUES, P. (1945) Sur l'existence et la nature de l'apport chimique dans certaines séries cristallophylliennes. Bull. Soc. géol. France, ser. 5, **15**, 255.

———. (1949) Contributions aux problèmes de l'apport dans le métamorphisme. *Ibid.*, **19**, 89.

LARSEN, E. S., and BRIDGMAN, P. W. (1938) Shearing experiments on some selected minerals and mineral combinations. Am. J. Sci., ser. 5, **36**, 81.

LAVES, F. (1950) The lattice and twinning of microcline and other potash feldspars. J. Geol., **58**, 548.

———. (1952) Phase relationships in alkali feldspars. *Ibid.*, **60**, 436.

LINDGREN, W. (1933) Mineral deposits. 4th ed. New York: McGraw-Hill Book Co., Inc.

MAGNUSSON, N. H. (1930) Långbans malmtrakt. Sveriges Geol. Undersökn., ser. c, No. 23.

MARSH, J. S. (1935) Principles of phase diagrams. New York and London: McGraw-Hill Book Co., Inc.

MICHEL-LÉVY, M. C. (1950) Reproduction artificielle de la sillimanite. Compt. rend. Acad. Sci., Paris, **230**, 2213.

MILCH, L. (1917) Über Adinolen und Adenolschiefer des Harzes. Ztschr. deutsch. geol. Gesellsch., **69**, 349.

MILLER, J. P. (1952) A portion of the system calcium carbonate-carbon dioxide–water, with geological implication. Am. J. Sci., **250**, 161.

MISCH, P. (1949) Metasomatic granitization of batholithic dimensions, Parts I, II, III. Am. J. Sci., **247**, 209, 372, 673.

MOSESMAN, M. A., and PITZER, K. S. (1941) Thermodynamic properties of the crystalline forms of silica. J. Am. Chem. Soc., ser. 2, **63**, 2348.

NEUMANN, F. (1925) Ueber die Stabilitätsverhältnisse der Modifikationen im polymorphen System Al_2SiO_5. Ztschr. anorg. Chem., **145,** 193.

NEWHOUSE, W. H. (1932) The composition of vein solutions as shown by liquid inclusions in minerals. Econ. Geol., **27,** 419.

———. (1933) The temperature of formation of the Mississippi Valley lead-zinc deposits. *Ibid.*, **28,** 744.

———. (1940) Openings due to movement along a curved or irregular fault-plane. *Ibid.*, **34,** 445.

———. (1941) The direction of flow of mineralizing solutions. *Ibid.*, **36,** 612.

NEWHOUSE, W. H.; HAGNER, A. F.; and DEVORE, G. W. (1949) Structural control in the formation of gneisses and metamorphic rocks. Science, **109,** No. 2825, 168.

NISSAN, A. H. (1946) Flow of water through very narrow channels. Nature, **157,** 267.

NOLL, W. (1932) Hydrothermalsynthese von Muskovit. Naturwissenschaften, **20,** 283.

NORTON, F. H. (1939) Hydrothermal formation of clay minerals in the laboratory. Am. Mineral., **24,** 1.

OFTEDAHL, C. (1948) Studies on the igneous rock complex of the Oslo region. IX. Skrifter. Norsk. Vidensk.-Akad. Oslo. I. Mat.-naturv. Kl., No. 3.

O'NEIL, T. F. (1948) The hydrothermal alteration of feldspar at 250° C. to 400° C. Econ. Geol., **43,** 167.

PAULING, L. (1949) The nature of the chemical bond. 2d ed. Ithaca N.Y.: Cornell University Press.

PERRIN, P., and ROUBAULT, M. (1949) On the granite problem. J. Geol., **57,** 357.

PETTIJOHN, F. J. (1949) Sedimentary rocks. New York: Harper & Bros.

PHEMISTER, T. C. (1945) The coast range batholith near Vancouver, British Columbia. Quart. J. Geol. Soc. London, **101,** 37.

POLDERVAART, A. (1947) The relationship of orthopyroxene to pigeonite. Mineral. Mag., **28,** 164.

POULTER, T. C., and WILSON, R. O. (1932) The permeability of glass and fused quartz to ether, alcohol, and water at high pressure. Phys. Rev., ser. 2, **40,** 877.

RAMBERG, H. (1944*a*) Petrological significance of sub-solidus phase transitions in mixed crystals. Norsk. geol. Tids., **24,** 42.

———. (1944*b*) Relation between external pressure and vapor tension of compounds, and some geological implications. Avhandl. Norske Videnskaps.-Akad. Oslo. I. Mat.-naturv. Kl., No. 3.

———. (1944*c*) The thermodynamics of the earth's crust. I. Norsk. geol. Tids., **24,** 98.

———. (1944*d*) En undersökelse av Vestrandens regionalmetamorf bergarter. *Ibid.*, **23,** 1.

———. (1945) The thermodynamics of the earth's crust. II. *Ibid.*, **25,** 307.

———. (1946) Kjemisk likevekt i gravitasjonsfeltet og dens betydning for jordskorpens differentiasjon. Medd. Dansk Geol. Foren., **11,** 13.

———. (1947) The force of crystallization as a well definable property of crystals. Geol. Fören. i. Stockholm Förh., **69,** 189.

———. (1948*a*) Titanic iron ore formed by dissociation of silicates in granulite facies. Econ. Geol., **43,** 553.

———. (1948*b*) Radial diffusion and chemical stability in the gravitational field. J. Geol., **56,** 448.

———. (1948*c*) On sapphirine-bearing rocks in the vicinity of Sukkertoppen, West Greenland. Medd. Grönland, **142,** No. 5, 1.

RAMBERG, H. (1949) The facies classification of rocks: a clue to the origin of quartzo-feldspathic massifs and veins. J. Geol., **57,** 17.

———. (1951) Remarks on the average chemical composition of granulite and amphibolite-to-epidote amphibolite facies gneisses in West Greenland. Medd. Dansk Geol. Foren., **12,** 27.

———. (1952) Chemical bonds and distribution of cations in silicates. J. Geol., **60,** 331.

RAMBERG, H., and DEVORE, G. (1951) The distribution of Fe^{++} and Mg^{++} in coexisting olivines and pyroxenes. J. Geol., **59,** 193.

RANKAMA, K. (1946) On the geochemical differentiation in the earth's crust. Bull. comm. géol. Finlande, Vol. **24,** No. 137.

RANKAMA, K., and SAHAMA, T. G. (1950) Geochemistry. Chicago: University of Chicago Press.

RATH, G. VAN. (1870) Geognostisch-mineralogische Fragmente aus Italien. III. Thiel. Ztschr. deutsch. geol. Gesellsch., **22,** 591.

READ, H. H. (1933) On quartz-kyanite-rocks in Unst, Shetland Islands, and their bearing on metamorphic differentiation. Mineral. Mag., **23,** 317.

REYNOLDS, D. L. (1946) The sequence of geochemical changes leading to granitization. Quart. J. Geol. Soc. London, **102,** 389.

———. (1947) The association of basic "fronts" with granitization. Science Progress, **35,** 205.

ROSENQUIST, I. T. (1942) Om öiegneisdannelse i fiellkjeder. Norsk. geol. Tids., **21,** 165.

———. (1949) Noen observasjoner og refleksjoner omkring Modum koboltgruver. *Ibid.*, **27,** 187.

ROY, R. (1949) Decomposition and resynthesis of the micas. J. Am. Ceram. Soc., **32,** 202.

SAHAMA, T. G., and TORGESON, D. R. (1949) Thermochemical study of the stability relation of geikielite and ilmenite in rocks. U.S. Dept. of the Interior, Bureau of Mines, Rept. Inv. No. 4407.

SAHLSTEIN, T. G. (1935) Petrographie der Eklogiteinschlüsse in den Gneisen des südwestlichen Liverpool-Landes in Ost-Grönland. Medd. Grönland, Vol. **95,** No. 5.

SANDER, B. (1930) Gefügekunde der Gesteine. Vienna: J. Springer.

———. (1948) Einführung in die Gefügekunde der geologischen Körper. 2 vols. Vienna and Innsbruck: J. Springer.

SCHLAEPFER, M., and NIGGLI, P. (1914) Neue Beiträge zur hydrothermalen Silicatbildung. Ztschr. anorg. Chem., **87,** 52.

SCHMIDT, W. (1932) Tektonik und Verformungslehre. Berlin: Gebr. Borntraeger.

SCHÜLLER, A. (1948) Petrogenetische Studien zum Granulitproblem an Gesteinen der Münchberger Masse. Heidelberg. Beitr. z. Mineral. u. Petrog., **1,** 269.

SCOTT, H. S. (1948) The decrepitation method applied to minerals with fluid inclusions. Econ. Geol., **43,** 637.

SEDERHOLM, J. J. (1899) Über eine archäische Sedimentformation im südwestlichen Finnland. Bull. comm. géol. Finlande, No. 6.

SEITH, W. (1933) Abhängigkeit der elektrolytischen Leitfähigkeit und der Selbstdiffusion in Kristallen von der kristallographischen Richtung. Ztschr. Electrochem., **39,** 538.

SENG, H. (1937) Für das Riecke'sche Prinzip! Neues Jahrb. Mineral. Geol., Beilage Bd. A, **73,** 239.

SLATER, J. C. (1939) Introduction to chemical physics. New York and London: McGraw-Hill Book Co., Inc.

SMEKAL, A. (1935) The structure-sensitive propeties of salt crystals. Internat. Conf. Physics, London, 1934, **2,** 93.

———. (1949) On the imitation of reactions between solid phases. Powder Met. Bull., **4,** No. 4, 120.

SMITH, F. G. (1950) A method for determining the direction of flow of hydrothermal solutions. Econ. Geol., **45,** 62.

SONDER, R. A. (1948) Gefügeregelung, Druckschieferung und Druckfarmung von Gesteinen. In: Festschrift Paul Niggli. Zurich: A. G. Gebr. Leemann & Co.; also in Schweiz. Min. Petr. Mitt., **28,** Heft 1, 140.

SOSMAN, R. B. (1927) The properties of silica. New York: Chemical Catalog Co., Inc.

STILLWELL, F. L. (1918) The metamorphic rocks of Adelie Land: Australasian Antarctic Exped. 1911–1914. Sci. Rept. ser. A, Vol. **3,** Part I.

SULLIVAN, C. J. (1948) Ore and granitization. Econ. Geol., **43,** 471.

SUNDIUS, N. (1916) Zur Frage der Albitisierung im Kirunagebiet. Geol. Fören. i Stockholm Förh., **38,** 446.

TALIAFERRO, N. L. (1943) Franciscan-Knoxville problem. Bull. Am. Assoc. Pet. Geol., **27,** 109.

THEILE, W. (1932) Temperaturabhängigkeit der Plastizität und Zeitfestigkeit von Steinsalzkristallen. Ztschr. Phys., **75,** 763.

TISELIUS, A. (1934) Die Diffusion von Wasser in einem Zeolithkristall. Ztschr. phys. Chem., **169** A, 425.

———. (1935) Sorption und Diffusion von Ammoniak in Analcim. *Ibid.*, **174** A, 401.

TILLEY, C. E. (1927) Vesuvianite and grossularite as products of regional metamorphism. Geol. Mag., **64,** 372.

———. (1937) Wollastonite solid solutions from Scawt Hill, Co. Antrim. Mineral. Mag., **24,** 569.

———. (1947) The gabbro-limestone contact zone of Camas Mòr, Muck, Inverness-Shire. Bull. comm. géol. Finlande, **24,** No. 140, 97.

TODD, S. S. (1949) Heat capacities at low temperatures and entropies of magnesium and calcium fluorides. J. Am. Chem. Soc., **71,** 4115.

TURNER, F. J. (1942) Current views on the origin and tectonic significance of schistosity. Trans. Proc. Roy. Soc. New Zealand, **72,** 119.

———. (1948) Mineralogical and structural evolution of the metamorphic rocks. Mem. Geol. Soc. Am. **30.**

TUTTLE, O. F. (1949*a*) Structural petrology of planes of liquid inclusions. J. Geol., **57,** 331.

———. (1949*b*) The variable inversion temperature of quartz as a possible geologic thermometer. Am. Mineral., **34,** 723.

TUTTLE, O. F., and BOWEN, N. L. (1949) High-temperature albite. Bull. Geol. Soc. Am., **60,** 1925.

VERHOOGEN, J. (1938) Thermodynamical calculation of the solubility of some important sulphides up to 400° C. Econ. Geol., **33,** 34, 775.

———. (1946) Volcanic heat. Am. J. Sci., **244,** 744.

———. (1949) Radial diffusion and chemical stability in the gravitational field: a discussion. J. Geol., **57,** 93.

VOGEL, R. C., and GIBSON, G. (1950) Migration of sodium ions through quartz plates in an electric field. J. Chem. Phys., **18,** 490.

Vogt, T. (1927) Geology and petrology of the Sulitjelma district. Norg. Geol. Undersökelse, No. 121.

Volmer, M. (1939) Kinetik der Phasenbildung. Dresden and Leipzig: T. S. Steinkopf.

Warren, B. E. (1930) The crystal structure and chemical composition of the monoclinic amphiboles. Ztschr. Kristallog. **72**, 493.

Wegmann, C. E. (1935) Zur Deutung der Migmatite. Geol. Rundschau, **26**, 305.

———. (1938) Geological investigation in southern Greenland. I. On the structural divisions of southern Greenland. Medd. Grönland, Vol. **113**, No. 2.

———. (1948*a*) Remarques sur le métamorphisme régional. Geol. Rundschau, **36**, 40.

———. (1948*b*) Zur Geschichte der Umwandlungsgesteine. *Ibid*., p. 48.

Weiser, H. B. (1939) Colloid chemistry. New York: John Wiley & Sons, Inc.; London: Chapman & Hall, Ltd.

Wells, A. F. (1945) Structural inorganic chemistry. 1st ed. Oxford: Clarendon Press.

White, D. E. (1947) Diagenetic origin of chert lenses in limestone at Soyatal, state of Queretaro, Mexico. Am. J. Sci., **245**, 49.

Wright, F. E., and Larsen, E. S. (1909) Quartz as a geologic thermometer. Am. J. Sci., **27**, 421.

Wright, W. I. (1938) The composition and occurrence of garnets. Am. Mineral., **23**, 436.

Yoder, H. S., Jr. (1950*a*) The jadeite problem. Parts I and II. Am. J. Sci. **248**, 225, 312.

———. (1950*b*) Stability relations of grossularite. J. Geol., **58**, 221.

———. (1952) The $MgO-Al_2O_3-SiO_2-H_2O$ system and the related metamorphic facies. Am. J. Sci. (in press).

Yoder, H. S., Jr., and Weir, C. E. (1951) Change of free energy with pressure of the reaction: nepheline and albite-2 jadeite. Am. J. Sci., **249**, 683.

Zener, C. (1950) Ring diffusion in metals. Acta Cryst., **3**, 346.

Zener, C., and Holloman, J. H. (1946) Problems in non-elastic deformation of metals. J. Appl. Phys., **17**, 69.

AUTHOR INDEX

SUBJECT INDEX

PRINTED IN U·S·A